# Data and Network Communications

## Online Services

**Delmar Online**
To access a wide variety of Delmar products
and services on the World Wide Web, point your
browser to:

      **http://www.delmar.com**
      or email: info@delmar.com

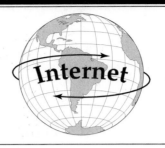

A service of I(T)P

# Data and Network Communications

## Michael A. Miller
### Senior Professor

DeVry Institute of Technology
Phoenix, AZ

Africa • Australia • Canada • Denmark • Japan • Mexico • New Zealand • Philippines
Puerto Rico • Singapore • Spain • United Kingdom • United States

## NOTICE TO THE READER

Cover Design: Nicole Reamer
Cover Photo: VCG/FPG International LLC

**Delmar Staff:**

Publisher: Alar Elken
Acquisitions Editor: Gregory L. Clayton
Developmental Editor: Michelle Ruelos Cannistraci
Executive Editor: Sandy Clark
Editorial Assistant: Amy E. Tucker

Production Manager: Larry Main
Senior Project Editor: Christopher Chien
Art Director: Nicole Reamer
Executive Marketing Manager: Maura Theriault
Marketing Coordinator: Paula Collins

COPYRIGHT © 2000
Delmar Thomson Learning

Printed in the United States of America

For more information contact:

**Delmar Publishers**
3 Columbia Circle,
Box 15015
Albany, New York USA 12212-5015

**International Thomson Publishing Europe**
Berkshire House
168-173 High Holborn
London, WC1V 7AA
United Kingdom

**Nelson ITP, Australia**
102 Dodds Street
South Melbourne, Victoria 3205
Australia

**Nelson Canada**
1120 Birchmont Road
Scarborough, Ontario
M1K 5G4, Canada

**International Thomson Publishing France**
Tour Maine-Montparnasse
33, Avenue du Maine
75755 Paris Cedex 15, France

**International Thomson Editors**
Seneca 53
Colonia Polanco
11560 D.F. Mexico

**International Thomson Publishing GmbH**
Konigswinterer Straße 418
53227 Bonn
Germany

**International Thomson Publishing Asia**
60 Albert Street
#15-01 Albert Complex
Singapore 189969

**International Thomson Publishing—Japan**
Hirakawacho Kyowa Building, 3F
2-2-1 Hirakawa-cho Chiyoda-ku
Tokyo 102 Japan

**ITE Spain/Parainfo**
Calle Magellanes, 25
28015-Madrid, Espana

1 2 3 4 5 6 7 8 9 10 XXX 05 04 03 02 01 00 99

**Library of Congress Cataloging-in-Publication Data**

Miller Michael A.
   Data and network communications / Michael A. Miller.
       p.     cm.
   ISBN 0-7668-1100-X
   1. Computer networks.   2. OSI (Computer network standard)
3. TCP/IP (Computer network protocol)     I. Title.
TK5105.5.M563   2000
004.6--dc21

99-36709
CIP

# Dedication

This book is dedicated to my wife, Ann, who had the patience
to provide me the freedom and time to write it.

# Contents

Preface   xi

## 1   Introduction to Data and Network Communications   1

1.1   Introduction   1
1.2   Data Communication System   6
1.3   Data Communication Links   10
1.4   Character Codes   12
1.5   Digital Data Rates   16
1.6   Serial Data Formats   17
1.7   Encoded Data Formats   20

Summary   24
Questions   25
Design Problems   26
Answers to Review Questions   27

## 2   The Telephone System   29

2.1   Introduction   29
2.2   The Telephone System   30
2.3   Specifications and Parameters   38
2.4   Line Impairments   52

Summary   56
Questions   57
Design Problems   58
Answers to Review Questions   59

## 3   Error Detection and Correction   61

3.1   Introduction   61
3.2   Asynchronous Data Methods   62
3.3   Synchronous Data Error Methods   72
3.4   Error-Testing Equipment   83

Summary   85
Questions   86
Design Problems   87
Answers to Review Questions   87

## 4   Open Systems Network Models   89

4.1   Introduction   89
4.2   Data Topologies   90
4.3   Data Switching   101
4.4   Types of Networks   105
4.5   The Open Systems Inteconnection (OSI) Architecture   108
4.6   Systems Network Architecture (SNA)   114
4.7   SNA Operating Sessions   121

Summary   122
Questions   123
Design Problems   125
Answers to Review Questions   125

## 5   OSI Physical Layer Components   127

5.1   Introduction   127
5.2   Units of a Communication Link   128
5.3   RS232C Interface   137
5.4   RS449 Interface Standard   145
5.5   RS422 and RS423 Interface Standards   147
5.6   FSK Modems   150
5.7   Additional Types of Modems   153
5.8   .34 and V.90 Modems   170

Summary   171

Questions   173

Research Assignments   175

Answers to Review Questions   175

## 6   Higher Capacity Data Communications   179

6.1   Introduction   179
6.2   Multiplexing Methods   181
6.3   Sampling Theorem   190
6.4   Quantization   197
6.5   Pulse Code Modulation   198
6.6   Delta Modulation   202
6.7   Digital T Carriers   206
6.8   Companding   213
6.9   CODECs   217

Summary   221

Questions   221

Design Problems   223

Answers to Review Questions   224

## 7   Data-Link Layer Protocols   227

7.1   Introduction   227
7.2   Data-Link Sections   228
7.3   Character-Oriented Protocols   230
7.4   Bit-Oriented Protocols   238
7.5   Protocol Analyzers   249

Summary   251

Questions   252

Answers to Review Questions   254

## 8   Network Architecture and Protocols   257

8.1   Introduction   257
8.2   Networks by Size   259
8.3   IEEE 802.3 and Ethernet   261
8.4   IEEE 802.4 Token Bus   274

8.5   IEEE 802.5 Token Ring   279
8.6   Network Interface Cards   283
8.7   Interconnecting LANs   285
8.8   IEEE 802.6 Metropolitan Area Network (MAN)   291
8.9   X.25 Packet Switch Protocol   293

Summary   296

Questions   296

Research Assignments   298

Answers to Review Questions   299

## 9   Integrated Services and Routing Protocols   301

9.1   Introduction   301
9.2   Integrating Services   302
9.3   Broadband ISDN   311
9.4   IEEE 802.9: Integrated Voice and Data Services   314
9.5   Digital Subscriber Line (DSL)   315
9.6   Private Branch Exchange   318
9.7   Asynchronous Transfer Mode (ATM)   324
9.8   Frame Relay   337

Summary   339

Questions   340

Research Assignments   343

Answers to Review Questions   343

## 10   The Internet and TCP/IP   347

10.1   Introduction   347
10.2   Internet History   348
10.3   Uses for the Internet   353
10.4   Accessing the Internet   358
10.5   Internet Addresses   361
10.6   Security on the Internet   365
10.7   Authentication   367
10.8   Firewalls   371
10.9   Intranets and Extranets   374
10.10   TCP/IP, the Technology behind the Internet   375

Summary    381

Questions    381

Research Assignments    384

Answers to Review Questions    384


## 11    Fiber Optic Communications    387

11.1    Introduction    387
11.2    Basic Concepts of Light Propagation    389
11.3    Fiber Cables    394
11.4    Light Sources    400
11.5    Optical Detectors    402
11.6    Fiber-Cable Losses    406
11.7    Wave Division Multiplexing    409
11.8    Fiber Distributed Data Interface    411
11.9    FDDI-II: Isochronous Traffic    418
11.10   The Fibre Channel    419
11.11   SONET    421

Summary    423

Questions    424

Research Assignments    427

Answers to Review Questions    427


## 12    Wireless Communication Systems    431

12.1    Introduction    431
12.2    Microwave Communications    432
12.3    Cellular Mobile Telephone Service    434
12.4    Personal Communications Systems    437
12.5    IEEE 802.11: Wireless LANs Using CSMA/CA    439
12.6    Cellular Digital Packet Network    442
12.7    Satellite Communications    443
12.8    Methods of Satellite Communications    452
12.9    Satellite Networking    457

Summary    461

Questions    461

Research Assignments    464

Answers to Review Questions    464

Appendix A—Acronyms and Key Terms    467

Appendix B—Complete Extended ASCII Code Set    479

Appendix C—Network Timeline    483

Appendix D—Facsimile    497

Appendix E—Answer Key to Odd-Numbered Questions    501

Index    517

# Preface

This text on the data and network communications field is intended to be a junior or senior level text in an Electronics Technology or Technician curriculum offering one or more courses on the topic. A community college or proprietary school with a Communications specialty could teach this text's material to sophomores after a course in basic communications. In order to do justice to the numerous facets of the data and network communications fields, no one area is treated with the kind of depth necessary to produce telecommunications engineering technologists. It is intended, rather, to provide the electronics technician and technologists with sufficient background in data and network communications technology for a solid, thorough understanding of what is in the field.

It is expected that the student has had, or is familiar with, the topics in courses that cover the following material:

a) AM and FM Radio
b) Basic Communications
c) Basic Electronic Circuits and Devices
d) Basic Digital Circuits and Devices
e) Microprocessors and Basic Computer Architectures.

Chapter 1 starts by giving the student a general overview of the data communications area using a point-to-point data-link model. It lays the basis for future chapters by establishing some necessary fundamentals.

Chapter 2 deals with the telephone system from the standpoint of data transfer usage. Electrical specifications and impairments that require consideration when using the telephone system facilities for networking are also covered.

Chapter 3 provides details on some common error detection and correction methods used in data, digital, and network communications systems.

Chapter 4 presents the two most common open system network layer models for data communications networks, the open systems interconnection (OSI) and systems network architecture (SNA) models.

Chapter 5 presents protocols and specifications at the physical layer of the OSI model. It deals with a look at the hardware for low- and medium-speed data communications as interfaced through the telephone system. This includes discussions on UARTs, modems, RS232C, and other physical interfaces. Modulation techniques involving frequency, phase, and amplitude changes are also explored in relationship to communications interfacing. The chapter finishes with a brief discussion on V.32 and V.90 modems used to send and receive data at 28.8 Kbps and 56 Kbps.

Chapter 6 introduces the student to multiplexing communications channels into a single entity and to higher capacity data channel techniques. Included are time division, frequency division multiplexing, and a discussion on the concepts of T1 digital lines and the equipment used to monitor and test them.

Chapter 7 takes the student to the next layer of the OSI model, the data-link layer. Basic types of data-link protocols, the asynchronous data-link protocol and BISYNC are introduced followed by an indepth look at the bit-oriented SDLC/HDLC protocol.

Chapter 8 covers a wide variety of communications networks including local area networks (LAN) and the IEEE 802 standard for networking, which are discussed in detail in this chapter.

Chapter 9 deals with integrated services digital networking and routing protocols. These include ISDN, digital subscriber line (DSL), ATM, and Frame Relay protocols. These services are provided for combined services such as voice, video, and data to be transmitted simultaneously across the public services telephone network.

Chapter 10 provides discussion of an application of both the network and transport layers of the OSI model. These are amply represented in the most widely used network today, the Internet. This chapter will supply an overview of the Internet including security issues, and then launch into the details of the TCP/IP protocol that makes its existence possible.

Chapter 11 discusses fiber optics and fiber-optic networks. In a text of this nature, a single chapter on fiber optics cannot cover this vast subject in its entirety, which usually requires a separate tome. However, it is the author's intent to provide enough material on the topic so that the student obtains a good background on the subject.

Chapter 12 brings us to the concepts and applications of various wireless communications systems. It starts with an overview of microwave communications, moves into cellular telephones, wireless networks, and ends with a thorough discussion on satellite networks.

Five appendixes are provided:

Appendix A is a list of all the abbreviations and acronyms used in this text and in the data and network communications field.

Appendix B shows the complete extended ASCII character set.

Appendix C provides a network timeline from a historical perspective.

Appendix D gives a brief overview of facsimile (fax) transmissions that are considered "optional" in a data communications course. They are presented here for those instructors who feel they are still of some value.

Appendix E is a list of solutions to the odd numbered questions at the end of each chapter.

The author wishes to show his appreciation to the editors and assistants at Delmar, Greg Clayton, Michelle Cannistraci, Amy Tucker, et al. for their guidance and help during the writing and reviewing of this text. Additional thanks are extended to the professional instructors who took time to thoroughly review the manuscript and make the many suggestions that were incorporated into the book to make it more complete and accurate. Specifically, those reviewers are:

Mike Awwad—DeVry Institute of Technology, North Brunswick, NJ
Don Custer—Western Iowa Tech Community College, Hornick, IA

Robert Diffenderfer—DeVry Institute of Technology, Kansas City, MO
Tom Diskin—College of San Mateo, San Mateo, CA
Everett Feight—Technical College of the Low Country, Beaufort, SC
John Giancola—DeVry Institute of Technology, Columbus, OH
Nawaz Khan—Iowa Western Community College, Council Bluffs, IA
Bruce Lampe—Pearl River Community College, Hattiesburg, MS
Harry Mendell Smith—Mt. San Antonio College, Walnut, CA
Roger Peterson—Northland Community and Technical College, Thief River Falls, MN
Bob Redler—Southeast Community College, Milford, NE
Lowell Tawney—DeVry Institute of Technology, Kansas City, MO
Jamie Zipay—DeVry Institute of Technology, Long Beach, CA

Michael A. Miller, Senior Professor
DeVry Institute of Technology
Phoenix, Arizona
mmiller@devry-phx.edu

# CHAPTER 1

# Introduction to Data and Network Communications

## OBJECTIVES

After completing this chapter, the student should be able to:
- discuss the history of data communications and networking.
- define basic data communications terminology.
- have an overview of a data communications system and its basic underlying characteristics.
- define the parts of a two-point communication model.
- realize the benefits of an open systems concept of communications modeling.
- define the character types represented in a binary character code.
- identify different data types, rates, and binary data formats.

## OUTLINE

1.1 Introduction
1.2 Data Communications System
1.3 Communication Links
1.4 Character Codes
1.5 Digital Data Rates
1.6 Serial Data Formats
1.7 Encoded Data Formats

## 1.1 INTRODUCTION

Now that we are fully absorbed by the Information Age and spending more time communicating and gathering information through the Internet, it has become necessary to have a working knowledge of the technology behind the scenes. We are faced with terms like baud rate, modems, cellular phones, TCP/IP, ATM, ISDN, etc., and trying to make decisions about our communications needs involving the systems that these terms apply to. In order to develop a useful working understanding of this technology requires you to have a good understanding of the background technology and basics of data communications.

## Historical Perspective

The transfer of data in digital form began around 1832 with the advent of Morse code, a systematic code that represents the printable characters of a language using a form of binary data. These characters are letters, numbers, or punctuation marks and are called **alphanumeric** as a class. Combinations of dashes (long signals) and dots (short signals) were used to code each character. A collection of these combinations is known as a **character code.** For Morse code to work, a signal or electrical current is placed onto an interconnecting line between a sender and a receiver when a switch of the key is closed by the sender. Short key closures created dots while longer closures produced dashes. The form of data modulation using a sine wave voltage as the signal is called **continuous wave keying (CWK)** because the signal itself is a continuous audio oscillation that is placed on or off the line by use of the key. The switching on and off of a direct current (DC) in place of the audio signal is also referred to as keying. In the latter case, an electromagnetic relay is energized in the presence of the current and released when the key or switch opens and removes current from the line. The length of time the relay is energized determines if a dash or dot has been sent.

Morse code served as a primary means of communicating information across vast distances for a considerable amount of time. With the invention of the telephone in 1876 and the creation of the telephone company system a year later, quicker means of transferring data evolved. By 1881, long distance trunk lines connected major cities in the eastern United States, eventually resulting in the birth of Atlantic Telephone and Telegraph (AT&T) in 1885. The early use of the telephone system and its principal use today is for voice communication. However, it was soon discovered that binary data could be converted to voice signals and sent over the telephone lines. Different methods had to be developed to generate the data and voice signals from the sender and interpret them at the receiving end. One of those developments was an early type of printer/keyboard, created by the Teletype Corporation, which employed electromechanical relays to replace the action of the key. Instead of continuing with dots and dashes, these teletype machines used the presence or absence of a 20 ma current to represent binary data. The presence of the 20 ma current signifies a logic high or "one" state often called a **mark** and the absence of current, a logic low or "zero" state called a **space.** The current either energized or released a relay as described above. These teletype machines combined with equipment that allowed them to be interfaced to the telephone system, provided the machinery for the development of Teletype and Telex systems in the 1930s.

Teletype machines were slow, noisy, and consumed large amounts of power. The Teletype system used in the United States sent and received data at a **baud rate** of 110 bits per second (bps). Baud rate is a measure of the rate at which binary data are transmitted and received. Part of the basis for a baud rate is the number of binary digits or bits are sent within one second. This measure is called the bit rate and the measurement is known as **bits per second** or **bps.** Differences between baud rate and bit rate occur because they

---

**alphanumeric**
printable characters in a character code, comprised of alphabet, numerical, and punctuation characters.

**character code**
binary code representing alphanumeric, formatting and data link characters.

**continuous wave keying (CWK)**
form of data transmission that uses the presence of a sine wave to represent a logic 1 and the absence, a logic zero. A form is used in Morse code where the duration of the signal represents a dot or a dash and the absence of the signal indicates no information is being sent.

**mark**
logic 1.

**space**
logic 0.

**baud rate**
digital information transfer rate.

**bits per second**
rate at which raw serial binary data is sent and received.

**bps**
bits per second.

**raw binary data**
digital data bits without any interpretation of meaning or use.

**throughput**
rate at which information is transferred from one point to another.

**symbol**
an electrical parameter used to represent one or more data bits.

define different but related information. The bit rate is a measure of **raw binary data,** which is the flow of binary bits with little concern to the actual content that they represent. Baud rate, on the other hand, is more of an information flow-through rate that factors in consideration about actual information data and noninformation overhead data.

Baud rate is a closer measure of information **throughput,** or the effective information data transfer rate from sender to receiver. Bit and baud rates are mathematically related with the result, for instance, that 110 bps baud rate actually translates to a binary bit rate of about 100 bps. It is easy to confuse baud and bit rate at lower data rates since they are both measured in bits per second.

As we get into higher data rates later on, we translate information rate into symbols per second or sps. A **symbol** is any element of an electrical signal that can be used to represent one or more binary data bits. The rate at which symbols are transmitted is the symbol rate in SPS. This rate may be represented as a systems baud rate in much the same manner that bits per second can be interpreted as a baud rate.

The tendency in studying data communications and, to a lesser extent, its application in the field, is to dismiss the differences between baud rate and bit or symbol rate. Many authors and working professionals use the terms interchangeably. For many low-speed applications the differences between them are insignificant. Thus 300 and 1200 bps modems originally used with personal computers were frequently referred to as 300 or 1200 baud modems. There is no problem here, since at these rates, one symbol is produced for each data bit, resulting in very similar numbers for bit, symbol, and baud rates. One note, some 1200 baud modems use a type of modulation scheme that produces two bits per symbol. In that case, there is certainly a difference between bit and symbol rates. Until the distinction becomes significant in a particular area under discussion, this text will consider bit and baud rate as being similar. This is not to minimize the differences, but a good many concepts are unaffected by them. Analysis of data transfer efficiency, rates, and bandwidth limitations is specified using baud rate. In those contexts the difference is critical.

The Telex system, used in Europe, was slower yet, ambling along at a mere 50 bps rate. These speeds were dictated by the need to operate the relays allowing them sufficient time to switch on and off. Additionally, since these machines were electromechanical in nature, they were highly prone to mechanical failures and constant adjustments and maintenance.

At about the same time, Marconi had invented radio transmissions and work was under way to establish wireless communications. The first amplitude modulated (AM) radio broadcasts as a commercial endeavor was radio station KDKA in Pittsburgh in 1921. By 1934 the United States government stepped in to regulate the growing additions to the nation's airwaves and the Federal Communications Commission (FCC) was born.

## The Computer and Data Communications

Parallel work moving communications toward today's information highway involved the computer, for without it we would still be keying in a lot of data by hand. The idea of doing math calculations using binary numbers had been toyed with by the telephone company's research arm, Bell Labs, for a number of years before a teletype machine was interfaced to one of these electronic calculators in 1940. What emerged from this marriage was the electronic computer. Data could be entered in from the teletype machine keyboard, processed by the calculator, and the results printed out on the teletype printer. Like the teletype machines, the calculator portion was derived from more relays because the binary system could easily be represented by a closure (logic 1) or open (logic 0) state of the relay. The first computers, the ENIAC Mark I and II were huge systems occupying many rooms and consuming large amounts of power to operate them.

The next big break came in 1947 with the invention of the transistor. Many functions of the relays had been replaced by vacuum tubes. These devices were improvements over relays since they no longer had movable parts, but they still consumed a lot of power. In addition, the vacuum tube, which required a filament element to produce heat to "agitate" electrons into movement, required air conditioned rooms to dissipate the heat they generated. The transistor provided all kinds of relief—no moving parts, much less heat created, small in size, and less expensive to make. It was what the computer world was waiting for.

Several different events in the 1950s impacted the future of data communications. One was the production of the first computer by International Business Machines (IBM)—big blue was launched. IBM became the leading producer of mainframe computers in the world, setting standards for many others to follow for years. The next big impact was heralded throughout the world as Russia launched the first satellite, Sputnik 1, in 1957. Today the greatest percentage of our daily communications is carried by communications satellites covering the globe. A less heralded event than Sputnik I, but one with as much influence on how we communicate, occurred in 1958 with the first coast-to-coast microwave radio link in Canada. Also that year, America entered the space race by launching Explorer 1, beginning many years of space insanity and rapid technological advancements.

Integrated circuits arrived in 1959 along with the first mini-computer, the PDP-1, based on the UNIX operating system. An operating system (OS) is a computer program that is used to configure the computer so that it can be used. Operating systems also provide utility programs that allow users to perform basic tasks by typing commands directly on the computer's command line. Communication satellites were launched into operation beginning in the 1960s. During this period small solid-state lasers are developed along with fiber optic cables that are used to carry the light generated by these small lasers leading to communications by light waves.

## The Telephone System

Up to the year 1968, if a vendor wanted to connect communications equipment to the telephone company's system, they had to rent the interface equipment from the telephone company. Many of these companies felt that this was unfair and led to a monopolizing of the phone system by AT&T. In response to the pressure from these companies, the government, through the Federal Communications Commission (FCC), produced Rule 67. This was a voluminous document, which clearly specified what a company had to do to be allowed to directly connect equipment to the phone company network. Despite the numerous rules and specifications detailed in that ruling, a number of manufacturers did proceed to develop and market modems for the purpose of allowing two computers to send and receive data over the telephone lines. A **modem** is a device that converts between the serial

**modem**
unit that converts between digital data and analog data.

digital data form produced by a computer to a form of analog signal that can be sent through the telephone voice circuits. On the receive side of the communications line, a modem reverses the process, returning the analog data back to serial digital data.

## Microprocessors and PCs

There is no doubt that the information age would never have emerged without the appearance of the microprocessor in the early 1970s. INTEL is credited with producing the first line of commercial microprocessors, starting with the 4004 and 8008. Over the years the industry has dutifully tracked INTEL's progress as each new microprocessor generation brought new and more powerful computers for our use. However, INTEL's processors were not the ones used to launch the personal computer into existence. Instead, an offshoot company, called ZILOG founded by three ex-INTEL engineers, developed the Z80 microprocessor used by an electronic hobby outfit called Radio Shack that resulted in the TRS-80 personal computer. The TRS-80 used a language called Basic, which was written by a pair of professors at Dartmouth College as a teaching language. Programs were entered by hand from the keyboard and later through an audio cassette interface.

## Networking

At about the same time as the TRS-80 was introducing the world to personal computers, the first data local area network (LAN), called ETHERNET was deployed to interconnect mainframes with terminals throughout a building. It was not long afterwards that Compuserve, one of the first of many bulletin board services, arrived to bring information into home owner's personal computers via telephone lines and modems.

In 1975 Bill Gates started Microsoft and in 1976 Steve Jobs and Stephen Wozniak began Apple. The computer industry was off and running. IBM entered the PC battles in 1981 and began to dominate the market as it had with mainframes. Apple remains its chief competitor, and in 1983, brought out the first graphic user interface (GUI) with the LISA computer. Microsoft responds with Windows 1.0 in 1985, which was a poor system in comparison. To round off the beginnings, add the inclusion of NETWARE 286 by NOVELL for interconnecting personal computers into a local area network (LAN).

Data rates for transmission have been on the rise. Early modems connected to personal computers ran at 300 bps. By 1987, 9600 bps modems were available. Networks, too, improved rapidly. Novell revised Netware for the 386 microprocessor-based IBM PC with NETWARE 386 in 1989. A fiber optic network, *fiber distributed data interface (FDDI),* came into being to handle faster data transfers. Networking across the telephone system gets a boost in 1990 from the Integrated Services Digital Network, or ISDN, which carries voice, binary data, and video information on the telephone lines.

Today, we have advanced versions of Windows that are adapted to network use and are vastly improved over the earlier versions. A number of companies are vying for the pieces of the network pie, making all kinds of super programs available to the endusers as a result. Not to be dismissed is the impact of the Internet on everyone's life. If nothing more, the amount of advertisement that ends with an Internet address is staggering. Every magazine has a "web site."

## The Internet

The Internet, which we have ignored to now, has been an ongoing system during this entire time period. Developed in the late 1950s as a network to share research information between military and university researchers. By 1984, it serviced approximately 1,000 users. As network technology improved, the number of users increased tenfold by 1988. When the system was opened wide for any users, its use escalated well beyond the projections anyone had to over one million users in 1992, up to 3 million in 1994 and 6 million a year later. What lies behind the popularity of the Internet is the access to all kinds of information at a reasonable cost to the user. You can shop on the Internet, make airline reservations, get educated, just plain chat, look up all kinds of information, entertain yourself with all kinds of amusements, and exchange ideas with people of like interests. The Internet is Highway 1 of the information super highway—welcome to the information age.

A more concise timeline list of events that impacted on the data communication and network industry is offered in Appendix F at the back of the book. A lot of additional items are included that were not discussed in the brief coverage in this section.

---

### Section 1.1  Review Questions

1. What is an early data communications system that uses continuous wave modulation?
2. What do mark and space tones represent?
3. What is used as a measure of serial data rate?
4. What function does a Modem provide?
5. What is the first data local area network (LAN)?

---

## 1.2  DATA COMMUNICATION SYSTEM

### Data Communications Link

**node**
entry point into a network.

**primary station**
controlling station in a network.

**remote** or **secondary station**
non-controlling station in a data link.

**line control unit**
controls the interface of peripheral devices to the data terminal.

The components of a basic communications link between two endpoints, or **nodes,** is illustrated in Figure 1-1. A node is any connection point to a communications link. For this two-point network, the node points are the **primary station** and the **remote** or **secondary station** at the other end of the communications link. Station refers to any section of hardware whose purpose is to communicate with another piece of communications hardware at a different location. Data link refers to the process of connecting or linking two stations together.

A primary station is responsible for establishing and maintaining the data link between it and a secondary station. Data sent from one station to another usually originates in parallel binary form from one or more peripheral devices connected to that station through a **line control unit.** This unit supplies the interface to the communications station and control of peripheral devices including, but not limited to, computer terminals, printers, keyboards, facsimile (FAX) machines, and data display terminals.

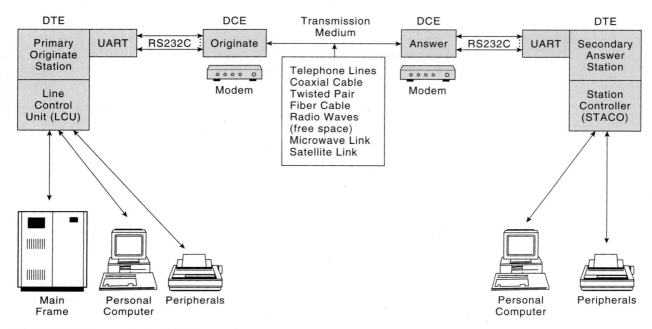

**FIGURE 1-1** Communications Link

**parallel data**
all the bits of a data word transferred at the same time.

**words**
fixed number of data bits.

**serial**
digital data transferred one bit at a time.

Communication stations may also be part of mainframe, personal workstation systems or access points (nodes) for networks. Note that the information supplied by the peripherals or networks could be anything from a series of keyboard characters to a stream of digitized video. This information is converted from its natural form into digital form by the peripheral and is presented as groups of parallel binary data to the system. **Parallel data** are a group of digital bits that are available at the same time, often referred to as digital or binary **words.**[1] An individual communications path is required for each bit, allowing data to move quickly, transferring a complete word each time. However, the need for multiple data paths is impractical and costly for long distance transfers. Instead, it is preferable to send data along a single data path between two stations. In order to do this, the parallel data need to be converted into **serial** form, with one bit of data sequentially following another. While parallel data allows many bits to be sent at once, serial data requires each bit to be sent separately.

[1]The conventional definition of word, as pplied to digital information, is a fixed number of binary bits indicating a selected size of information. Thus a computer system using a 32-bit data bus operates using 32-bit words. With the advent of 16-bit and 32-bit microprocessors, manufacturers have taken liberties with the term word. For example, Motorola's user manuals for the 68000 family microprocessor specifically defines word as having 16 bits of data and a long word as containing 32 bits. However, as it develops, Motorola, like other makers of microprocessors, is not consistent. In the user manual for its newer processors, word is redefined as containing 32 bits! In addition, INTEL, the manufacturer of the X86 line, uses double word to denote a 32-bit binary word. This is mentioned so that students who may be studying microprocessors concurrently with a course on data communications are made aware of the conflicting use of the term word. In this text the term refers to its original meaning, that is, binary data in a fixed number of bits.

**EXAMPLE 1-1**

Compare the time it takes to send the following short message from one station to another, first as an 8-bit or byte parallel data and then as serial data. The transfer rate is 1 ms per transfer.

2C3B in hexadecimal, which is 0010110000111011 in binary

**SOLUTION**

As byte parallel data, there are two groups of 8-bit words for this message, 2C or 00101100 and 3B or 00111011. It takes two transfers, or 2 ms, to send the data in that form. In comparison, sending the 16 bits of data serial at 1 ms per transfer, results in 16 ms of total time required to complete the transmission.

## UARTS

Devices that perform the parallel-to-serial conversion (and vice versa at the receiving station) are the **universal asynchronous receiver transmitter (UART)** and the **universal synchronous/asynchronous receiver transmitter (USART).** The conversion process and the rates at which parallel data are sent to the UART or USART and the rate at which serial data are sent and received are controlled by the computer system to which the UART or USART is connected. UARTs and USARTs, which are discussed in detail in Chapter 5, are produced in medium scale integrated circuit packages. They perform other tasks besides parallel-to-serial conversion that are required for the successful transfer of information.

## DTE and DCE Equipment

The computer system, communications station, UART, and line control unit (LCU) are grouped together and classified as **data terminal equipment (DTE),** as shown in Figure 1-2. The computing system of the DTE contains software needed to establish and control the communications link between the primary and secondary stations. An applications program used by the DTE, called a **protocol,** defines a set of rules that determine the requirements for the successful establishment of a data link and the transfer of actual information between the stations. Protocols exist at many levels of an overall network or communications system. While they can become quite sophisticated, their basic premise never changes. They are always used to set the requirements for moving information between two or more node points within a network or system. Protocols at the lowest level are discussed in Chapter 4, while higher-level protocols are presented in several later chapters.

Applications programs also direct control information to the line control unit and UART to allow data flow from the peripheral currently serviced by the LCU to the UART and out to the **data communications equipment,** or **data circuit terminating equipment,** both known by the acronym **DCE.** In the two point system of Figure 1-1, the DCE is a **modulator-demodulator (modem).** This device is used to convert the serial data stream into a form that can be used by the connecting **medium** to transfer data over

**universal asynchronous receiver transmitter (UART) and universal synchronous/ asynchronous receiver transmitter (USART)**
devices used to interface between DTE (parallel) and DCE (serial) data.

**data terminal equipment (DTE)**
the hardware responsible for controlling communications.

**protocol**
a set of rules for successful data transfers.

**data communications equipment (DCE) or data circuit terminating equipment (DCE)**
data communication equipment or data circuit terminating equipment.

**modulator-demodulator (modem)**
unit that converts between digital data and analog data.

**medium**
the transmission path for data.

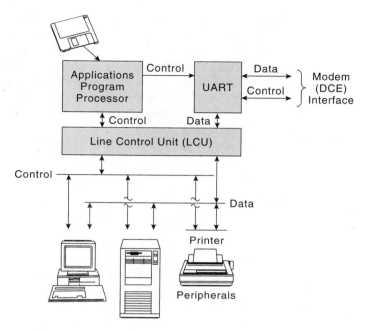

**FIGURE 1-2** Data Terminal Equipment (DTE)

long distances. One common medium is the existing telephone lines that normally carry voice calls. In order to be able to use the telephone system, the serial data needs to be converted to audio range signals. Several types of modems, discussed in detail in Chapter 4, perform this conversion using a number of methods dictated by the rate at which the data needs to be transmitted. One such method changes logic 1 (mark) and logic zero (space) levels to audio signals of two different frequencies. Thus when a logic 1 is sent, the mark signal or tone is sent and a logic 0 generates the space tone. The receiving modem at the secondary station converts these tones back to binary logic states, which are, in turn, sent to the UART or USART for conversion to parallel data used by the receiver's computer system.

The medium between the primary and secondary stations can be as simple as a coaxial or twisted pair cable as used by local area communications networks. Telephone lines are another example of the use of twisted pairs of wires for local connections between phone users and phone switching stations. The telephone company, as well as other carriers, have long adopted the use of radio transmission at many levels to complete communication links. These include microwave, cellular, and satellite transmissions.

## Hardware Interfaces

Interconnecting data terminal equipment to data communication equipment so they will work harmoniously is complicated because different manufacturers produced varied types of devices. A need for a standard interface between the DTE and DCE units is crucial. One example of a commonly used interface standard, the RS232C, was written by communications engineers for the *Electronics Industries Association (EIA),* which is one of many organizations responsible for establishing standards for a variety of electronics and electrical applications. *RS* stands for recommended standard, which means there is no enforcing authority to assure the proper use or complete compliance with the specifications of the standard. Enforcement of its use falls to the consumer, who benefits by not purchasing those systems that do not include the standard and buying those that do. Encompassed in the standard are functional, electrical, and physical specifications for users wishing to connect DTE to DCE equipment. This standard and others are examined in detail in Chapter 4.

### Section 1.2  Review Questions

1. Which DTE block interfaces the communication function with peripheral devices?
2. What is the term that denotes a fixed group of binary bits?
3. Which system block converts parallel data to serial data form?
4. What is the overall function of an RS232C interface?
5. What functional unit makes up the DCE?

## 1.3  DATA COMMUNICATION LINKS

Data communications links are configured to satisfy particular requirements for a given system. The simplest link is the one used in the previous discussion. It contains a single primary and a single secondary station connected to node end points of the link. The primary station initiates (or originates) the communication link and maintains control over that link until all data transfers are completed. The answering station is the secondary. Both stations must use the same protocol, data rates, and data codes for data to be correctly sent and received. The actual method of sending and receiving data is further divided into three types—simplex, half duplex, and full duplex illustrated in Figure 1-3.

A system or a particular data transmission can be configured to send and receive data in one direction only (from primary to secondary, for example as shown in Figure 1-3a). This transmission is referred to as a **simplex** transmission. This type of data transfer link can be performed by the system of Figure 1-1, if, for example, the primary would always transmit data to the secondary and the secondary would not be required to respond or send anything in return. Simplex transmissions are useful in environments where large quantities of data are sent without acknowledgement of the reception. The data can be sent fast and continuously. An example application of a simplex system would be an inventory dump from a warehouse to a company's central offices. Once the inventory was taken and entered into a computer, the files containing the inventory data can be sent to

**simplex**
transmission of data in one direction only.

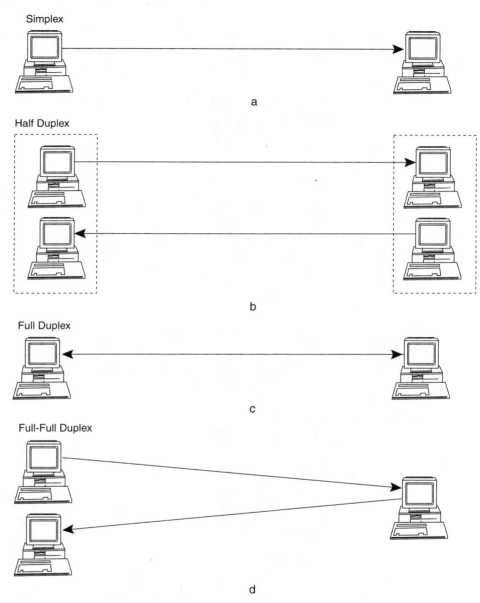

**FIGURE 1-3** Communicaton Directions

the accounting department. This data can be transferred or "dumped" late at night when the system would be less occupied. Since no one is working in the accounting department at the time, there would be no immediate acknowledgement of receipt of the inventory data. Ideal use for a simplex transmission. Of course, if an error should occur, there would be no way for the secondary to inform the primary that it happened.

**half duplex**
ability to send and receive data but not at the same time.

**full duplex**
ability to transmit and receive data simultaneously.

There are two basic methods of bidirectional data transfer, **half duplex** and **full duplex.** Half duplex allows transmission of data in both directions between primary and secondary stations, but restricts these transfers to one direction at a time as shown in Figure 1-3b. The primary might begin by sending a message from its transmitting circuits to the secondary's receiving units, which receives and stores the data. After the message is completely sent, the secondary can reply with a message of its own from its transmitter to the primary's receiver.

Full duplex differs from half duplex in that both stations can transmit to each other and receive each other's messages simultaneously as illustrated in Figure 1-3c. One way to achieve this is for the primary and secondary to transmit binary data using two different sets of audio tones for logic 1 and 0. For instance, the primary might send data using 1250 Hz and 1750 Hz signals for mark and space while the secondary would use 2050 Hz and 2750 Hz. The receive portions of both stations' modems would be tuned to the appropriate set of tones in order to differentiate between those it sent and those it received.

**full-full duplex**
ability to transmit data to one station and receive data from a different station simultaneously.

A fourth, but lesser-used form of communication is known as **full-full duplex.** Like full duplex, data can be sent and received simultaneously. The difference is that in a full-full duplex system, data is sent to one location and received from a different location. See Figure 1-3d as an illustration of the process.

---

### Section 1.3 Practice Questions

1. Which type of transmission sends data in one direction only?
2. Which type of transmission sends and receives data simultaneously?

---

## 1.4 CHARACTER CODES

Basic to every system deployed and used today is the transfer of information via a digital code. Text characters—alphabetic, numeric, punctuation, formatting, attribute, etc.—use selected combinations of 1s and 0s in a fixed binary word size, which is tabulated into a set known as a **character code.** Besides data communications applications, binary character codes are used wherever a digital processor or computer is used to interpret letters, numbers, and directed functions. A common example most students are familiar with is the personal computer keyboard and display screen. As the user types on the keyboard, a processor within the keyboard converts the key press into a binary code and signals the computer. When text is to be sent to the screen, the text is converted from the characters we read and are familiar with to the same binary code used by the keyboard.

Coding became more sophisticated by replacing the series of dots and dashes used in the Morse code with combinations of binary data bits formulated into a character code. One of the earliest of these character codes is a 5-bit word code known as the *Baudot code,* named for the same engineer who is responsible for developing baud rate as a measure for digital data transfers. This code was used in the European Telex communications system in the late 1940s through the early 1970s. It is presented here for historical perspective, but since it is no longer in use, we shall bypass details about it.

## ASCII

A parallel communication system, known as the *teletype* system performed similar functions in North America. The character code used by this system is the more familiar ASCII code. *ASCII* is an acronym for *American standard code for information interchange.* The original ASCII code used a 7-bit binary word to represent the printable characters, formatting characters, and data-linking, or control code, characters used by data communication systems today. Somewhere in the 1980s, the code was expanded to an 8-bit word size. Known as **extended ASCII,** this code also includes the Greek symbols used in mathematics, special alphabet characters used by different languages, such as the umlaut on German vowels, and other speciality characters (smiling faces, rudimentary graphics, etc.). The complete extended ASCII chart is presented at the back of the book in Appendix B. Its size makes it inappropriate to reproduce in this chapter. Instead, Table 1-1 lists the original ASCII set to allow us to explore some of its components.

**extended ASCII**
8-bit character code representing alphanumeric graphic, and link control characters.

The table shows the characters (ASCII) and hexadecimal (HEX) representation of the 7-bit binary codes for each ASCII character. The table in Appendix B also includes the decimal equivalent of each ASCII hexadecimal code. By looking on the table, you can find the **alphanumeric** characters you would expect—alphabetic, numeric, and punctuation characters. The remaining characters are classed into two groups—**formatting** and **data-linking** or **control characters.** Formatting characters are responsible for how text appears on the page and includes line feed (LNFD), carriage return (CRET), horizontal and vertical tabs (HTAB, VTAB), form feed (FMFD), etc.

**formatting characters**
characters responsible for the appearance of text, like line feed, carriage return, etc.

**data-linking** or **control characters**
characters used to establish and maintain a communications link between two stations.

Data-linking or control characters are those used by protocols to establish and maintain a data link. They include characters for indicating the beginning of a transmission (STX—start of text), ending a transmission (EOT—end of transmission), acknowledge (ACK), delimiter (DLE), device control (DC1–DC4), etc.

A further extension to the ASCII code set has been formulated under the name of **UNICODE.** Besides the basic extended ASCII set, this 16-bit character code embraces the use of foreign letters and special symbols such as the German umlaut. Since the code includes a total of 65,536 possible characters, a table for it is *not* included with this text.

**UNICODE**
16-bit extension of the ASCII character code.

## EBCDIC

The chief character code competition for ASCII came from Big Blue—IBM (International Business Machines, which has its own character code used, primarily, for its mainframe computers. This code, known as EBCDIC (pronounced EB—CE—DIC) for **extended binary coded decimal interchange code,** is listed in Table 1-2. EBCDIC is an 8-bit binary code that represents the same set of characters that the original 7-bit ASCII code did, but in a different sequence. This text will use the ASCII code throughout in examples, discussion, questions, and problems.

**extended binary coded decimal interchange code**
8-bit IBM character code.

---

**TABLE 1-1**

## ASCII Character Code

*Binary codes are shown in their hexadecimal (HEX) equivalent*

| HEX | ASCII | HEX | ASCII | HEX | ASCII | HEC | ASCII |
|-----|-------|-----|-------|-----|-------|-----|-------|
| 00 | NULL | 20 | Space | 40 | @ | 60 | ' |
| 01 | SOH | 21 | ! | 41 | A | 61 | a |
| 02 | STX | 22 | " | 42 | B | 62 | b |
| 03 | ETX | 23 | # | 43 | C | 63 | c |
| 04 | EOT | 24 | $ | 44 | D | 64 | d |
| 05 | ENQ | 25 | % | 45 | E | 65 | e |
| 06 | ACK | 26 | & | 46 | F | 66 | f |
| 07 | BELL | 27 | ' | 47 | G | 67 | g |
| 08 | BKSP | 28 | ( | 48 | H | 68 | h |
| 09 | HTAB | 29 | ) | 49 | I | 69 | i |
| 0A | LNFD | 2A | * | 4A | J | 6A | j |
| 0B | VTAB | 2B | + | 4B | K | 6B | k |
| 0C | FMFD | 2C | ' | 4C | L | 6C | l |
| 0D | CRET | 2D | – | 4D | M | 6D | m |
| 0E | SHOUT | 2E | . | 4E | N | 6E | n |
| 0F | SHIN | 2F | / | 4F | O | 6F | o |
| 10 | DLE | 30 | 0 | 50 | P | 70 | p |
| 11 | DC1 | 31 | 1 | 51 | Q | 71 | q |
| 12 | DC2 | 32 | 2 | 52 | R | 72 | r |
| 13 | DC3 | 33 | 3 | 53 | S | 73 | s |
| 14 | DC4 | 34 | 4 | 54 | T | 74 | t |
| 15 | NACK | 35 | 5 | 55 | U | 75 | u |
| 16 | SYNC | 36 | 6 | 56 | V | 76 | v |
| 17 | ETB | 37 | 7 | 57 | W | 77 | w |
| 18 | CAN | 38 | 8 | 58 | X | 78 | x |
| 19 | ENDM | 39 | 9 | 59 | Y | 79 | y |
| 1A | SUB | 3A | : | 5A | Z | 7A | z |
| 1B | ESC | 3B | ; | 5B | [ | 7B | { |
| 1C | FLSP | 3C | < | 5C | \ | 7C | : |
| 1D | GPSP | 3D | = | 5D | ] | 7D | } |
| 1E | RDSP | 3E | > | 5E | ^ | 7E | ~ |
| 1F | UNSP | 3F | ? | 5F | — | 7F | DEL |

## TABLE 1-2

## EBCDIC Character Code

*Binary codes are shown as their hexadecimal (HEX) equivalents.*

| HEX | EBCDIC | HEX | EBCDIC | HEX | EBCDIC | HEX | EBCDIC | HEX | EBCDIC |
|-----|--------|-----|--------|-----|--------|-----|--------|-----|--------|
| 00 | NULL | 20 | DIGSEL | 50 | & | 91 | j | D0 | } |
| 01 | SOH | 21 | STSIG | 5A | ! | 92 | k | D1 | J |
| 02 | STX | 22 | FLSEP | 5B | $ | 93 | l | D2 | K |
| 03 | ETX | 24 | BYPASS | 5C | * | 94 | m | D3 | L |
| 04 | PNOFF | 25 | LNFD | 5D | ) | 95 | n | D4 | M |
| 05 | HZTAB | 26 | ENDBLK | 5E | ; | 96 | o | D5 | N |
| 06 | LWRCASE | 27 | ESC | 5F |  | 97 | p | D6 | O |
| 07 | DELETE | 2A | STMESS | 60 | – | 98 | q | D7 | P |
| 09 | RLF | 2D | ENQR | 61 | / | 99 | r | D8 | Q |
| 0A | Repeat | 2E | ACK |  |  | A1 | ~ | D9 | R |
| 0B | VERTAB | 2F | BELL |  |  | A2 | s | E2 | S |
| 0C | FMFD | 32 | SYNC | 6B | , | A3 | t | E3 | T |
| 0D | CARET | 34 | PNON | 6C | % | A4 | u | E4 | U |
| 0E | SHFTOUT | 35 | RCDSEP | 6D | – | A5 | v | E5 | V |
| 0F | SHFTIN | 36 | UPCASE | 6E | > | A6 | w | E6 | W |
| 10 | DLE | 37 | EOT | 6F | ? | A7 | x | E7 | X |
| 11 | DC1 | 3C | DC4 | 7A | : | A8 | y | E8 | Y |
| 12 | DC2 | 3D | NACK | 7B | # | A9 | z | E9 | Z |
| 13 | DC3 | 3F | SUB | 7C | @ | C0 | { | F0 | 0 |
| 14 | Restore | 40 | Space | 7D | ' | C1 | A | F1 | 1 |
| 15 | NewLine | 4A |  | 7E | = | C2 | B | F2 | 2 |
| 16 | BckSp | 4B | . | 7F | " | C3 | C | F3 | 3 |
| 17 | Idle | 4C | < | 81 | a | C4 | D | F4 | 4 |
| 18 | Cancel | 4D | ( | 82 | b | C5 | E | F5 | 5 |
| 19 | EndMed | 4E | + | 83 | c | C6 | F | F6 | 6 |
| 1A | UnitBkSp | 4F | | | 84 | d | C7 | G | F7 | 7 |
| 1C | IntF1Sep |  |  | 85 | e | C8 | H | F8 | 8 |
| 1D | IntGpSep |  |  | 86 | f | C9 | I | F9 | 9 |
| 1E | IntRcSep |  |  | 87 | g |  |  |  |  |
| 1F | IntUnSep |  |  | 88 | h |  |  |  |  |
|  |  |  |  | 89 | i |  |  |  |  |

*Note:* Unused codes have been omitted from this chart.

---

**EXAMPLE 1-2**

Compare the ASCII and EBCDIC codes for the capital letter E.

**SOLUTION**

From Table 1-1, we find the ASCII code for the letter E is a hexadecimal 45, which is 1000101 as a 7-bit binary code. The EBCDIC code for E, from Table 1-2, is C5, which is 11000101 as an 8-bit binary code.

*NOTE*

The extended ASCII code for E has the same hexadecimal value as the standard ASCII code, but the binary equivalent contains an additional 8th bit: 01000101.

---

**Section 1.4 Review Questions**

1. What are the three types of characters represented by the ASCII character code?
2. What are the word sizes for ASCII and extend ASCII codes?
3. What is the extended ASCII code for OK?

---

# 1.5 DIGITAL DATA RATES

Digital information in serial form is transferred at a distinct data rate. It takes time to send information, one bit at a time, from one place to another. Data are sent serially to reduce the number of transmitting lines or paths to a single pair (electrically active and a return line). As mentioned earlier in the chapter, the rate at which digital information is sent or received is called the bit rate, whose unit is bits per second (bps).

When other methods are used to send data using groups of bits in each transmission, a different measure of data rate is required. This measure is known as symbols per second (sps), where a symbol represents the bit group. Symbols can be either in binary or any other format. Sometimes they are an analog signal that uses different frequencies or phases to represent groups of data bits. What is most significant for the field of data communications is that a symbol can be created through various processes from a group of binary bits. As an example, the ASCII code, instead of being represented as a 7-bit code, could be set up to use a different voltage level for each character. In other words, 1.2113 volts could indicate the letter C and 1.2114 volts a letter D. We are not concerned with the practicality of such a system, but since there is a unique 7-bit binary code for each of the characters in the ASCII set, there could also be a unique voltage to represent each character. There would then be a correlation between the ASCII binary codes and those voltage levels as well. That is, 1000011, the binary ASCII code for the letter C is equivalent to 1.2113 volts and 1000100 for a D is equivalent to a 1.2114 volt level. The bottom line is that 7 bits can be used to generate a single symbol (in this case a voltage level) for each character. The single symbol can be sent as one entity, which is faster than sending the equivalent 7 individual bits.

Given the same rate of transmission, 1,000 bits per second and 1,000 symbols per second, we can compare the time it would take to send a single ASCII character using the voltage example in the preceding paragraph. The time of transmission in both cases is the reciprocal of the rate, or 1 millisecond (ms) per bit or per symbol. Sending the data in digital form requires sending seven consecutive bits for a single character, which means it takes a total of 7 ms to send the letter D in binary. When the seven bits are used to form a single voltage symbol, then only that one symbol is sent, taking only 1 ms time.

---

### EXAMPLE 1-3

Compare the rate of transmitting the letter C as both a 7-bit binary stream transmitted at 4900 bps and as an equivalent 7-bit group symbol.

### SOLUTION

Sent as serial binary data, the letter C is transmitted at 4900 bps. As a symbol, the rate of transmission is one seventh of that data rate, or 700 sps, to achieve the same transfer rate of 4900 bps. Conversely, if the 4900 bps rate is maintained to send the symbols at 4900 sps, seven times as much data can be sent in the same time it takes to send data at 4900 bps.

---

### Section 1.5  Review Questions

1. What is the symbol rate for a 2400 bps second system that generates one symbol for every 3 bits?
2. What is the advantage of using symbols rather than bits to send data?

---

## 1.6  SERIAL DATA FORMATS

Whether data are sent as bits or symbols, it is transmitted serially in one of two forms, **synchronous** or **asynchronous.**

**synchronous**
serial data that requires a synchronizing clock signal between sender and receiver.

**asynchronous**
serial data that does not require a synchronizing clock or signal between sender and receiver.

### Synchronous Data

Synchronous data require a coherent clocking signal between transmitter and receiver, called a *data clock,* to synchronize the interpretation of the data sent and received. The data clock is extracted from the serial data stream at the receiver by special circuits called *clock recovery circuits.*

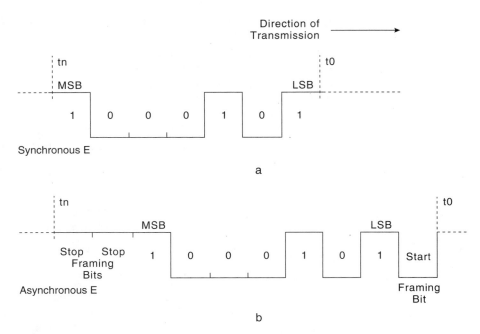

**FIGURE 1-4** Comparing Data Formats for ASCII E

Once the clock is recovered at the receiving end, bit and character synchronization can be established. *Bit synchronization* requires that the high and low condition of the binary data sent matches that received and is not in an inverted state. *Character synchronization* implies that the beginning and end of a character word is established so that these characters can be decoded and defined. Overall synchronization is maintained by the clock recovered from the message stream itself.

Figure 1-4a shows how a synchronous binary transmission would send the ASCII character E (hex 45 or 1000101). The least significant bit (LSB) is transmitted first, followed by the remaining bits of the character. There are no additional bits added to the transmission.

## Asynchronous Data

**framing bits**
bits that denote the beginning
and end of a character.

Asynchronous data formats incorporate the use of **framing bits** to establish the beginning (start bit) and ending (stop bit) of a data character word as shown in Figure 1-4b. A clocking signal is not recovered from the data stream, although the internal clocks of the transmitter and receiver must be the same frequency for data to be correctly received. To understand the format of an asynchronous character, it is first necessary to be aware of the state of the transmission line when it is idle and no data is being sent. The idle condition results from the transmission line being held at a logic 1, high state, or mark condition. The receiver responds to a change in the state of the line as an indication that data has been sent to it. This change of state is indicated by the line going low or logic 0, caused by the transmission of a start bit at the beginning of the character transmission as shown in Figure 1-4b. Data bits representing the code of the character being sent follow next ending with one or two stop bits. The stop bits actually specify the minimum time the line must return to a logic 1 condition before the receiver can detect the next start bit of the next character.

## Transmission Efficiency

Notice that the synchronous data uses just the seven bits required for the E character's code while the asynchronous stream needs 10 bits (one start, seven data, and two stop bits). The synchronous stream is more efficient than the asynchronous because it does not require the overhead (framing) bits that the asynchronous stream needs. Efficiency is a mark of performance calculated as a ratio of data or information bits to total bits as shown in Equation 1-1.

$$\text{efficiency} = \frac{\text{data bits}}{\text{total bits}} \times 100\% \qquad (1\text{-}1)$$

---

**EXAMPLE 1-4**

Compare the efficiency of sending the synchronous and asynchronous E character.

**SOLUTION**

Since all of the synchronous bits are data bits, the efficiency of sending the synchronous E is 7/7 x 100% = 100%.

The asynchronous, on the other hand, has a much less efficiency: $7/10 \times 100\%$ = 70%.

---

A more efficient stream of data takes less time to be transmitted simply because there are less bits to be sent. However, the overall efficiency of a transmission relies on more than the efficiency of individual characters within a message. For asynchronous data, the entire message will retain a 70% efficiency because no additional bits or overhead are required to send the data. Bit and character synchronization are built into the framing bits. Synchronous data, on the other hand, requires a **preamble** message, which is a set pattern of binary ones and zeros used to facilitate clock recovery, so the data to be bit and character synchronized before data can be correctly received. This adds additional bits to be sent and reduces the overall efficiency of the transmission. Despite this added burden, synchronous transmissions remain more efficient than asynchronous ones.

**preamble**
used by synchronous data streams to establish character synchronization.

---

**EXAMPLE 1-5**

Compare the efficiency of sending a 256-byte message synchronously and asynchronously. The synchronous transmission requires an 8-byte preamble for clock recovery.

**SOLUTION**

Using the same asynchronous configuration as in Example 1-6, the efficiency of transmitting the 256 bytes asynchronously remains 70%.

The total number of bytes in the synchronous message is 256 data plus 8 preamble, or 264 bytes. Thus the efficiency of the synchronous messages becomes:

$$\frac{256 \text{ data}}{264 \text{ total}} \times 100\% = 97\%$$

The synchronous transmission is still more efficient than the asynchronous one despite the required additional overhead of the 8-byte preamble.

---

**Section 1.6  Review Questions**

1. What is used by asynchronous data streams to establish bit and character synchronization?
2. Which serial data type is more efficient?
3. What is used by synchronous data streams to establish character synchronization?

---

# 1.7  ENCODED DATA FORMATS

Besides different character codes and data types (synchronous and asynchronous), digital data can be transmitted or coded into different electrical signal formats. Each of the following forms has its own advantage and/or use. They are illustrated in Figure 1-5. Each data signal format, sent as a serial stream of data may be interpreted as generating a square wave whose frequency varies according to the changing bit pattern. Depending on the signal format type, the frequency of the "square wave" usually gets lower as the number of consecutive ones or zeros increases. The highest frequency condition generally occurs with alternating ones and zeros. This is referred to as the worse case condition, because it produces the highest fundamental frequency for the square wave and determines the bandwidth required of the system that sends the data using a particular format.

## Raw Data Form

**unipolar non-return-to-zero (UNZ)**
basic raw digital data.

For instance, the first data form is technically known as **unipolar non-return-to-zero (UNZ).** It is your basic raw serial binary ones and zeros. Unipolar indicates that the logic 1 level is always one polarity (as shown in Figure 1-5a, this polarity is +V). Non-return-to-zero indicates that the logic 1 does not return to 0V midway through the bit time as some others will. By observing the waveform created by the data, it appears as a square wave with a varying time period (and, hence, frequency). Notice that the shortest time period occurs when the data is made up of alternating ones and zeros (on the left side of the

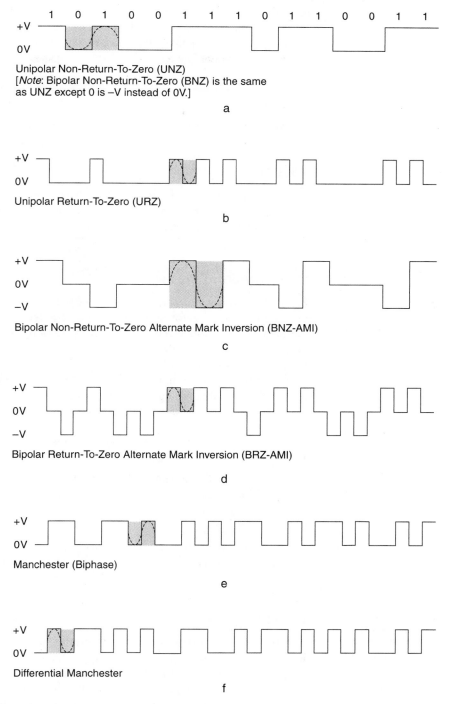

Unipolar Non-Return-To-Zero (UNZ)
[*Note*: Bipolar Non-Return-To-Zero (BNZ) is the same
as UNZ except 0 is −V instead of 0V.]

a

Unipolar Return-To-Zero (URZ)

b

Bipolar Non-Return-To-Zero Alternate Mark Inversion (BNZ-AMI)

c

Bipolar Return-To-Zero Alternate Mark Inversion (BRZ-AMI)

d

Manchester (Biphase)

e

Differential Manchester

f

**FIGURE 1-5** Binary Data Forms

data stream). The fundamental frequency of the "square wave" is shown as a dotted sine wave. Since it takes two bits of data to form the "square wave," you can see that the frequency of the fundamental sine wave of the worst case (alternating ones and zeros) is one-half the bit rate. The remaining formats are shown in relation to this UNZ form.

---

**EXAMPLE 1-6**

What is the fundamental frequency of the sine wave of the worst case state of a UNZ serial data stream sent at 1,000 bits per second (bps)?

**SOLUTION**

The time period for one bit at 1,000 bps is 1 ms. Two bits use a total of 2 ms. The worse case square wave occupies two-bit periods as shown in Figure 1-5a. The frequency of the fundamental sine wave of that square wave is 1/(2 ms) or 500 Hz.

---

## Return-to-Zero Form

**unipolar return-to-zero (URZ)**
binary code format.

Figure 1-5b shows the same data stream coded as a **unipolar return-to-zero (URZ)** format. With return-to-zero forms, the logic one is returned to a 0V level midway through the data period. Notice that worst case condition now occurs with consecutive ones (about midway through the data stream in the figure). Notice also that the fundamental frequency is twice that of the UNZ format. The advantage, for synchronous data streams of the return-to-zero form, is that there are transitions or changes of state for consecutive logic 1 data bits that are not there in non-return-to-zero formats. These transitions aid in clock recovery for synchronous receivers.

---

**EXAMPLE 1-7**

Compare the fundamental frequencies for the worst case conditions of the UNZ and URZ formats.

**SOLUTION**

As illustrated by the dotted sine waves in Figure 1-5, the fundamental sine wave for the URZ is twice the frequency as that for the UNZ format.

---

**bipolar non-return-to-zero alternate mark inversion (BNZ-AMI)**
binary code using alternating voltage levels for logic 1 data.

**polarity violation**
data error control for alternate mark inversion data streams when two consecutive data bits appear with the same polarity.

## Bipolar Forms

The **bipolar non-return-to-zero alternate mark inversion (BNZ-AMI)** shown in Figure 1-5c is encoded with the polarity of the logic 1s alternating between +V and −V. This format adds a built-in automatic error checking capability into the transmission of data. As long as the logic 1 levels continue to alternate polarity, it is assumed that the data is good. If two consecutive logic 1 bits appear at the same polarity, then a **polarity violation** has occurred and the data is believed to be corrupted.

Figure 1-5d takes the bipolar convention one step further by returning to zero at the center of every logic 1 bit period. This form is called *bipolar return-to-zero alternate mark inversion (BRZ-AMI)*. The addition of the capability of returning to zero adds more transitions between +V, −V, and 0V levels and aids in recovering clock signals for synchronous data transmissions.

## Manchester Encoded Forms

The last two coded formats are special cases specified in networking protocols. They are known as Manchester and differential Manchester encoding. They both add a built-in clock capability at the center of each data bit period. Here's how they are formulated. For Manchester encoding, a data bit is split into two parts. The first half of the data bit period is the inverse of the data level—a one appears as a zero level and a zero appears as a one level. At the midway point the level is inverted. This midway transition occurs at the middle of each data bit time and becomes the recovered clock signal at the receiver. Manchester encoding has the advantage of facilitating easy clock recovery. One problem with Manchester encoding is that the interpretation of the data is dependent on the level of the first-half of the data period.

Differential Manchester moves the detection of the data level to the leading edge of a data bit time. It accomplishes this by using a comparison of the data levels at the beginning of a data bit to determine if that bit is a one or a zero. If there is a change of state at the beginning of the data bit period, then the data bit is a zero. If there is not a change of state, then the data bit is a one.

Compare the Manchester formats in Figure 1-5. Use the UNZ of Figure 1-5a as the reference data stream for each of the other coded formats. Notice for Manchester how each first-half of the data period is the inverse of the data level. Check the differential Manchester at the beginning of each data bit. Notice that a logic zero causes the differential data to change state while a logic 1 in the UNZ form does not cause a change of state in the differential code. Both forms provide a clock transition at the midpoint of each data bit.

---

### EXAMPLE 1-8

Show the Manchester and differential Manchester codes for the bitstream of the extended ASCII character message for OK.

**SOLUTION**

The ASCII code for OK (hex 4F4B), starting with the LSB of O on the left is:

1111001011010010

Manchester inverts the first half and changes state midway through the bit time (see Figure 1-6).

Differential Manchester causes the state to change at the beginning of bit if it's a zero and not change if it's a one and then changes state midway through the bit time (see Figure 1-7).

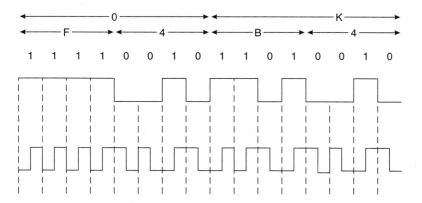

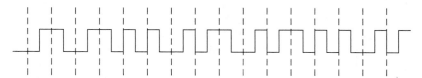

**FIGURE 1-6** Manchester Encoding OK

**FIGURE 1-7** Differential Manchester Encoding OK

**Section 1.7 Review Questions**

1. How does return-to-zero and non-return-to-zero binary data forms differ?
2. How can bipolar alternate mark inverted data be used to detect corrupted data?
3. What is the purpose for the logic transition in the middle of every Manchester encoded bit?

# SUMMARY

This chapter began by following the history of the development of data communications from Morse code to the Information Highway. Along the way we saw the development of networking and its influence on today's communication means. After the historical perspective, we examined a basic communication model to get a feel for a two-point communications link. Basic terminology such as DTE, UART, modem, medium, duplex, etc., was defined in relationship to this model.

Binary data comes in many formats and since the bulk of data communications involves serial data transfers, we explored the difference between asynchronous and synchronous data as well as several forms of digital encoding used by networks.

Much of this preliminary information will reappear in subsequent chapters that deal with many aspects of data and network communications.

# QUESTIONS

### Section 1.1

1. List five significant events that affected the development of network communications.
2. Explain how Morse code can be considered a binary code.
3. List four companies that are at the forefront of network communications.
4. Define in your own terms what the information highway means to you.

### Section 1.2

5. Which of the following units are part of the DTE and which are part of the DCE?
   a) UART   b) modem   c) line control unit   d) computer
6. Give the basic purpose of each of the following units
   a) line control unit   b) UART   c) RS232C   d) modem
7. Explain the difference between a primary and secondary station.
8. What is a medium? Give three examples of a medium.

### Section 1.3

9. What type of system transfers:
   a) data in one direction only?
   b) both directions simultaneously?
   c) both directions one at a time?
   d) two directions simultaneously but from different stations?

### Section 1.4

10. Define character codes.
11. What types of characters are represented in a character code? Define each character type.
12. Give an example of each type of character represented by a character code.
13. What are the ASCII and EBCDIC hexadecimal equivalent of the binary codes for each of the characters below:
    a) B        c) !         e) STX     g) carriage return   i) SYNC
    b) f        d) line feed   f) N)      h) EOT

### Section 1.5

14. A system converts EBCDIC characters into symbols and transmits the symbols at 8848 sps. Each symbol represents a single EBCDIC character. What is the bit rate of the stream?

15. Compare the time it takes to transfer 256 bytes of data using an 8-bit parallel system versus a serial system. Both transfer data at a rate of 1200 transfers per second.

**Section 1.6**

16. Describe the parts of an asynchronous character data stream.
17. How does synchronous data differ from asynchronous data? Give a benefit for each data type.
18. Show the binary data stream for the asynchronous ASCII message Do it! using 1 start and 1 stop per character for framing. Place least significant bits of each character on the left and begin the message with the first character (D).
19. Show the binary synchronous EBCDIC data stream for the message This time? using two SYNC characters as a preamble. Place least significant bits of each character on the left and begin with the first character (T).
20. What is the efficiency of a 256-ASCII character message using asynchronous data with two (2) stop bits?
21. What is the efficiency of a 256-EBCDIC character message using synchronous data with two SYNC characters for a preamble?

**Section 1.7**

22. What is the difference between UNZ and URZ data formats?
23. What function can be done directly using bipolar alternate mark inversion that can not be done with unipolar data formats?
24. What is the fundamental sine wave frequency for a data transmission using UNZ data format? Compare this to the fundamental sine wave of a BRZ-AMI format.
25. Show the binary format for the ASCII synchronous message GO with no preamble, using Manchester and differential Manchester formats.

# DESIGN PROBLEMS

1. Another type of character code is the *automatic request for retransmission (ARQ),* which uses 7 bits per character. It is unique in that each code contains exactly three (3) logic 1 bits and four (4) logic 0 bits. The code is used to aid in error detection. Design such a code that represents uppercase letters, numbers, and basic punctuation (?,.!'"), and space characters. How does this code actual assist in detecting transmission errors?

2. It is desirable to transmit messages at a rate of 9600 bps onto a system whose bandwidth is 50 to 2700 Hz. Select the parameters for such a system. They should include the type of binary data format; synchronous or asynchronous transmission; number of bits per symbol; simplex, half duplex, or full duplex; character code used; and any other information you feel is pertinent. Justify each of your choices.

# ANSWERS TO REVIEW QUESTIONS

### Section 1.1
1. Morse Code
2. Logic 1 and 0
3. Baud rate
4. convert digital data to analog data
5. Ethernet

### Section 1.2
1. Line control unit (LCU)
2. word
3. UART or USART
4. interconnect DTE and DCE
5. modem

### Section 1.3
1. simplex
2. full duplex

### Section 1.4
1. alphanumeric, formatting, data link
2. ASCII: 7-bits
   Extended ASCII: 8-bit
3. 4F48

### Section 1.5
1. 800 sps
2. faster transfer rate

### Section 1.6
1. framing or start bit
2. synchronous
3. preamble

### Section 1.7
1. Return to zero data causes a logic 1 data bit to be returned to a zero state at the midpoint. Non-return-to-zero data does not.
2. Alternate mark inversion uses polarity violation to detect the possibility of corrupted data.
3. The transition in the center of Manchester encoded data is used as a clock signal.

# CHAPTER 2

# The Telephone System

## OBJECTIVES

After reading this chapter, the student will be able to:
- define the different parts of the telephone system.
- understand the specifications of the telephone system.
- recognize and understand telephone system impairments that affect data communications systems.
- determine how and when to use line conditioning to assure more dependable data communications.

## OUTLINE

1. Introduction
2. The Telephone System
3. Specifications and Parameters
4. Line Impairments

## 2.1 INTRODUCTION

One of the first considerations for designers of network systems was what to use as the medium to transfer digital data from source to destination. Much of their choices hinged on the size of the network being designed. Once the decision was made as to the medium to be used, other design elements evolved based on the electrical and functional characteristics of that medium. Additionally, the medium and the system selected had to be cost-effective. For early long distance communication, the telephone system became the choice simply because it was there. Early network designers decided to accept the parameters and limitations of the telephone system and design their networks around them. The choice was to either design circuits and equipment to work on the existing telephone system or develop a completely new interconnecting network to handle the transfer of digital information. Time and money dictated using the existing telephone network, although the decision was not without problems—the telephone company was pretty autonomous in those days. It could dictate requirements and costs for the use of its lines and equipment. On the other hand, those in control of the telephone company wisely detected a source of revenue that would only grow and elected to work with the emerging data communications field.

---

**Section 2.1 Review Question**

1. Give reasons why the telephone system was selected to carry digital information in early networks.

---

## 2.2  THE TELEPHONE SYSTEM

**TELCO**
telephone company.

**POTS**
plain old telephone system.

**RBOC**
regional bell operating company.

An early acronym for the telephone company was **TELCO,** which is still being used to a fair extent today. It is being gradually replaced by a newer one—**POTS** (Plain Old Telephone System). Either one is easier to use than spelling out telephone system or telephone company. Since the divesture of AT&T in 1984, another acronym has emerged to refer to some of the local telephone companies, **RBOC** (Regional Bell Operating Company). These companies quickly came to be known as the "Baby Bells." In this chapter, I will use TELCO, and later on switch to POTS, where it is more applicable. Figure 2-1 will be used as a reference for the following discussion on the basic telephone system.

The most common and familiar use of TELCO begins when a user, known as a subscriber, picks up the telephone handset and initiates a call by going off hook. In the telephone set, a multiganged switch activates with release of the handset from its cradle or hook. This switch completes a direct current path between the subscriber and the local switch station.

**local loop**
Telephone lines between subscriber and local switch station.

The current developed from a –48V battery is sent through this **local loop** and is sensed by the user as a dial tone, which the local switch sends when it senses the current flow. The user, upon hearing this dial tone, begins to dial the desired number of the party she wishes to communicate with. The term *dial* reflects the original way a telephone number was entered into the local switch by a subscriber. The user would literally dial the number using a rotary dial. The number on the dial is selected and the dial rotated to a stop. Upon releasing the dial, it would return to its original position at a given rate. A switch opens and closes for each position the dial passes on its return, generating a number of pulses equivalent to the number selected. The fixed rate of return of the dial assured a fixed pulse rate for these pulses. Figure 2-2 shows the general specifications and timing diagram for the rotary dial. The local switch station then interprets the pulses and makes the correct connection to the destination subscriber.

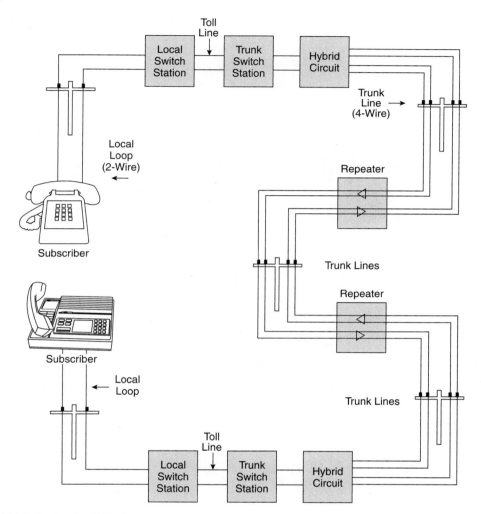

**FIGURE 2-1** Telephone System Elements

## Touch Tone Pad

**dual tone multiple frequency (DTMF)**
telephone touch tone pad for dialing numbers.

The dial has long since been replaced by the Touch Tone (registered trademark of AT&T) keypad, which is laid out as blocknumbered push buttons. Each button, when selected and pushed, sends a pair of tones to the local switching office, which interprets the tones as the number associated with the button pushed. The name for this method of sending and detecting phone numbers is **dual tone multiple frequency (DTMF).** The Touch Tone pad and the tonal frequencies associated with each row and column of the pad are shown in Figure 2-3. Pressing one of the buttons on the pad sends the two tones associated with that button's row and column.

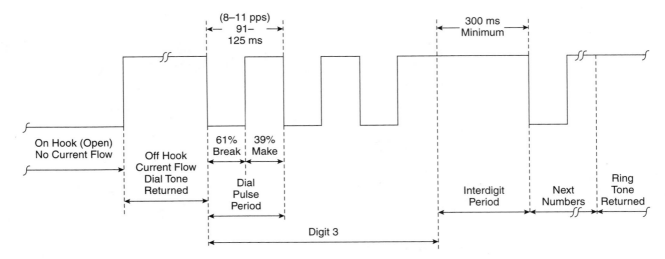

**FIGURE 2-2** Line Characteristic During Dialing

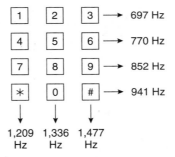

**FIGURE 2-3** Dual Tone Multiple Frequency (DTMF) Touch Tone Pad and Associated Frequencies

---

**EXAMPLE 2-1**

What is the series of tones that are sent when dialing the number 911?

**SOLUTION**

Pressing the button marked 9 sends an 852-Hz and a 1,477-Hz tone. Button 1 sends both 697-Hz and 1,209-Hz tones. The sequence for 911 then is 852/1,477-Hz followed by two signals at 697/1,209-Hz.

---

The specifications for DTMF are listed in Table 2-1 for both the sending subscriber and receiving local switch station. The power specifications include minimum single frequency power and maximum frequency pair power as well as the difference of power for

**TABLE 2-1**

**DTMF Specifications**

| Transmitter | Parameter | Receiver |
|---|---|---|
| −10 dBm | Minimum Power Level (Single Frequency) | −25 dBm −55 dBm Optional |
| +2 dBm | Maximum Power Level (Frequency Pair) | 0 dBm |
| 4 dB | Maximum Power Difference Between Frequencies in a Pair | +4 dB −8 dB |
| 50 ms | Minimum Duration of Digit | 40 ms 23 ms Optional |
| 45 ms | Minimum Interdigit Period | 40 ms 23 ms Optional |
| 3 s | Maximum Interdigit Period | 3 s |
| ±1.5% | Frequency Deviation Maximum | ±1.5% |
| N/A | Maximum Echo Level Below Primary Frequency Level | −10 dB |
| N/A | Maximum Echo Delay | 20 ms |

any two tones sent and received. The minimum duration of a digit specifies the minimum times that a tonal pair must remain on for the number associated with that pair to be recognized by the local switch station.

The interdigit times specify the minimum and maximum time between tones. A "fast-fingered" caller cannot push keypad buttons fast enough to beat the 40 ms minimum time. If, however, while dialing a number, a caller forgets a digit and has to check for it, he or she had best find that digit and press the button associated with it within 3 seconds of the last number pressed or redial the entire number.

To correctly interpret the numbers being sent, the frequency of the tones from the telephone set originating the call are required to be within ±1.5% of their rated value. Echoing tones back from the switch station need to be at a level considerably lower than those being sent by the subscriber. If their power is at too high a level, the echoing tones will be reflected back along with the new tones being sent or in the interdigit time between tones and could cause a misinterpretation of the new number or cause the local station to believe that an additional number was sent. These echoes should also not occur too long after the original tones were sent, again to avoid mistakes in interpreting phone numbers. DTMF specifies the level of the echo to be no more than −10 dB below the tones sent by the subscriber and to be delayed by no more than 20 ms.

Pop music has occasionally taken advantage of the tonal quality of the touch tone phone to spark their current release. Some tunes have also been written with the telephone as the central theme. A successful example of this application is ELO's *Telephone Line*. The record uses sounds of pulse phones and touch-tone signals, particularly in the instrumental interlude between vocals.

## Local Loop

If the call being initiated is to be connected to another subscriber within the same local area, then the local switching station routes the call to that destination or to a second local switch station, which in turn, completes the connection to the destination subscriber. These type of calls remain on the local loop and are connected using the red and green twisted pair of telephone wires. The type of cable used for these local loop connections is called *Unshielded Twisted Pair (UTP)*. As shown in Figure 2-4, they are unshielded, meaning that there is no braided wire or metal jacket surrounding each twisted pair. The wires are twisted around each other to reduce the effect of cross-coupling signals from other pairs of wires within a bundle of telephone wires. For a single subscriber line, there are four wires present: red, green, yellow, and black. Of the four, only the red and green are required for regular voice communications and connection.

The switch station senses whether the called party is in use (the handset is off hook) or is free. Upon detecting an off-hook condition at the called number, a busy tone, which is a mixture of 480-Hz and 620-Hz tones, is returned to the calling party. It is sent altering half a second on and half a second off. This tone is the local busy tone. When long distance lines are clogged, a different form of busy signal is sent to the caller. The effect is the same—your call cannot be completed.

If the line is free, a ring signal is sent to the called station. It consists of a 20-Hz sine wave signal at a voltage between 90 and 120 Vrms. This amount of voltage was necessary to operate the electromechanical ringing relay through its ringer coil in older telephone sets. The ringing signal is on for two seconds and off for four seconds. Today's ringing circuits are solid-state units that drive a small speaker. Even though they could have been designed to operate at lower voltages, they were not. Realizing that many older telephone sets are and will remain in use, the newer ringer circuits had to be designed to handle this ringing signal.

Ever have the experience of calling a party and having them come online before you ever hear the ringing signal in your handset? That is, you dial the party expecting to hear their ring, but instead you get a "Hello" from them. This occurs because the ring signal you hear is not the same ring signal sent to the called party. Instead, you hear a 480-Hz tone returned by the switching station that indicates it is sending the 20-Hz ring signal to the called party. These two signals are not synchronized with each other and often result in the phenomenon just described. Another effect of these two separate signals is that just be-

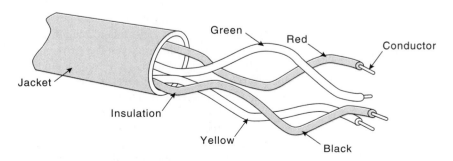

**FIGURE 2-4** Telephone UTP cable

cause you hear the 480-Hz ring signal does not guarantee that a phone is actually ringing at the called party's location. If the phone lines are down or the phone set is disconnected at the called party, they won't ring, but you will still hear the ring signal in your handset.

## Long Distance Lines

**trunk line**
long distance telephone lines.

**toll station**
long distance switch station.

**tandem switch**
switch station that interconnects several local switch stations and/or toll stations.

For long distance or message unit calls, the call is routed through **trunk lines** via **toll stations.** Toll stations are switching stations used to select which long distance trunk lines are to be used to route your call. These trunk lines are terminated in another toll station, which may connect the call to another trunk line or to a local switch station, depending on the call's destination. Longer calls may require longer trunk lines and additional intermediate switching stations. The hierarchy of the telephone company switching stations, shown in Figure 2-5, begins with the local switch station, which has direct connections to the local subscribers. Many calls are completed at the local switch station. Other calls require connections through higher levels of switching. Local switch stations are classified as class 5 stations. They are connected to each other and in clusters to a **tandem switch,** which is also classified as a class 5 station. The subscribers, tandem, and local switch stations are considered as the local loop.

Tandem stations, in turn, may connect the incoming call to other tandem stations, which pass the call on to a local switch station to be connected to the called subscriber. Tandem stations are also the beginning of the long distance connection. They are connected through class 4 toll stations, which route the calls through trunk lines to another toll station. In the event that the call is not within a specific primary area (message unit call), it is routed up through a class 3 primary center for the beginning of a toll call. Several of these primary centers are further interconnected through class 2 sectional centers and finally to class 1 regional centers.

In the United States and Canada, this hierarchy takes the form of a tree with twelve regional stations at the top, connecting to a number of sectional stations, which, in turn, connect to several primary stations, each of which connect to a number of toll stations that service a group of tandem stations, which finally connect to the over 25,000 total local stations at the root of the tree. It should be noted that the switching procedure is such that the minimum number of connections is made to complete a call. As shown in Figure 2-5, there are parallel connections as well as cross connections between levels to satisfy this requirement.

The TELCO term for interfacing calls through the switching station hierarchy has been termed BORSHT, the letters of which are initials for the following elements of the TELCO system:

*Battery,* signifying the –48V battery used to supply the direct current for off hook and dial detection.

*Overvoltage protection,* which is builtin protection against voltage surges and induced voltages from electrical storms.

*Ringing,* designating the 20–Hz ringing signal.

*Supervision control,* which is the response to on-/off-hook, etc.

*Hybrid circuit,* that interfaces two-wire and four-wire lines.

*Testing,* for purposes of checking the system.

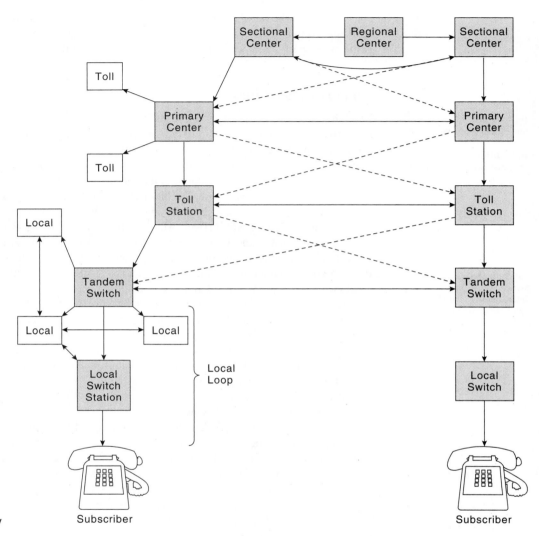

**FIGURE 2-5**
Switching Hierarchy

Toll calls are often routed through many switch stations to find a clear path from caller to called party. Busy times of the day may route a call between Phoenix and Tucson, Arizona, through Sante Fe, New Mexico, and El Paso, Texas. However, on one particular day each year the peak period is so jammed with calls that all routes are blocked. Happy Mother's Day!

**repeaters**
long distance line amplifier used to reamplify and reshape degraded signals.

Most upper-level switching stations employ **repeaters** in the trunk lines, as illustrated in Figure 2-1. This devices are basically regenerating amplifiers that boost and reshape weak or distorted signals that have traveled considerable distance on the trunk lines. The UTP wiring used for the local loop to handle full duplex conversations, tend to attenuate these calls over long distance.

## LATA Structure

**local access and transport area (LATAs)**
telephone system hierarchy.

Since the divesture of AT&T in 1984, the telephone system has been restructured into **local access and transport areas (LATAs).** There are 181 such areas in the continental United States, one in Alaska, Hawaii, and Puerto Rico. Of the 181 LATAs, 163 are local RBOCs and others are independents. These companies handle local calls while calls between LATAs are managed by inter-LATA carriers or long distance companies. The RBOCs supply local exchange traffic and intra-LATA toll calls. They must allow access by subscribers to inter-LATA or long distance carriers.

The size of a LATA, though somewhat keyed to geographic areas, is more dependent on the number of subscribers within the area. LATA hierarchal structure remains closely mirrored to the previous AT&T structure. The classes of switch stations are the same and are summarized here:

*Class 1:* Ten regional centers in the United States and two in Canada.
*Class 2:* Sectional centers.
*Class 3:* Primary centers.
*Class 4:* Toll centers.
*Class 5:* End or local offices.

Each LATA must contain at least one class 4 switch center to manage inter-LATA and intra-LATA connections. While connections to complete a call generally follow the hierarchy, cross connections between same and different classes (including skipping classes) are made when traffic volume or efficiency dictates it.

## Two-Wire and Four-Wire Interfaces

A circuit called a *hybrid circuit* is used to interface the two-wire UTP lines to a four-wire trunk line that transports the conversation in a half duplex mode. The hybrid circuit shown in Figure 2-6 receives the call signals on a pair of wires, sending them through the primary windings of transformers T1 and T2 and impedance matching components, R1, R2, and C. The calls are coupled to the transformer secondaries, amplified, and sent out on one pair of wires. The same signals appear at the output of a second amplifier, but being in phase, have no potential difference and are canceled.

Conversations returning are amplified through the second amplifier and fed, out of phase, to the junction of T1 and T2 and to one side of a two-wire connection. The signal traveling through the primaries of T1 and T2 are now out of phase and are coupled across to the secondaries and canceled at the input of amplifier 1. Hence, the signal returns along the two-wire path but is not echoed back along the four-wire line.

### Section 2.2  Review Questions

1. What components of the telephone system make up the local loop?
2. What are the two names for long distance switching stations?
3. What is the purpose of repeaters in long distance trunk lines?
4. What function does a hybrid circuit perform?

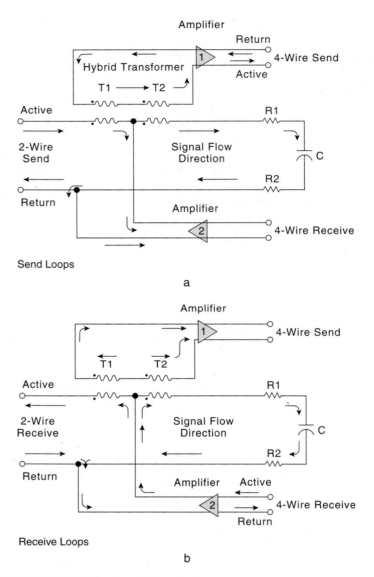

**FIGURE 2-6** Hybrid Two-Wire/Four-Wire Interface

## 2.3 SPECIFICATIONS AND PARAMETERS

Many of the specifications, parameters, and impairments that apply to the telephone system are critical when considering using this system for transporting digital information. One primary specification of major concern is the bandwidth limitation of the system.

## Bandwidth

Bandwidth limits the maximum frequency and switching speeds that can be applied to TELCO lines. Numerous studies were conducted on a wide variety of people to determine what the TELCO bandwidth should be. An example of these early studies was the one used to develop the **C-message** curve of Figure 2-7. People of varying ages, backgrounds, and sex were asked to listen to a set of tones and to place a mark on whether they: a) just heard the sound; b) heard it comfortably; or c) heard it loudly. The absence of a mark indicated that the tone was not heard at all. Each participant donned a set of headphones and were subjected to a number of tones at different frequencies. These frequencies were applied in random order and the results were compiled and graphed into the C-message curve that became the basis for the bandwidth specification that the telephone company adopted.

**C message**
Voice bandwidth between 300 and 3 Khz.

Ideally, the voice bandwidth of the telephone system is 0 to 4 kHz, but a more practical range of 300 to 3 kHz is used in the design of telephone and data communications equipment. Another result of the test, which did not come as a great surprise to anyone, was that women hear higher tones that men do not. Most of the rolloff on the high end of the band was due to this disparity.

Additional tests and inputs from psychologists suggested a fascinating phenomenon about silence and the human mind. Human beings do not like silence, particularly on the telephone line. When we pick up a telephone handset and hear a "dead," or quiet line, our first impulse is to believe that there is something wrong. This assumption, in the case of the telephone system, is absolutely correct, even if the silence occurs after the call has been connected. Because of our dislike of pure silence, TELCO is designed so that a small amount of noise is always present on the line, even when no one is talking. This noise is sensed in speaker side of the handset so that the user will feel confident that there is activity on the line.

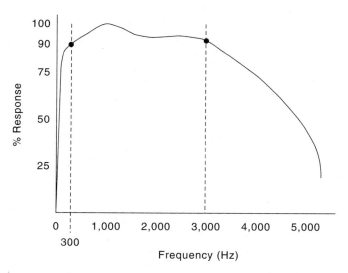

**FIGURE 2-7** Telephone System C-Message Response Bandwidth

## Line Impedance

Another important specification is the line impedance used as a standard for the telephone lines and equipment. TELCO specifies the impedance to be 600 ohms at both sending and terminating ends. Some telephone lines have been in existence since the telephone company first began. Others are recent additions and some are no longer twisted-pair copper, but instead are now fiber optic lines. Most telephone switching equipment has been converted to solid-state devices and integrated circuits, replacing relays and older vacuum tubes. The end effect is that line and equipment impedances can fluctuate from lows of 100 ohms to highs above 5,000 ohms.

## Types of Line Service

**direct distance dialing (DDD)**

**leased** or **dedicated lines**
private line between two subscribers.

Complications concerning impedance arise as a result of the routing calls take to be completed. Physical routes are established at the time a call is initiated. This type of routing is known as connectionless service, which is performed through the telephone company's **direct distance dialing (DDD)** network. The advantage of this service is that lines are free when not in use. In contrast, a connection-oriented service is one in which a permanent pathway is established between two communicating stations. These connections are established through **leased** or **dedicated lines.** Anytime communication needs to occur, the users simply begin on this permanent connection. The downside of connection-oriented service is that resources are permanently allocated even when they are not used. The up side is that the circuit is the same each time it used. Later in the chapter we will see how this advantage is exploited to allow a link between two entities to have stable and predictable specifications.

One problem with connectionless service is that each time a call is made, a different route is established to complete the call. This results in varying specifications on the connection each time it is established. You may have experienced this change by sensing a change in the quality of a voice connection. There have been times when you cannot hear a caller well and suggest that they hang up and redial the call. Immediately, the call comes through and is ten times better. Consider the implication of this when trying to send digital data over these same lines.

---

### EXAMPLE 2-2

Illustrate the possible routing of a call between Phoenix and Tucson, Arizona.

### SOLUTION

An earlier statement referred to the difficulty of completing a call on Mother's Day. In a more general sense, this is what can happen when trying to establish a call between Phoenix and Tucson, which are cities within the same state and located within a four-hour drive.

---

An individual in Phoenix desires to place a call to Tucson. These two cities are used for illustration because of their close proximity to each other. There are also few inhabited areas between the two cities. One would expect calls from one to the other to be direct and simple. Most of the time they are as shown in Figure 2-8a. There are long distance trunk lines and switching stations that connect these two locations. However, during peak hours (middle of the work day, for instance), these lines may become busy due to heavy traffic. If this were to prevent calls from being completed, the telephone company would never have reached the level of success that it has.

The telephone switch network will try to route the call along the shortest and most direct route. If that cannot be done, a different route will be sought. The call, in this example, might first be routed to a switching station in Albuquerque, New Mexico, then down to El Paso, Texas, and finally on to Tucson before the connection is completed, as shown in Figure 2-8b. The caller does not detect anything, making the whole process transparent to the user. Furthermore, the caller probably is not concerned about the route taken as long as the connection is made and is billed only for the distance between Phoenix and Tucson.

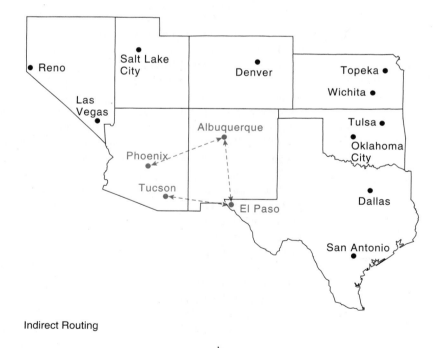

Direct Connection

a

Indirect Routing

b

**FIGURE 2-8** Long-Distance Routing Examples

The scenario in Example 2-2 is a common occurrence and is not limited in scope to minor rerouting. Calls have been known to take some long roundabout routes to be completed. It is only on certain days, such as Mother's Day, that the telephone system becomes so overcrowded that a call cannot be completed through some form of routing. Additionally, a call placed between the same two numbers within minutes of each other may not take the same path to be completed.

## Line Conditioning

Dedicated or leased lines, because of their permanent status, can be conditioned to tighten telephone company parameters so that they can transfer digital data more dependably. The parameters that can be affected by conditioning are gain or attenuation fluctuations, propagation delay variables, signal-to-noise ratios, and harmonic distortion. Initial specifications for these parameters are listed by the telephone company as **basic conditioning.** These specifications are what the telephone company has designed their lines and equipment to satisfy.

**basic conditioning**
Standard telephone company specifications for attenuation distortion and propagation delay variances. Also known as c/conditioning.

Changes in a signal's strength are dependent on many factors, not the least of which is the frequency of the signal. The reactance of cables and equipment changes with respect to the signal's frequency, causing the amplitude of the signal to vary with respect to its frequency. Gain or attenuation variations throughout the bandwidth of the telephone system are measured in reference to power of a standard signal frequency of 1004 Hz applied to the telephone lines. The actual power received is measured and given a reference value of 0 dB. **Decibel (dB)** is a measure of a power ratio between one power level and another used as a reference level. When the reference for the ratio is 1 mW, then the level of the power is in **dBm.** When the reference is any other power level then the difference between that level and a newer one is in dBs. A change of +3 dB is equivalent to doubling the power while a change of −3 dB represents a drop to half the power of the reference level. The reference level is usually designated 0 dB since there is no change between the reference and another signal measured having the same power level.

**decibel ratio (dB)**
gain ratio of powerout over power in.

**dBm**
power in decibels referenced to 1 milliwatt (mW).

When the telephone company set up their specification for gain fluctuations across the bandwidth, they started with a 1,004-Hz signal generated by a sending station. The power at the receiving station was measured and that level designated as 0 dB. The frequency at the transmitter was altered in steps, holding the sending output level constant. Received power was measured and compared to the 1,004-Hz power reference. From these measurements, a set of standards for basic line conditioning were formulated. An example of results that could appear is shown in Figure 2-9. Here a reference of 0 dB is established at a power level of 12 dBm, or 12 dB above the 1 mW power level. As the frequency of the signal is altered, the power fluctuates. In the graph shown, it rises to a maximum level of 15 dBm, which is 3 dB above the 0-dB reference, and attenuates to −12 dB below the reference. These measurements can show how well the system performs over the range of the 300–3 kHz telephone bandwidth.

A graph for the specifications for gain/attenuation variation for basic conditioning is shown in Figure 2-10. A typical response curve for a line is included for purposes of illustration. On the graph the 0-dB reference point is at 1,004 Hz. Note that the sample response curve always goes through this point. The straight lines on the graph show the

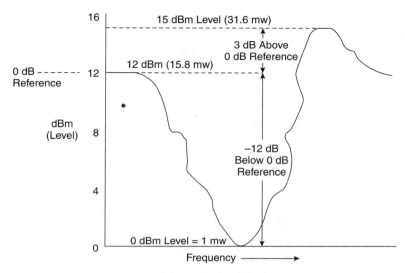

**FIGURE 2-9** Relationship of Level (dBm) and Gain/Attenuation (dB)

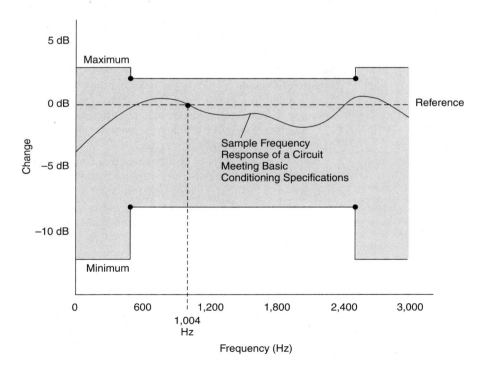

**FIGURE 2-10** Basic Conditioning Limits Graph

minimum and maximum changes in power allowed for basic conditioning. For the entire bandwidth, the signal's level must remain within +3 dB and −12 dB of the reference level at 1004 Hz. Additionally, within the frequency range of 500 to 2,500 Hz, the variation limits are reduced to +2 dB and −8 dB. What this translates to is that for equipment and lines within the telephone system to operate satisfactorily, signals traveling on that system will not experience changes in their amplitudes beyond the limits specified with reference to levels at 1004 Hz.

**Basic conditioning,** also known as C1 conditioning, is for voice-grade quality communications meaning that they are sufficient to guarantee successful voice communications. Other conditioning levels, known as C1 through C5, offer more stringent guidelines for the gain variations required for digital and data communication networks.

### Envelope Delay Distortion

In addition to specifying gain variation limits, C conditioning also details requirements concerning fluctuations in propagation delay throughout the 300 to 3 KHz bandwidth. To grasp the implication of propagation delay, consider a cable connected between two points. The cable has specific DC characteristics, such as resistance, and specific AC characteristics, such as impedance. The DC characteristics are those that are the same regardless of the frequency of the signal applied to the line. In contrast, AC characteristics are dependent on, and distinctly different for, each signal frequency value. Line impedance is the effective resistance and capacitance or inductive reactance load presented to the transmitter by the cable in response to the signal frequency. These effects are created by the inherent reactive quality of the cable created by its physical makeup. One result of this reactive quality is the changes in gain or attenuation just described. A second effect involves the time it takes for a signal to propagate down a line.

Figure 2-11 shows a graph comparing the time of arrival of three signals denoted f1, f2, and f3. The three signals, each at a different frequency, leave point A at the same time, $t0$. All three propagate through the same medium, a pair of telephone wires for instance. Because of the changes in reactance each signal feels, they arrive at point B at different times. The graph shows that when signal f2 arrives at point B, signals f1 and f3 still have a way to go. It is interesting to note that the longer propagation delays may occur at both higher and lower frequencies from the signal with the shortest propagation time.

One effect of propagation delay, besides arrival time, is pulse spreading. Consider the square wave signal shown in Figure 2-12 possibly created by sending a number of alternating 1s and 0s using non-return-to-zero (NRZ) format. A square wave is composed of a fundamental sine wave whose frequency is one-half that of the square wave, and numerous higher odd harmonics with frequencies that are an odd multiple of the fundamental. As this square wave propagates down the line, those sine waves are delayed at different rates. The ones that experience shorter delays arrive at the receiving end first, while the ones having longer delays arrive measurably later. The effect is to spread the shape of the square wave so that it becomes difficult to decipher the logic 1 and 0 levels. Another effect of spreading is to reduce the additive effects of the sine waves, causing the amplitude to reduce. If the spreading is severe enough, part of the wave shape can become totally lost, which, for digital data, can result in the loss of one or more bits.

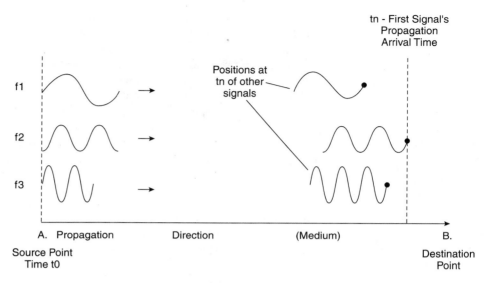

**FIGURE 2-11** Effects of Propagation Delay on Three Different Signals

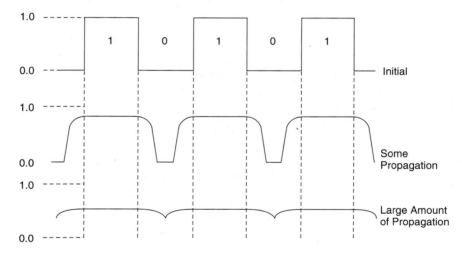

**FIGURE 2-12** Pulse-Spreading Effect of Propagation Delay

Measuring propagation time is not an easy task. It would be convenient if we could start a signal down a line, signal the other end that we did so, and measure how long it took the signal to reach the far end. The problem is that it also takes time for the start signal to get to the other end. Another way might be to echo the signal back—that is, to send it down the cable and then back to the originating source and measure the round-trip delay. This delay then can be divided in half to determine the propagation time in one direction. There are two problems with this method: First, whatever is used to echo the signal takes time to perform the turnaround, receiving and then sending the signal back. Second, how long do you send the signal before stopping to measure the return signal? This type of propagation measurement, if it could be made, would be an absolute propagation time measurement. The value obtained would be the actual propagation time. Fortunately, absolute propagation time is not a crucial quantity. If every signal arrived with the same propagation time, the overall effect would be a uniform delay in the communication. The true concern is the relative delay between signals that could cause pulse spreading to occur. We need to measure and quantify this relative delay.

**envelope delay distortion**
method used to test and measure the variance in propagation delay of different signals within the voice band.

The method used to measure propagation time is called **envelope delay distortion,** which is a clever process used to make this measurement. The signals that are to be measured are used as carrier signals that are modulated with a low sine wave signal as shown in Figure 2-13a. The frequency and amplitude of the modulating sine wave is held constant for each carrier signal sent. This modulating signal is known as the *envelope* that surrounds the carrier frequency. As each carrier signal that is modulated and sent is received, the time between the peaks ($T_{pk}$) of the envelope is measured (Figure 2-13b). After all the carrier signals have been sent and measured, the $T_{pk}$ times of each are compared to each other with the resulting difference ($T_d$) becoming a measure of the relative difference in the propagation time between any two signals (Figure 2-13b). For basic and C conditioning, the carrier frequency that results in the shortest time between envelope peaks is used as a reference (usually occurring at 1,800 Hz for telephone lines). Specifications for maximum relative delay are then listed for each level of conditioning.

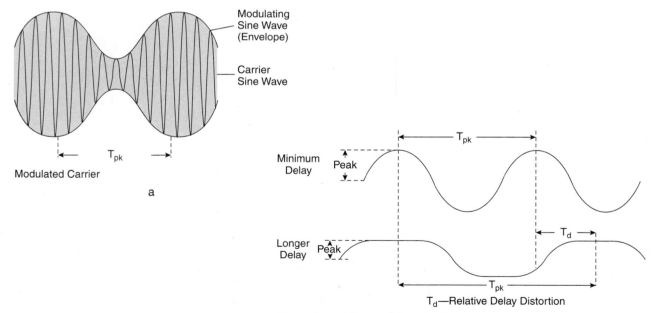

Modulating Sine Wave (Envelope)

Carrier Sine Wave

$T_{pk}$

Modulated Carrier

a

Minimum Delay   Peak

$T_{pk}$

Longer Delay   Peak

$T_d$

$T_{pk}$

$T_d$—Relative Delay Distortion

Comparison of Envelope Signals Affected By Delay of Carrier Signals

b

**FIGURE 2-13** Propagation Delay Distortion

**TABLE 2-2**

## Telephone Line Conditioning Parameters

| Condition Level | Gain Limitations | | | Propagation Delay | |
|---|---|---|---|---|---|
| | | Limits (dB) | | | Limit (ms) |
| | Frequency (Hz) | Max | Min | Frequency (Hz) | Max |
| Basic | 500–2,500 | +2 | −8 | 800–2,600 | 1.75 |
| Voice | 300–499 | +3 | −12 | | |
| Grade | 2,501–3,000 | +3 | −12 | | |
| C1 | 1,000–2,400 | +1 | −1 | 1,000–2,400 | 1.00 |
| Station to One | 300–999 | +2 | −6 | 800–999 | 1.75 |
| or More Stations | 2,401–2,700 | +2 | −6 | 2,401–2,600 | 1.75 |
| C2 | 500–2,800 | +3 | −3 | 1,000–2,600 | 0.50 |
| Same as C1 | 300–499 | +2 | −6 | 600–999 | 1.50 |
| | 2,801–3,000 | +2 | −6 | 500–599 | 3.00 |
| | | | | 2,601–2,800 | 3.00 |
| C3 | 500–2,800 | +.5 | −1.5 | 1,000–2,600 | 0.11 |
| Local | 300–499 | +.8 | −3 | 600–999 | 0.30 |
| Private Line | 2,801–3,000 | +.8 | −3 | 500–599 | 0.65 |
| | | | | 2,601–2,800 | 0.65 |
| C3 | 500–2,800 | +.5 | −1 | 1,000–2,600 | 0.08 |
| Trunk | 300–499 | +.8 | −2 | 600–999 | 0.26 |
| Private Line | 2,801–3,000 | +.8 | −2 | 500–599 | 0.50 |
| | | | | 2,601–2,800 | 0.50 |
| C4 | 500–3,000 | +2 | −3 | 1,000–2,600 | 0.30 |
| Maximum of | 300–499 | +2 | −6 | 800–999 | 0.50 |
| Four Interconnected | 3,001–3,200 | +2 | −6 | 2,601–2,800 | 0.50 |
| Stations | | | | 600–799 | 1.50 |
| | | | | 2,801–3,000 | 1.50 |
| | | | | 500–599 | 3.00 |
| C5 | 500–2,800 | +.5 | −1.5 | 1,000–2,600 | 0.10 |
| Station-to- | 300–499 | +1 | −3 | 600–999 | 0.30 |
| Station Only | 2,801–3,000 | +1 | −3 | 500–599 | 0.60 |
| | | | | 2,601–2,800 | 0.60 |

Table 2–2 summarizes the specifications for basic and C conditioning for both gain or attenuation variations and envelope delay distortion. Column 1 in the table denotes the type of conditioning. Column 2 holds the frequency range in hertz (Hz) for the limits shown. Column 3 is the gain/attenuation variations allowed in dB. Column 4 is the frequency range for envelope delay measurements. And column 5 contains the actual delay variations in milliseconds (ms) using a modulating sine wave of 100 Hz.

## EXAMPLE 2-3

What are the parameters for a communication system that interconnects three stations? Illustrate the bandwidth response for gain/attenuation variations and envelope delay distortion for these parameters.

## SOLUTION

According to Table 2-2, C4 conditioning is required to interconnect up to four stations. C4 has the following gain/attenuation variations, which are shown on Figure 2-14a:

1. +2 dB maximum and –3 dB minimum variation from 1004 Hz between 500 and 3 kHz.
2. +2dB maximum and –6 dB minimum from 300 to 499 Hz and 3,001 to 3,200 Hz.

Envelope delay distortion for C4 conditioning, shown in Figure 2-14b, has the following limitations:

1. 0.3 ms relative delay to 1,800 Hz between 1,000 and 2,600 Hz.
2. 0.5 ms relative delay from 800 to 999 Hz and from 2,601 and 2,800 Hz.
3. 1.5 ms relative delay from 600 to 799 Hz and 2,801 to 3,000 Hz.
4. Lastly, between 500 and 599 Hz, the allowable relative delay is 3.0 ms.

## Noise and Harmonic Conditioning

C conditioning specifies limits dealing with signal strength and propagation delay, which can be tightened to accommodate the needs of transferring data up to 4800 bps on dedicated telephone circuits. Additional factors concerning signal-to-noise ratio and harmonic distortion affect the manner in which higher data rates are transmitted using the telephone system. The symbols used to transport data at these rates are produced by modulating the phase and amplitude of a constant frequency analog signal. Since amplitude modulation is part of the process, the resulting signals are affected by a system's signal-to-noise ratio.

Detecting changes in the phase of these signals can be corrupted by the existence of harmonics of the original signal since these harmonics would appear to the receiver as changes in the signal's phase as well as in its frequency.

D conditioning addresses these concerns by tightening the telephone company's basic specifications for both signal-to-noise and harmonic distortion. TELCO specifies the acceptable signal-to-noise ratio (SNR) of their system to be 24 dB. That is the noise level is 24 dB below the desired signal level. D conditioning improves this specification by 4 dB to 28 dB.

SNR measurements can be done in one of two ways: unweighted, which is a measure of the existing noise on a line without any active signal on it, or weighted, which measures noise with a signal present. Unweighted noise is known as **inherent** or **white noise** and is the noise generated by circuits "at rest." A signal noise ration figure can be determined by dividing the intended signal level by the unweighted noise level. It becomes ap-

**inherent or white noise** steady low-level noise generated by equipment.

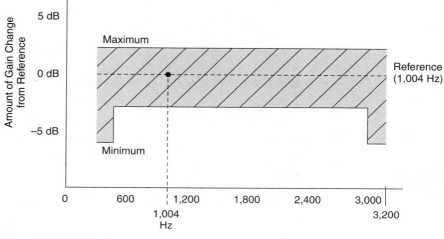

Attenuation/Gain Graph

a

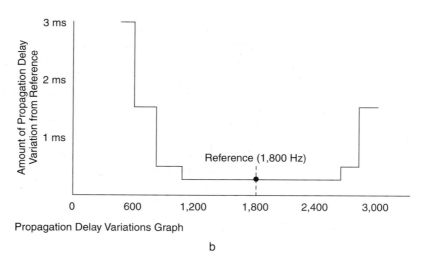

Propagation Delay Variations Graph

b

**FIGURE 2-14** C4 Conditioning Graphs

parent that this measure is not entirely accurate. The SNR value can be artificially inflated by applying enough signal power to present a high number. However, what is not taken into account, is that when the signal is applied and the circuits become active, they generate additional noise because of the increased activity. When a circuit is fed a signal, the semiconductor and resistive devices, which are part of the circuit, become agitated. There is more electron activity within these devices because of the energy brought by the applied signal. As a result, these semiconductors and resisters generate additional noise above the inherent or white noise. Remove the signal and after a short period,

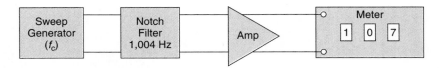

**FIGURE 2-15** Weighted Noise Measurement

**weighted noise**
noise generated when a signal is applied to a circuit.

**notch filter**
A narrow bandstop filter.

these devices settle back to their ambient, or at-rest, state and the noise is reduced. The additional noise generated by applying a signal to a circuit or system is called **weighted noise.**

To measure weighted noise, a signal must be applied and then removed and the existing noise measured quickly before the circuits have time to return to their ambient state. The C-notched test system shown in block form in Figure 2-15 is designed to make the weighted noise measurement. The sweep generator supplies a sine wave signal whose frequency ($f_c$) constantly changes while holding its amplitude steady. A **notch filter** between the sweep generator and the circuit under test (represented by the amplifier (AMP) in Figure 2-14) allows all the sine wave signals generated by the sweep generator to pass through except one. The signal that is absent is gone for a very short period of time. This is because the notch filter has a high $Q$ and very narrow bandwidth. As a result, for a brief period, the signal is not present at the circuit's output, but the noise generated by the circuit activity is. A power meter connected to the output monitors the level of the noise by making and latching the power level when the signal is not present (or notched out). A true signal-to-noise measurement can now be determined by dividing the applied signal power by the C-notched noise level.

A schematic and frequency response curve for an active notched filter is shown in Figure 2-16. Notch filters are characterized by a high $Q$ resulting in a narrow bandstop. Equations 2-1 through 2-3 summarize the relationship between notch filter components, notch frequency ($fn$), and the bandstop (or inverted bandwidth, $-BW$).

$$fn = \frac{1}{2\pi C\sqrt{(R_1)(R_2)}} \tag{2-1}$$

$$Q = 0.5\sqrt{\frac{R_2}{R_1}} \tag{2-2}$$

$$BW = \frac{fn}{Q} \tag{2-3}$$

**EXAMPLE 2-5**

Select reasonable values for the resistor and capacitor components of a notch filter with a notch frequency of 1,004 Hz and a bandstop of approximately 50 Hz.

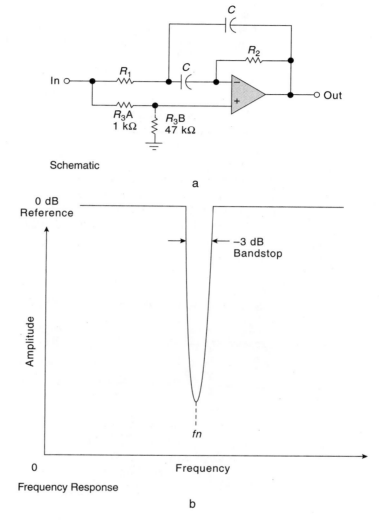

Schematic

a

Frequency Response

b

**FIGURE 2-16** Active Notch Filter

**SOLUTION**

Begin by working backwards using equations 2-1 through 2-3. Determine the Q from *fn* and *BW:*

$$Q = fn/BW = 1,004 \text{ Hz}/50 \text{ Hz} = 20.08$$

Next use Equation 2-2 to find the ratio of $R_2$ to $R_1$:

$$Q = 0.5\sqrt{R_2 / R_1} \qquad R_2 / R_1 = 1,613$$

Then select a value of C, say 0.01μf, as a starting point and solve for one of the resistor values. Use the $R_2/R_1$ ratio to determine the other resistor value. If any of the component values are not reasonable, change the value of C and try again.

$$fn = \frac{1}{2(3.14)(0.01\mu f)\sqrt{1,613(R_1)-R_1}} = 1,004\ \text{Hz}$$

$$R_1 = 395\ n \quad R_2 = 1,613R_1 = 637\ k n$$

The other parameter that D conditioning affects is harmonic distortion. The presence of a second harmonic signal (one that is twice the frequency of the desired or fundamental frequency) must not exceed a level of –35 dB below the level of the fundamental signal to meet D-conditioning. Third harmonic signals (those three times the frequency of the fundamental signal) are required to be an additional –5 dB lower, for a total of –40 dB below the desired signal level.

**Section 2.3 Review Questions**

1. What is the advantage of connectionless service?
2. What is the benefit to digital data links in using dedicated lines?
3. What is the TELCO bandwidth and specified line impedance?
4. What type of telephone lines can be conditioned?
5. What are the reference frequencies for gain variation and envelope delay measurements?
6. What are the tightest limits for conditioning that can be applied to telephone lines? What frequencies do these limits apply to?
7. What two electrical parameters are specified using D-conditioning?
8. What are the characteristics of a notch filter?
9. What is the difference between weighted and unweighted noise?

## 2.4 LINE IMPAIRMENTS

**impulse impairment**
change in a signal's amplitude or phase for a very short time period.

Once a line is conditioned, normal signal variations due to telephone equipment is minimized. However, sudden and unexpected changes in signal strength, caused by external factors, can result in creating numerous data errors. These changes, which are random and usually last for short periods of time, are termed **impulse impairments** because of their sudden and short (impulse) disturbance of line activity (impairment). Examples of impairment causes are induced currents into the lines from lightening strikes, a power surge from the power station, or magnetic disturbances from a solar flare. An extreme cause is an atomic explosion, which would cause numerous spiking on any lines that sur-

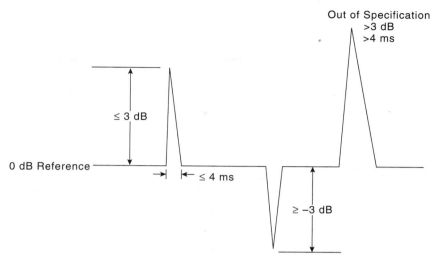

**FIGURE 2-17** Impulse Gain Hits

vive the explosion. For regular voice communications, they are a source of annoyance and not much more, but for data communications they can cause data errors to occur.

## Impulse Gain Hit

One type of impulse impairment is an impulse gain variation known as an impulse gain hit, which the telephone company specifies as a change in signal gain of no more than ±3 dB, lasting no more than four ms as illustrated in Figure 2-17. These variations are in addition to the gain variations specified in the conditioning tables. The telephone company considers changes within the four-mS time slot to be random and not in violation of the conditioning limits.

## Impulse Phase Hit

Electrical disturbances, such as those from electrical storms, also can cause changes in the phase of a signal for a short duration. Recall that a symbol is a single electrical parameter that represents a group of binary bits and that variations in symbols can be based on amplitude, frequency, or the phase of a signal. On a system that bases the formation of symbols on the signal's phase, a sudden change in the phase, known as an impulse phase hit, which lasts for a short period, can result in data errors. To minimize this effect, system designers must adhere to the telephone company specification for impulse phase changes. These specifications limit impulse phase shifts to 20° in phase (positive or negative, as shown in Figure 2-18) as permissible on telephone lines. Again, this is for a short period of time, but it is a concern for reliable communications that depend on phase differences between signals to supply digital information.

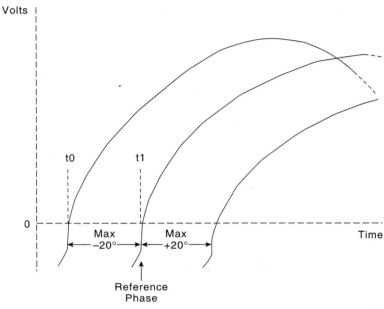

FIGURE 2-18 Impulse Hits

### Cross Talk

**cross talk**
coupling of one signal to a neighboring line.

An additional difficulty that can cause problems in using telephone lines is **cross talk** between neighboring lines. Most of us have experienced this problem on long distance calls at one time or another. While talking with your party, you can distinctly hear, albeit at a very low volume, another phone conversation going on. What you are experiencing is cross talk, in which one line is inducing current into the other in the same manner as a transformer inducing current from its primary to secondary winding. Once again, for voice conversations, this is a small annoyance, but for data communications, it is a possible source of data errors. There are no telephone company specifications addressing lines that cross talk.

The occurrence of cross talking should not be confused with the old telephone system of using party lines. In the beginning, the demand for this new method of communicating by telephone far outdistanced the availability of telephone numbers, which were only five digits long. To compensate for this need, the telephone company devised the concept of a party line in which two or more subscribers would share the same telephone number. This temporarily alleviated the backlog of customers waiting for telephone service. However, this solution did cause some not-so-amusing problems. Individuals with party service could eavesdrop on others sharing their number. Also, if one of the callers was a bit long-winded, the other user had to have a lot of patience. Add to this the problem that when the phone rang, neither of the party-line users would know for which user the call was intended. But in spite of all this, one did have telephone service.

## Impedance Mismatch

Telephone equipment is constantly updated and sections of old lines are replaced with newer cables, or in many instances, with fiber optic cables. This all leads to improved telephone service for voice communications. However, the merging of older systems with newer ones has a tendency to create impedance and performance mismatches throughout the system that may effect data communications traffic. We detect instances of drastic degradation on the system when after a call has been completed, we have difficulty hearing the party on the other end of the line. This problem is quickly alleviated by hanging up and redialing the call. This causes different connections to be made in completing the call, hopefully using a better connection.

Other problems with mismatched equipment result in signal attenuation, distortion of signal shapes, and phase shifts. Some of these consequences can be minimized by conditioning the line as previously discussed. Since line conditioning is only performed on dedicated or leased lines, the user is guaranteed that the parameters for the conditioning are maintained.

One impairment resulting from mismatches on the line that is not improved by conditioning is called **standing waves.** A standing wave is created when a signal reaches a termination point that has an impedance mismatch with the line. Because of the mismatch, some of the signal is not accepted by the receiver, but is echoed back along the line toward the sending end as seen in Figure 2-19. This echo causes the original to be attenuated and distorted. Standing waves are minimized by adjusting the impedances of the transmitting and receiving ends to match the line and each other. The closer the match, the less standing wave occurs.

**standing waves**
Echo created by impedance mismatch between sending and receiving terminations.

## Echoes

Another form of echo is produced on purpose on local switch lines by the telephone company. We have an aversion to silence, particularly on the telephone. If the line sounds too quiet, we immediately believe there is something wrong with the line. In order to ease our fears, the telephone company causes a small amount of the voice transmission to be echoed back to the talker. This way, the caller perceives that there is some activity on the line while they are talking or when the line is idle. In the case of an idle period, where neither party is talking, the noise on the line is echoed so the level of the noise is high enough to be heard.

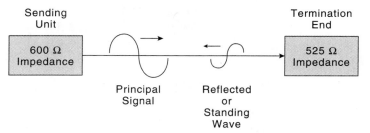

**FIGURE 2-19** Standing Wave Created by Impedance Mismatch.

When trunk lines are used to complete long distance calls, repeaters are used to reamplify and reshape signals that have become degraded by traveling. These repeaters not only do this reamplifying and reshaping, but they also supply the echo signal and control the direction of voice communication. When a voice is detected, the amplifier on that side of the line defeats the repeater amplifier on the other pair of wires carrying a return signal. This defeat stays in place until that first amplifier detects a lack of voice. At that point, the other side, if it detects a replying voice can now come on and turn the other line's amplifier off.

How do we know this is going on? Have you ever noticed how difficult it is to hold a conversation on the telephone with someone who talks constantly? This is because the constant sound of their voice causes the repeater amplifier on their side of the line to defeat the one on yours, so you cannot respond until something makes the other person stop talking long enough for his or her repeater amplifier to detect the absence of a voice.

**echo suppressors**
circuit that disables amplifiers in repeaters to prevent transmitted signals from being echoed back.

We can be lighthearted about the consequences of **echo suppressors** in repeaters that are responsible for disabling the amplifiers and generating the low-echo signal, when we consider voice calls. However, if we are to achieve full-duplex-data uninterrupted transfers using long distance lines, we have to be able to defeat these echo suppressors. These are defeated by sending a tone in the frequency range between 2,010 and 2,240 Hz for 400 ms on the line. This defeats the echo suppressors and full-duplex operation can commence and be maintained as long as signals within the voice bandwidth remain on the line. The loss of a signal for as short a time as 5 ms will allow the suppressors to be enabled again.

## Jitter

**jitter**
small constant change in a signal's amplitude or phase.

One last impairment to consider has direct impact on systems that use the phase or amplitude of signals to carry data symbols. Amplitude and phase **jitter** shown in Figure 2-20 is a small, constantly changing swing in a signal's phase, amplitude, or both. It differs from impulse-type impairments by its constant nature. Because the change is usually small in size and predictable, this type of problem is easy to resolve or compensate for. This doesn't mean jitter can be ignored—it must be dealt with. It is just easier to deal with a problem that is always present rather than one that only occurs occasionally, like impulse hits.

### Section 2.4  Review Questions

1.  What are the two types of impulse impairments and what are the telephone company's acceptable specification for them?
2.  Why are echo suppressors detrimental to full-duplex operation?
3.  How do jitter problems differ from impulse problems?

## SUMMARY

The very availability of the telephone company system provided the early designers of data communication systems and networks to adopt it as a leading medium for transferring data. Once it was decided to do this, the specifications, limits, and impairments expe-

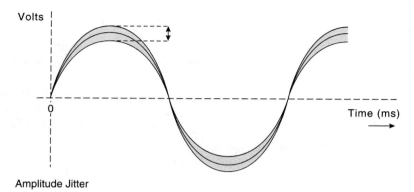

Amplitude Jitter

**FIGURE 2-20** Examples of Jitter

rienced by the telephone system and defined by the telephone company had to be taken into consideration during the design process.

To alleviate some of the concerns about the telephone system, dedicated or leased lines could be used, which may be conditioned to more exacting specifications for gain and delay variations throughout the voice bandwidth. This did not remove all concern over impairments that effect the telephone system. These impairments included impulse hits, cross talk, echoes, and jitter. All of these elements had to be addressed and resolved for dependable data communications to go forward.

# QUESTIONS

### Section 2.2

1. A call connecting two subscribers on the same city block is routed through _____.
2. What type of lines are long distance calls carried on?
3. What is the purpose of a toll station? a tandem station?
4. What type of circuit converts two-wire local loop lines to four-wire trunk lines?
5. What is the purpose of a repeater on long distance lines?

### Section 2.3

6. What is the voice bandwidth of the telephone system?
7. Why is the telephone system referred to as a connectionless system?
8. What is the difference between weighted and unweighted noise?
9. What frequency is "notched" when taking C-notched weighted noise measurements?
10. What two parameters are specified by C conditioning?
11. What does basic C conditioning specify?
12. Describe the limits for a C-2 conditioned line. Draw the response graphs for attenuation/gain and envelope delay distortion.
13. What type of telephone lines can be conditioned?

14. What is the reference frequency for attenuation/gain and propagation delay variance measurements?
15. Explain how envelope delay distortion is measured.
16. Is envelope delay distortion an absolute or relative delay test?
17. What parameters are specified for D conditioning?
18. How much are lines improved by using D conditioning?
19. A circuit has a weighted noise level of -35 dBm when a 100-dBm signal is applied. What is the signal-to-noise ratio (SNR) of the circuit?

**Section 2.4**

20. Define impulse hit.
21. What types of impulse hits are specified by the telephone company? What are their acceptable limits?
22. What is the cause of cross talk between telephone lines?
23. What are the causes of signal interference in telephone lines?
24. What is a standing wave and what creates them in the telephone system?
25. What is the purpose of echo suppressors?
26. What type of data communications transfer do echo suppressors undermine?
27. How is echo suppression defeated?
28. What is jitter?
29. What type of jitter may be experienced by telephone system communication?
30. Contrast jitter and impulse hits.
31. Which types of impairments are caused by or result in the following activity?
    a) Electrical storm
    b) "Ghost" voice on the line
    c) Constant line "static"
    d) Gain set to high on a repeater
    e) Line impedance mismatch
    f) Atomic explosion
    g) Ham radio transmitting antenna close to telephone lines

# DESIGN PROBLEMS

1. A system is experiencing a phase jitter of ±5°. Design a circuit that can be used to remove or minimize this jitter. The input and output impedance of the system that produces the jitter is 600 $n$.

2. Design a circuit to decode a touch tone (DTMF) keypad. Use either two sine-wave generators to generate the tones or a Touch Tone telephone set to test your circuit.

3. Design a trunk line repeater circuit that has the following capabilities:
    a) Amplify four-wire signals
    b) When a signal in one direction is being amplified, the responding line should be disabled to suppress echoes.
    c) Option: add a circuit to defeat the echo suppression function of part b in response to a steady 1,200-Hz signal for a minimum of 400 ms. Echo suppres-

sion remains defeated as long as the line remains in use. It is restored when the line is quiet for one second.

4. Design an active notch filter for a notch frequency of 2,200 Hz and a bandstop of 100 Hz.

## ANSWERS TO REVIEW QUESTIONS

### Section 2.1
Telephone lines already existed and were a reliable means of communication.

### Section 2.2
1. subscribers, local switch stations, interconnecting lines
2. toll and tandem stations
3. regenerate and reshape voice signals
4. convert between two-wire local lines and four-wire trunk lines

### Section 2.3
1. conserve resources—makes circuits available to other users
2. a) same line characteristics each time it is used
   b) guaranteed connection anytime you want to use it
3. 300 –3 KHz at 600 $n$
4. dedicated, leased, or private lines
5. gain variance: 1,004 Hz
   envelope delay distortion: 1,800 Hz
6. C3 conditioned lines:
   + .5 dB, –1 dB 500 – 2,800 Hz
   0.8 ms delay 1,000 – 2,600 Hz
7. signal-to-noise ratio (SNR) and harmonic distortion
8. high Q and narrow bandstop (bandwidth)
9. Weighted noise occurs when a signal is applied to a circuit. Unweighted noise is the circuit's inherent noise without a signal applied.

### Section 2.4
1. impulse gain hit: ±3 dB for up to 4 ms
   phase hit of up to 20° for up to 4 ms
2. Echo suppressors disable one of the two-wire pairs in a four-wire trunk line when a signal is detected on the other pair.
3. Jitter is a constant small variance in a signal's amplitude or phase, while an impulse change is a larger change for a short duration of time.

# CHAPTER 3

# Error Detection and Correction

## OBJECTIVES

After reading this chapter, the student will be able to:
- determine how errors in asynchronous digital data transmissions are detected using parity.
- determine how errors in asynchronous digital data transmissions are corrected using LRC/VRC.
- determine how errors in synchronous digital data transmissions are detected using checksum and CRC.
- determine how errors in synchronous digital data transmissions are corrected using Hamming Code.
- determine how errors are corrected using automatic request for retransmission.

## OUTLINE

3.1   Introduction
3.2   Asynchronous Data Error Methods
3.3   Synchronous Data Error Methods
3.4   Error-Testing Equipment

## 3.1  INTRODUCTION

The occurrence of a data bit error in a serial stream of digital data is an infrequent occurrence. Even less frequent is the experience of numerous errors within the transmission of a single message. Usually if a number of errors occur then it can be presumed that either a significant interference occurred effecting the transmission line or that there is a major failure in the communications path. Largely because of the extremely low bit-error rates in data transmissions, most error detection methods and algorithms are designed to address the detection or correction of a single bit error. However, as we shall soon see, many of these methods will also detect multiple errors. Error correction, though, will remain a one-bit error concern.

# 3.2 ASYNCHRONOUS DATA ERROR METHODS

**parity**
error-detection (parity) and error-correction (VRC) techniques based on the odd or even count of logic 1's in a transmitted character.

Probably the most common and oldest method of error detection is the use of **parity.** While parity is used in both asynchronous and synchronous data streams, it seems to find greater use in low-speed asynchronous transmission applications, however, its use is not exclusive to this.

## Parity Error Detection

Parity works by adding an additional bit to each character word transmitted. The state of this bit is determined by a combination of factors, the first of which is the type of parity system employed. The two types are even and odd parity. The second factor is the number of logic 1 bits in the data character. In an even parity system, the parity bit is set to a low state if the number of logic 1s in the data word is even. If the count is odd, then the parity bit is set high. For an odd parity system, the state of the parity bit is reversed. For an odd count, the bit is set low, and for an even count, it is set high.

---

**EXAMPLE 3-1**

What is the state of the parity bit for both an odd and an even parity system for the extended ASCII character B?

**SOLUTION**

The extended ASCII character B has a bit pattern of 01000010 (42 H). The number of logic 1s in that pattern is two, which is an even count. For an even parity system, the parity bit would be set low and for an odd parity system, it would be set high.

---

To detect data errors, each character word that is sent has a parity bit computed for it and appended after the last bit of each character is sent as illustrated in Figure 3-1. At the receiving site, parity bits are recalculated for each received character. The parity bits sent with each character are compared to the parity bits the receiver computes. If their states do not match, then an error has occurred. If the states do match, then the character *may* be errorfree.

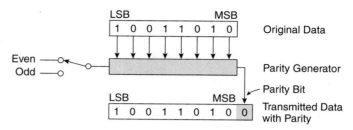

**FIGURE 3-1** Appending Parity Bit

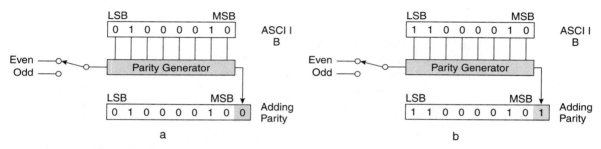

**FIGURE 3-2** Even Parity for ASCII B (a), Parity for Bad ASCII B (b)

---

**EXAMPLE 3-2**

The ASCII character B is transmitted with an even-parity bit appended to it. Illustrate how the receiver would detect an error.

**SOLUTION**

As shown in Figure 3-2a, the state of the even-parity bit for the ASCII B is low, so the complete data stream for the character sent, starting with the least significant bit (LSB) is: 010000100. Notice there are now nine bits—eight bits for the extended ASCII character B and one for the parity bit. The breakdown of the data stream is:

| LSB | | | | | | MSB | Parity |
|---|---|---|---|---|---|---|---|
| 0 | 1 | 0 | 0 | 0 | 0 | 1 | 0 | 0 |

Suppose that the LSB becomes corrupted during transmission. The receiver receives the character as: 110000100. When the receiver computes a parity bit for the character data, it results in a high state of the parity, bit as shown in Figure 3-2b. This is compared with the transmitted parity, which is a low state. Since they do not agree, the receiver determines that an error has occurred. Note that the receiver cannot determine which bit is bad, only that one of them is wrong.

---

A match between transmitted parity and receiver-calculated parity does not guarantee that the data has not been corrupted. Indeed, if an even number of errors occur in a

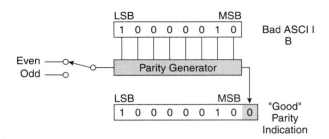

**FIGURE 3-3** Multiple Errors in ASCII B

single character, then the parity for the corrupted data will be the same state as the good data. For instance, suppose the two lowest bits in the character B were bad as seen in Figure 3-3. The total number of ones in the data stream would still be an even count and the parity bit calculated at the receiver would be a low state and would match the one transmitted. This does not present a major problem, since the occurrence of two errors in an eight-bit character is excessive and usually indicates a major problem in the system. Such a problem would cause errors to occur in other characters and one of them would eventually be detected. Since the occurrence of errors is extremely low, parity is successful in detecting more than 95% of the errors that occur.

## Parity Generation

The hardware circuit used to generate the state of the parity bit is composed of a number of exclusive OR gates as shown in Figure 3-4. To understand how the exclusive OR gate can be used to determine the even or odd count of the data bits, we first must examine its truth table.

$A$ and $B$ are used to designate the two inputs to an exclusive OR gate and $Y$ is its output state. An exclusive OR is a device whose output is low if the inputs are the same (or an even number of ones) and a high if they are opposite (or an odd number of ones). The exclusive OR gate can be used for a number of functions, including binary addition and subtraction, controlled inversion, and bit comparison.

The output of an exclusive OR for binary addition is used to represent the sum of its two inputs. Notice that $0 + 0 = 0$, $0 + 1 = 1$, $1 + 0 = 1$, and $1 + 1 = 0$. Of course, the last addition, in true binary form is 10, but since there is no way for a single exclusive OR gate to show the carry over, only the 0 will show on the truth table.

Subtraction can be viewed in a similar manner. $0 - 0 = 0$, $0 - 1 = 1$ (after borrowing), $1 - 0 = 1$, and $1 - 1 = 0$. Once again, the exclusive OR indicates the difference result as long as we ignore the need to borrow to subtract 1 from 0.

As a bit comparator, the exclusive OR's output ($Y$) is low when both inputs ($A$ and $B$) are the same logic level and high when they are different.

When considering the exclusive OR for inverting a bit under controlled conditions, in Table 3-1, treat input $A$ as a control input and input $B$ as a data bit input. Notice that

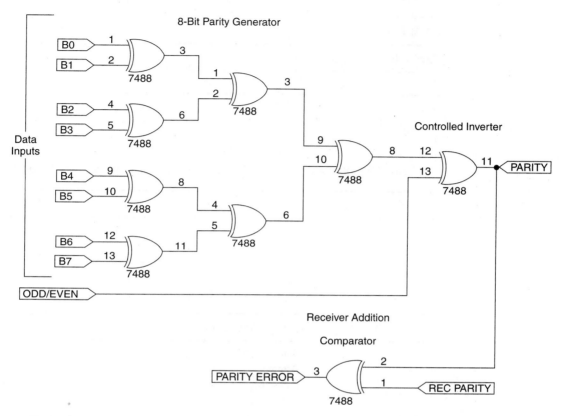

**FIGURE 3-4** Parity Generator

when the control input $A$ is low (0), the output $Y$ is the same state as the data input $B$. When the control input $A$ is low, output $Y$ is the opposite state of input $B$.

In this chapter, we will use all these functions of the exclusive OR, starting with the parity-bit generator and checker.

In Figure 3-4, each of the bits in the data-word inputs are summed together using the exclusive OR gates. The final sum's LSB result will be the even parity state. Since we need to be able to select either an even-parity or odd-parity system application, there has to be a way to produce one or the other. What is most interesting for the generation of a parity bit, is that when the LSB of a binary sum is low, the sum is even and when the sum is odd, the LSB is high. This defines an even parity result directly from the output of the exclusive OR summers. To make the output reflect an odd-parity system, it is only a matter of inverting that output to get the odd-parity state. However, we want to do this when an odd-parity system is desired. As long as an even-parity system is used, we do not want to invert the output. The solution to this problem is to use another exclusive OR as a controlled inverter at the output of the generator circuit as shown in Figure 3-4.

The final step, this time at the receiver, is to compare the parity sent with the one the receiver computes. In the receiver, another parity generator is employed to generate

**TABLE 3-1**

**Exclusive OR Truth Table**

| A | B | Y = A + B or A − B |
|---|---|---|
| 0 | 0 | 0 |
| 0 | 1 | 1 |
| 1 | 0 | 1 |
| 1 | 1 | 0 |

*Exclusive OR Truth Table Used for Sum or Difference Discarding Carry Out or Borrow*

| Control | Data Input | Data Output (Y) | |
|---|---|---|---|
| 0 | 0 | 0 | Non-Inverted Output |
| 0 | 1 | 1 | |
| 1 | 0 | 1 | Inverted Output |
| 1 | 1 | 0 | |

*Exclusive OR Truth Table Controlled Inverter Application*

| A | B | Comparison Results (Y) |
|---|---|---|
| 0 | 0 | 0 – Same Logic Level |
| 0 | 1 | 1 – Different |
| 1 | 0 | 1 – Different |
| 1 | 1 | 0 – Same Logic Level |

*Exclusive OR as Digital Comparator*

a received parity bit. This bit is compared to the transmitted parity bit to determine if an error has occurred.

The parity circuit is identical at both the transmitter and receiver sites with the exception of the use of the comparator. The transmitter does not require its use while the receiver does. Parity generators are produced as medium-scale integrated circuits (MSIC), like the Signetics 8262 shown in Figure 3-5. This chip uses two NOR gates on the output to provide a means for disabling parity using the inhibit input for systems that do not use parity for error detection. By using two NOR gates both odd- and even-parity states are available on separate output pins. Comparisons by the receiver using this chip require an external exclusive OR for the comparator function.

---

**EXAMPLE 3-3**

In the received asynchronous ASCII data stream using two-stop bits, shown below, there is an error. First, determine which parity system, odd or even, is being used. Then determine which character is bad. Lastly decode the message using the ASCII

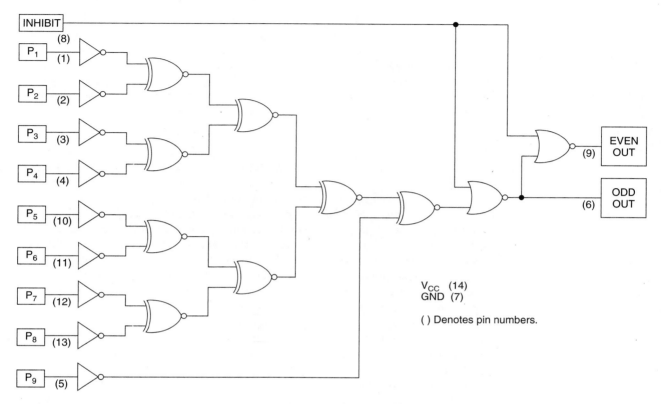

**FIGURE 3-5** IC Parity Generator

chart from Chapter 1. From this verify that the bad character is indeed bad. What was the intended message? (Spaces between each 4-bits are included for readability)

0000  1001  0110  1010  0110  1100  0010  1101  1000  0011  1111  0100
0010  0010  011

**SOLUTION**

Each character consists of seven ASCII bits, a start bit and two-stop bits plus the parity bit. The start bit is the leftmost bit of each character, followed by the LSB data bit through the MSB bit, then the parity bit, and finally, the two-stop bits. This means that each character contains eleven total bits. So step one is to define each character:

0000  1001  011    0101  0011  011    0000  1011  011
0000  0111  111    0100  0010  011

Next strip out the start and stop bits:

0001  0010    1010  0110    0001  0110    0000  1111    1000  0100

Now count the ones in the first seven bits of each word and determine the even- and odd-parity states for each. From this we can surmise that even-parity was used— all the characters except the middle one satisfy even-parity results. This means that the middle character probably has the error in it.

By using the ASCII chart in Chapter 1, the message spells out "Hehp!" This message should probably be "Help!" verifying that the middle character is incorrect.

## ARQ

To correct errors using parity, the receiving station can only request that the message containing the error be retransmitted. A system that is capable of requesting retransmission of a bad message automatically in response to detecting an error has *Automatic Request for Retransmission* or *Automatic Repeat Request (ARQ)* processing within its communications software package. ARQ was originally designed to be used with a special type of character code that used a seven-bit character size. The uniqueness of that code was that each character code contained three bits that were high and four that were low. If any character received is detected with more or less than three high bits in it, it is flagged as bad and the receive station automatically requests that the character be retransmitted. The automatic request process has been incorporated into other error detection software such as those which respond to parity errors with any available character code.

Not all systems using parity have ARQ functions included in their programs. Some use a parity error flag contained within a status register, which, when read by an application program, causes a message to be sent to a terminal to inform a user that a data error has occurred. It is then up to that user to determine if it is necessary to request the message be retransmitted. The advantage of doing this is that some errors are less critical than others and do not require taking up communication time for a retransmission of the data. For instance, an error in a plain text message may be obvious enough that the user can easily determine the correct character in the text. In that case, there is no need to have the text resent. The drawback to this method is that the display of the message must be interrupted to inform the user that an error has occurred and that it does require some action on the user's part.

## Data Correction Using LRC/VRC

**vertical redundancy check (VRC)**

**longitudinal redundancy check (LRC)**
error-correction method that uses parity and bit summing.

As described in the preceding section, parity is primarily used for detecting errors in a serial data character. A bad parity match indicates a logic error has occurred in one of the character's data bits. The use of parity called a **vertical redundancy check (VRC)** can be extended to allow single-bit error correction to take place in a received data stream. By having the ability to correct an error, a receiver would not require a message to be retransmitted, but could do the correction itself. The trade-off in using an error-correction scheme is that an additional character has to be sent with the message and additional software and/or hardware must be used to create and interpret that character. For asynchronous data transmission, that character is known as the **longitudinal redundancy check (LRC)** character.

Using a VRC/LRC system, the message is sent with each character containing the regular even-parity bit known as the VRC bit. As with error-detection schemes, any mismatch between transmitted and received VRCs indicates that the character contains a bad data bit. In order to correct the bad bit, what is left to be done is to determine which of the character's bits is the bad one. This is where LRC comes in. It is used to create a cross-matrix type of configuration where the VRC bit denotes the row (character) and the LRC, the column (bit position) of the message's bad bit. At the sending site, each of the data bits of each character is exclusive ORed with the bits of all the other data bits. This is best illustrated by example.

---

### EXAMPLE 3-4

Determine the states of the LRC bits for the asynchronous ASCII message "Help!"

**SOLUTION**

The first step in understanding the process is to list each of the message's characters with their ASCII code and even VRC parity bit:

| LSB | | | | | | MSB | VRC | CHARACTER |
|---|---|---|---|---|---|---|---|---|
| 0 | 0 | 0 | 1 | 0 | 0 | 1 | 0 | H |
| 1 | 0 | 1 | 0 | 0 | 1 | 1 | 0 | e |
| 0 | 0 | 1 | 1 | 0 | 1 | 1 | 0 | l |
| 0 | 0 | 0 | 0 | 1 | 1 | 1 | 1 | p |
| 1 | 0 | 0 | 0 | 0 | 1 | 0 | 0 | ! |

Next, for each vertical column, find the LRC bit by applying the exclusive OR function. To make this process easier, you can consider the results of the exclusive OR process as being low or zero (0), if the number of ones (1) are even, and one (1) if the count is odd. For instance, in the LSB column, there are two 1's, so the LRC bit for that column is a 0. And for the rest:

| LSB | | | | | | MSB | VRC | CHARACTER |
|---|---|---|---|---|---|---|---|---|
| 0 | 0 | 0 | 1 | 0 | 0 | 1 | 0 | H |
| 1 | 0 | 1 | 0 | 0 | 1 | 1 | 0 | e |
| 0 | 0 | 1 | 1 | 0 | 1 | 1 | 0 | l |
| 0 | 0 | 0 | 0 | 1 | 1 | 1 | 1 | p |
| 1 | 0 | 0 | 0 | 0 | 1 | 0 | 0 | ! |
| 0 | 0 | 0 | 0 | 1 | 0 | 0 | 1 | LRC |

When the message is transmitted, the LRC character is sent following the last character of the message. The receiver reads in the message and duplicates the process, including the LRC character in the exclusive OR process. If there were no errors, then the VRC's would all match and the resulting LRC would be 0.

**EXAMPLE 3-5**

Show how a good message would produce an LRC of 0 at the receiver.

**SOLUTION**

Repeat the process as before, but include the LRC character this time. Note that the number of 1s in each column are always even if there are no errors present:

| LSB | | | | | | MSB | VRC | CHARACTER |
|---|---|---|---|---|---|---|---|---|
| 0 | 0 | 0 | 1 | 0 | 0 | 1 | 0 | H |
| 1 | 0 | 1 | 0 | 0 | 1 | 1 | 0 | e |
| 0 | 0 | 1 | 1 | 0 | 1 | 1 | 0 | l |
| 0 | 0 | 0 | 0 | 1 | 1 | 1 | 1 | p |
| 1 | 0 | 0 | 0 | 0 | 1 | 0 | 0 | ! |
| 0 | 0 | 0 | 0 | 1 | 0 | 0 | 1 | LRC |
| 0 | 0 | 0 | 0 | 0 | 0 | 0 | 0 | Receiver LRC |

So far, the LRC/VRC combination seems to aid in detecting errors. If an error occurs, one of the VRCs won't match and the LRC would not be zero. The question is how is it used to pinpoint the bit in the bad character. Again the best way to see how this works is to continue our example.

**EXAMPLE 3-6**

Illustrate how LRC/VRC is used to correct a bad bit.

**SOLUTION**

We will use the same message, but by placing an error in the received data would cause the l character to print as an *h*. You can compare the data with the good example to satisfy yourself as to which bit is bad and confirm that the LRC process does indeed pick out the same bit.

| LSB | | | | | | MSB | VRC | CHARACTER |
|---|---|---|---|---|---|---|---|---|
| 0 | 0 | 0 | 1 | 0 | 0 | 1 | 0 | H |
| 1 | 0 | 1 | 0 | 0 | 1 | 1 | 0 | e |
| 0 | 0 | 0 | 1 | 0 | 1 | 1 | 0 | h |
| 0 | 0 | 0 | 0 | 1 | 1 | 1 | 1 | p |
| 1 | 0 | 0 | 0 | 0 | 1 | 0 | 0 | ! |
| 0 | 0 | 0 | 0 | 1 | 0 | 0 | 1 | LRC |
| 0 | 0 | 1 | 0 | 0 | 0 | 0 | 1 | Received LRC |

The first thing that we note is that the VRC for the character *h* will not match the one the receiver will generate, which will be a 1 since the number of ones in *h* are odd and the number of ones in l was even. The received LRC will have two indications that something is no longer right. Its own VRC bit will not match the transmitted VRC bit and, secondly, its value is not zero. The state of the VRC doesn't matter since any non-zero result will indicate that there is a problem. This is how the system uses the receive LRC to determine which bit is bad: The receiver already knows which is the bad character by checking the VRC (parity) bits. By inspecting the receive LRC, the answer comes jumping off the page. The one bit in the LRC that is not a zero is in the same bit position as the bad bit in the *h/l* character! Now that the bad character and the bad bit position are discovered, the receiver needs only to invert that bad bit to make the character correct.

---

We can illustrate the matrix concept behind LRC/VRC by circling the bad VRC and LRC bit and seeing where they intersect in the message matrix as shown for our example here:

| LSB | | | | | MSB | VRC | CHARACTER |
|---|---|---|---|---|---|---|---|
| 0 | 0 | 0 | 1 | 0 | 0 | 1 | 0 | H |
| 1 | 0 | 1 | 0 | 0 | 1 | 1 | 0 | e |
| 0 | 0 | 0<——1——0——1——1——1 | | | | | | h/l |
| | | ? | | | | | | |
| 0 | 0 | 0 | 0 | 1 | 1 | 1 | 1 | p |
| | | &#124; | | | | | | |
| 1 | 0 | 0 | 0 | 0 | 1 | 0 | 0 | ! |
| | | &#124; | | | | | | |
| 0 | 0 | 0 | 0 | 1 | 0 | 0 | 1 | LRC |
| | | &#124; | | | | | | |
| 0 | 0 | 1 | 0 | 0 | 0 | 0 | 1 | Received LRC |

**forward error correction (FEC)**
method of correction that occurs as messages are received.

This error-correction method and others which are similar, are known as **forward error correction (FEC)** because errors are corrected as the message is received. There is no requirement to retransmit the message as long as the errors remain infrequent. If more than one error occurs in a message, then more than one LRC and one VRC bit will be bad and there is no way to determine which LRC bit goes with which VRC character. In this case, the excessive number of bit errors is indicative of a severe or catastrophic condition. Once the cause of the problem is resolved, the message will have to be retransmitted in its entirety.

**Section 3.2  Review Questions**

1. What is the state of the parity bit using an odd-parity system for the extended ASCII character M?
2. Why doesn't a good match between transmitted and received parity bits guarantee that the character is good?
3. What is the chief advantage of the ARQ function?

4. List the functions that an exclusive OR logic gate can be used for.
5. Which type of parity system is used for the VRC bits in the LRC/VRC error-correcting scheme?
6. How many errors in a message can be corrected using LRC?
7. What does an incorrect VRC bit indicate?
8. What does a one in a bit position in a LRC character indicate?

# 3.3  SYNCHRONOUS DATA ERROR METHODS

**overhead**
Any non-data bits or characters sent with a transmission.

Synchronous data are transmitted at higher data rates in as an efficient manner as possible. Start-and-stop framing and parity bits are omitted from the data stream to reduce overhead. **Overhead** is defined as any bits sent that do not contain actual data information. This includes framing bits, preambles, error-detection characters, or bits, etc. It should be noted that in some synchronous data systems, parity is occasionally employed for error detection. Most high-speed synchronous transmissions, however, do not follow this practice. The reason is that most errors in high-speed transmissions occur in bursts, which could render parity-error detection less effective. These error bursts result from some external interference or other effect on the line that causes several bits to be corrupted at once. Single-bit errors occur less frequently. Because of this, and the desire to reduce the overhead in synchronous transmissions, error-detection methods have evolved to detect single and multiple errors within a data stream.

Synchronous error detection works by creating an additional character to be sent with the data stream. At the receive site, the process is duplicated and the two error-detection characters are compared similarly to comparing two parity bits. If the characters match then the data received has no errors. If they do not match, an error has occurred and the message has to be retransmitted. Note that one major difference between using error-detection characters versus single-parity bits, is that if the transmitted and received characters match, then the data is good. Using parity, matching parity bits does not guarantee that the character received was good. The computation of error characters is carried on quickly to support the higher data rates of transmission.

## CRC

**cyclic redundancy check (CRC)**
error-detection method that uses a pseudo-division-process.

One of the most frequently used error-detection methods for synchronous data transmissions is **cyclic redundancy check (CRC)** developed by IBM. This method uses a pseudo-binary-division process to create the error or CRC character, which is appended to the end of the message. The hardware circuitry that generates the CRC character at the transmitter is duplicated at the receiver. This circuitry is incorporated into the transmit-and-receive shift registers that send and receive the actual message. We will begin by exploring the method used to create and check the CRC character and then view the circuitry that performs the operation.

In the original specification for CRC, IBM specified a 16-bit CRC character designated as CRC-16. The process uses a constant "divisor" to perform the "division" process, which appears in binary for CRC-16 as:

$$1000 \quad 1000 \quad 0001 \quad 0000 \quad 1$$

Again, spaces are used for clarity. In actuality, there are no spaces. Now that we have a "divisor" we need something for it to be divided into. This is the message. The process is begun by adding 16 zeros (one less than the number of bits in the "divisor") after the last bit of the message. These 16-bits will eventually be replaced by the CRC character. The "divisor" is then exclusive ORed with the seventeen most significant bits of the message. Enough additional message bits are appended to the result of the exclusive OR process to fill out 17-bit positions starting with the first logic 1 of the exclusive OR result. The process is repeated until the last bits of the message (including the added zeros) are used. The result from the last exclusive OR process becomes the CRC-character that replaces the 16-zero bits originally added to the message (leading zeros are added as needed). This process is best viewed by example, which will use a smaller CRC "divisor" to shorten the process.

---

### EXAMPLE 3-7

Compute the CRC-4 character for the following message using a "divisor" constant of 10011:

$$1100 \quad 0110 \quad 1011 \quad 01$$

**SOLUTION**

CRC-4 is used for illustration purposes since an example using CRC-16 looks cumbersome on paper and is difficult to follow. However, the principle is the same. Notice that the "divisor" is 5-bits, one more than the number indicated by the CRC type (CRC-4). The same was true for CRC-16, which had a 17-bit "divisor." We start the process by adding four zeros to the data stream and removing the spaces we have been using for convenience:

$$1100011010110110000$$

Next, set up the problem to appear as a division problem:

$$10011 \quad \overline{)1100011010110110000}$$

Start the "division" process by exclusive OR the "divisor" with the first five bits of the message:

$$
\begin{array}{r}
10011 \quad \overline{)1100011010110110000} \\
\underline{10011} \\
1011
\end{array}
$$

Now bring "down" one bit so that the result of the exclusive OR process is filled out to the "divisor" size and repeat the process:

$$
\begin{array}{r}
10011 \quad \overline{)110001101011010000} \\
\underline{10011} \\
10111 \\
\underline{10011} \\
100
\end{array}
$$

Continue with the process until all of the bits in the message plus the added four zeros are used up:

$$
\begin{array}{r}
10011 \quad \overline{)110001101011010000} \\
\underline{10011} \\
10111 \\
\underline{10011} \\
10010 \\
\underline{10011} \\
11011 \\
\underline{10011} \\
10000 \\
\underline{10011} \\
11100 \\
\underline{10011} \\
11110 \\
\underline{10011} \\
11010 \\
\underline{10011} \\
1001 = \text{CRC character}
\end{array}
$$

The CRC character is appended onto the end of the message and transmitted. At the receiver, the process is repeated, except that there are no zeros added to the message. Instead, the CRC character fills up those positions. If the result of the process at the receiver produces zero then no errors occurred. If any bit or combination of bits are wrong, then the receiver will yield a non-zero result.

### EXAMPLE 3-8

Demonstrate how a receiver detects a good message and one with several errors in it.

### SOLUTION

Redo the process of EXAMPLE 3-7, but this time use the CRC character in place of the extra zeros:

```
10011    )110001101011011001
         10011
         10111
         10011
           10010
           10011
             11011
             10011
             10000
             10011
               11110
               10011
               11010
               10011
                 10011
                 10011
                 0000 = CRC character is zero
```

The changes in the bits brought down are highlighted. Notice how they produce different results from EXAMPLE 3-7. This eventually results in a CRC of 0000 if everything is correct. Now let's suppose there are three errors in the message that will also be highlighted. Follow the problem through to see how the CRC will be non-zero:

```
                        bad bits
10011    )110001110111011001
         10011
         10111
         10011
           10011
           10011
               011101
               10011
               11101
               10011
               11100
               10011
                 11110
                 10011
                 11011
                 10011
                 1000 = CRC character is non-zero
```

*NOTE*

Given the small size (CRC-4) of this example, there could easily be error combinations that would produce a zero CRC result. This is the reason that most CRC systems today are either CRC-16, CRC-32 or CRC-64.

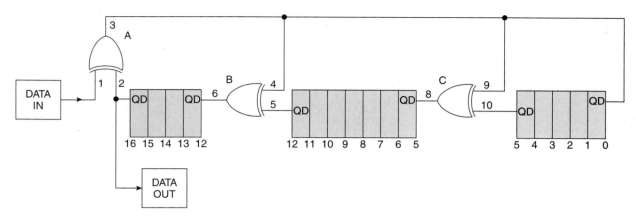

**FIGURE 3-6** CRC-16 Block Diagram

A block diagram of the shift register circuit that implements CRC-16 is shown in Figure 3-6. Each rectangle represents a data flip-flop with the D-input on the right and the Q output on the left of the block. The numbering on the bottom of the flip-flop register reflects a design process used to develop the circuit. We shall see how that process works shortly. Exclusive OR gates are placed in between certain sections of the shift register determined by the "divisor." To facilitate the design of the CRC-16 or any other CRC circuit, a form of quadratic expression is used to represent the "divisor." This expression is developed by using the powers of 2 for each bit position of the "divisor" that contains a logic 1. For CRC-16, those positions are:

$$16 \qquad 12 \qquad 5 \qquad 0$$
$$1000100000010000 1$$

which, as a quadratic expression, is written as:

$$G(X) = X^{16} + X^{12} + X^5 + 1 \qquad \text{note: } X^0 = 1$$

The flip-flops in the register are numbered by similar bit position numbers. Exclusive OR gates are inserted at the indicated bit positions in the quadratic expression. The Q output of the last flip-flop drives one input of an exclusive OR gate.

The other input is supplied from the serial data stream to be transmitted or received. The Q output of the flip-flop numbered 12 drives the input of another exclusive OR gate as does the one numbered 5. The second inputs of these gates are driven by the output of the exclusive OR connect to bit 16 and the data input. This same line is returned to the input of flip-flop number 0. The output data stream is taken from the last flip-flop (number 15/16). To see how this design is developed, let's do one for the CRC-4 circuit used in Example 3-7 and 3-8.

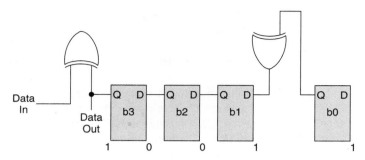

**FIGURE 3-7** Exclusive OR Placements

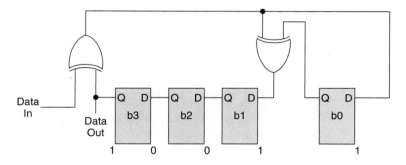

**FIGURE 3-8** CRC-4 Circuit

---

**EXAMPLE 3-9**

Develop the block diagram for the CRC-4 circuit used in Examples 3-7 and 3-8.

**SOLUTION**

First create the quadratic expression from the "divisor":

$$\begin{array}{cccc} 4 & & 1 & 0 \\ 1 & 0 & 0 & 1 \end{array} \ 1 = X^4 + X^1 + 1$$

Since this is a CRC-4 error-detection circuit, there will be four flip flops in the circuit. An exclusive OR is connected to the last flip and between the first and second one as seen in Figure 3-7.

To complete the circuit, connect the data in to the other input of the exclusive OR connected to the last flip-flop. Take its output and connect it to the other exclusive OR and the D input of the first flip-flop. There you have it—one CRC-4 circuit as shown in Figure 3-8.

---

When using the circuit, it is first reset to all zeros (in effect adding zeros to the message). The input data stream is fed into the circuit and shifted through. The output is sent to the transmitting circuit. After the message is fully shifted through the register, its contents will contain the CRC character, which is then shifted out attaching itself to the end of the message. As it is shifted out, zeros are shifted in, initializing the register for its next use.

The receiver directs the incoming data stream to the data input of its CRC shift register and the process is repeated. The last set of bits received and shifted through the register is the transmitted CRC character. If all is well, then the register will contain all zeros after everything has been shifted in and through the circuit.

To reduce possible confusion between shifting in zeros to clear the register and detecting a CRC result of zero, some CRC systems are set up so that the receiver's "divisor" is the complement of the transmitters. The result of this setup is that when the receiver is finished, instead of having all zeros in the CRC register if the message is good, it will have all ones.

## Checksum Error Detection

**checksum**
error-detection process that uses the sum of the data stream in bytes.

Another method of error detection uses a process known as **checksum** to generate an error-detection character. The character results from summing all the bytes of a message together, discarding and carry-over from the addition. Again, the process is repeated at the receiver and the two checksums are compared. A match between receiver checksum and transmitted checksum indicates good data. A mismatch indicates an error has occurred.

This method, like CRC, is capable of detecting single or multiple errors in the message. The major advantage of checksum is that it is simple to implement in either hardware or software. The drawback to checksum is that, unless you use a fairly large checksum (16- or 32-bit instead of 8-bit), there are several data-bit patterns that could produce the same checksum result, thereby decreasing its effectiveness. It is possible that if enough errors occur in a message that a checksum could be produced that would be the same as a good message. This is why both checksum and CRC error-detection methods do not catch 100% of the errors that *could* occur, they both come pretty close.

---

### EXAMPLE 3-10

What is the checksum value for the extended ASCII message "Help!"?

### SOLUTION

The checksum value is found by adding up the bytes representing the Help! characters:

| | |
|---|---|
| 01001000 | H |
| 01100101 | e |
| 01101100 | l |
| 01110000 | p |
| 00100001 | ! |
| 00010000 | Checksum |

---

The hardware solution relies once more on exclusive OR gates, which perform binary-bit addition. Each 8-bits of data are exclusive ORed with the accumulated total of all previous 8-bit groups. The final accumulated total is the checksum character.

## Error Correction

Error detection is an acceptable method of handling data errors in lan-based networks because retransmission of most messages result in a short delay and a little extra use of bandwidth resources. Imagine a satellite orbiting around Jupiter or Saturn, transmitting critical visual data as binary stream information. The time it takes for those transmissions to reach Earth is measured in hours. During this time, the satellite has adjusted its orbit and is soaring across new territory and sending additional data. Correcting errors in these messages cannot be done by retransmission. A request for that retransmission takes as long to get to the satellite as the original message took to get to Earth. Then consider the time it would take to retransmit the message. What would the satellite do with new data, reach it while it tries to handle the retransmitting of old data? The memory needed to hold the old data in case it would need to be resent is astronomical to say the least. Instead, an error-correcting method such as the Hamming code is used so that errors can be corrected as they are detected.

## Hamming Code

**Hamming code**
error-correction method based on the number of logic 1 states in a message.

For synchronous data streams, a error-correcting process called **Hamming code** is commonly used. This method is fairly complex from the standpoint of creating and interpreting the error bits. It is implemented in software algorithms and relies on a lot of preliminary conditions agreed upon by the sender and receiver.

Error bits, called Hamming bits, are inserted into the message at random locations. It is believed that the randomness of their locations reduces the statistical odds that these Hamming bits themselves would be in error. This is based on a mathematical assumption that because there are so many more messages bits compared to Hamming bits, that there is a greater chance for a message bit to be in error than for a Hamming bit to be wrong. Another school of thought disputes this, claiming that each and every bit in the message, including the Hamming bits, has the same chance of being corrupted as any other bit. Be that as it may, Hamming bits are inserted into the data stream randomly. The only crucial point in the selection of their locations is that both the sender and receiver are aware of where they actually are.

The first step in the process is to determine how many Hamming bits (**H**) are to be inserted between the message (**M**) bits. Then their actual placement is selected. The number of bits in the message (M) are counted and used to solve the following equation to determine the number of Hamming (H) bits:

$$2^H \geq M + H - 1 \qquad (3\text{-}1)$$

Once the number of Hamming bits is determined, the actual placement of the bits into the message is performed. It is important to note that despite the random nature of the Hamming bit placements, the exact same placements must be known and used by both the transmitter and the receiver. This is necessary so that the receiver can remove the Hamming bits from the message sent by the transmitter and compare them with a similar set of bits generated at the receiver.

---

**EXAMPLE 3-11**

How many Hamming bits are required when using the Hamming code with the extended ASCII synchronous message "Help!" ?

**SOLUTION**

The total number of bits in the message is:

$$M = 8\text{-bits/character} \times 5 \text{ characters} = 40 \text{ bits}$$

This number is used in Equation 3–1 to determine the number of Hamming bits:

$$2^H \geq 40 + H + 1$$

The closest value to try is 6 bits for $H$, since $2^6 = 64$, which is greater than $40 + 6 + 1 = 47$. This satisfies the equation.

---

Once the Hamming bits are inserted into their positions within the message, their states (high or low) need to be determined. Starting with the least significant bit (LSB) as bit 1, the binary equivalent of each message-bit position with a high (1) state is exclusive ORed with every other bit position containing a 1. The result of the exclusive OR process is the states of the Hamming bits. Once again, as with previous error detection- and correction-processes, it is best to view how the Hamming code works by using an example.

---

**EXAMPLE 3-12**

Determine the states of the six Hamming bits inserted into the message "Help!" at every other bit position starting with the LSB.

**SOLUTION**

In the last example, we determined that six Hamming bits were required for the "Help!" message. For simplicity, we shall insert the Hamming bits a little less randomly:

```
   H        e        l        p              !
01001000011001010110110001110000001H0H0H0H0H0H1H
```

Starting from the LSB on the right, the first 1 is encountered in bit position 2, the next in position 12 and so forth:

| Bit Position | Equivalent Binary |
|:---:|:---:|
| 2 | 0 0 0 0 1 0 |
| 12 | 0 0 1 1 0 0 |
| 19 | 0 1 0 0 1 1 |
| 20 | 0 1 0 1 0 0 |
| 21 | 0 1 0 1 0 1 |
| 25 | 0 1 1 0 0 1 |
| 26 | 0 1 1 0 1 0 |
| 28 | 0 1 1 1 0 0 |
| 29 | 0 1 1 1 0 1 |
| 31 | 0 1 1 1 1 1 |
| 33 | 1 0 0 0 0 1 |
| 36 | 1 0 0 1 0 0 |
| 37 | 1 0 0 1 0 1 |
| 42 | 1 0 1 0 1 0 |
| 45 | 1 0 1 1 0 1 |
| H   = | 1 0 0 1 1 0 |

All these binary values are exclusive ORed together—an odd number of ones produces a 1, and an even count, a 0—to create the Hamming bits values. These values are substituted for the H-bits in the message. The entire thing is then transmitted and the process repeated at the receiver:

0100100001100101011011000111000000110000010110

If the message was received without any errors, then the Hamming-bit states produced at the receiver will match the ones sent. If an error in one bit did occur during transmission, then the difference between the transmitted Hamming bits and the receiver results will be the bit position of the bad bit. This bit is then inverted to its correct state.

The limitation imposed by the Hamming code is twofold. First, it works only for single-bit errors, and secondly, if one of the Hamming bits becomes corrupted, then the receiver will actually invert a correct bit and place an error in the message stream.

---

## EXAMPLE 3-13

Demonstrate how the Hamming code is used to correct a single-bit error in the data stream.

## SOLUTION

During the transmission of the message, bit 19 experiences a noise spike that causes it to be received as a 0 instead of 1. The receiver goes through the process of determining the states of the Hamming code, resulting in this calculation:

```
2            0 0 0 0 1 0
12           0 0 1 1 0 0
20           0 1 0 1 0 0
21           0 1 0 1 0 1
25           0 1 1 0 0 1
26           0 1 1 0 1 0
28           0 1 1 1 0 0
29           0 1 1 1 0 1
31           0 1 1 1 1 1
33           1 0 0 0 0 1
36           1 0 0 1 0 0
37           1 0 0 1 0 1
42           1 0 1 0 1 0
45           1 0 1 1 0 1
H      =     1 1 0 1 0 1
```

Notice that bit 19 is not included in the list since it was received as a low-state instead of a high-state. Now we compare the Hamming code transmitted to this one the receiver just derived:

```
Transmitted code:   1 0 0 1 1 0
Receiver code:      1 1 0 1 0 1
                    0 1 0 0 1 1 = bit 19
```

There is no "black magic" mystery to why the Hamming code works. The originally transmitted codes are formulated by adding binary bits together (the exclusive OR process), ignoring carries. A similar process occurs at the receiver. If a bit has changed, then the two sums will be different and the difference between them will be the bit position number that was not added at either the transmitter or the receiver. By comparing the two Hamming codes using exclusive OR gates, the numbers are effectively being subtracted from one another (another function of the exclusive OR gate) and the difference is the bad bit position.

### Section 3.3 Review Questions

1. Why is CRC-16 preferred to parity for error detection in synchronous data systems?
2. Why are larger CRC "divisors" preferred over shorter ones?
3. What is the 16-bit checksum value for the extended ASCII message "That's a 10-4"? Do not forget space characters!
4. Why are larger checksums preferred over shorter ones?

5. How many Hamming bits are required for the extended ASCII message "Now is the time for all good men. . . "? Do not forget to count all the space and . characters!
6. What happens if a Hamming bit is corrupted during data transfer?
7. What is the stated purpose of placing Hamming bits randomly throughout a message?

# 3.4 ERROR TESTING EQUIPMENT

There are many different types of equipment used for testing the effect of errors on a data communication link. Two types of equipment used to check for error rate occurrences in communications systems are the **bit-error rate tester (BERT)** and the **error-free second (EFS) test box.**

**bit-error rate tester (BERT)**
instrument for testing errors in a bit stream.

**error free seconds (EFS)**
The number of seconds data is transmitted without errors.

**bit-error rate (BER)**
Measure of the number of errors in a stream of data.

## Bit-Error Rate

**Bit-error Rate (BER)** is the measure of the occurrence of an incorrect bit in a stream of data. It is classified as one error in so many bits transmitted. For example, one bit in $10^6$ bits is a bit-error rate of one in a million bits transmitted. Frequently, this specification is shortened to a bit-error rate of $10^{-6}$ since the one-bit error is understood. The exponent is negated, indicating the infrequency of the error occurrence.

Another measure of bit-error rate is by percentage and is calculated using this formula:

$$\text{bit - error rate } = \frac{\text{number of bad bits}}{\text{total number of bits sent}} \times 100\% \tag{3-2}$$

Bit-error-rate testers are available that can test a data link for a number of different types of error occurrences. Chief among these is the bit-error rate, but also included are parity- and framing-error testing. The tester can be used singularly by tapping into the line between the terminal and the modem as shown in Figure 3-9. In this placement, the tester can be used to monitor the line or to inject test data sequences into the line.

**loop-back**
Circuit that returns transmitted data to the source for the purposes of testing the line.

Generally, the receiving end is terminated in a **loop-back** arrangement at some point. Loop-backs take the received data and return them to the sending station. This is accomplished by connecting the transmit and receive data lines together at the originating point (point A in Figure 3-9) to test the local sending loop; the remote end of the telephone line (point B) to test the telephone line connection; or point C, for testing the secondary's receive modem.

A known pattern of data is generated by the bit-error tester and sent down the line. One common and familiar pattern is "Quick brown fox jumps over the lazy dog's back,"

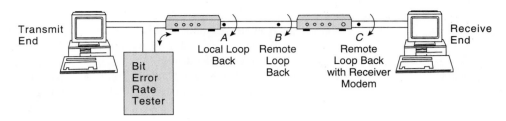

**FIGURE 3-9** Line Test Using a Bit Error Rate Tester

which contains all the letters of the alphabet. Other patterns include alternating ones and zeros and repeating single characters. When the data reaches the loop-back, they are echoed back to the sender. The bit-error tester monitors the returned data and checks for errors. A counter in the tester keeps track of how many errors occurred to determine the bit-error rate. Most bit-error testers are capable of operating at a wide range of data rates and can be used to test asynchronous and synchronous data systems.

## Error-Free Seconds

Another measure of bit errors called *error-free seconds (EFS)*, is used as a measure of error occurrences for data transmissions from 2.4 Kbps to 2.5 Mbps. Instead of measuring bit errors occurring in a specified number of transmitted bits, EFS is a measure of the number of seconds of transmission time that contain at least one error. An error-free second percentage is computed using this formula:

$$\text{error-free seconds} = 100\% - \frac{\text{seconds with an error}}{\text{total transmission seconds}} \times 100\% \qquad (3\text{-}3)$$

Error-free second testers are used in the same manner as bit-error rate testers and serve a similar purpose for digital data systems.

---

### EXAMPLE 3-14

Contrast the specifications, bit-error rate, and error-free seconds for a system that experienced five errors in 25 Mbytes of data transmitted in 20 seconds. Two errors occurred in the seventh second of transmission, one error in the twelfth second and the last two errors in the last second of transmission.

### SOLUTION

The bit-error rate is found by using Equation 3-2:

$$\text{bit-error rate} = (5 \times 100\%)/(25\,M \times 8) = 2.5 \times 10^{-6}\%$$

While error-free seconds uses Equation 3-3:

$$\text{error-free seconds} = 100\% - (3 \text{ error seconds})/(20) \times 100\% = 85\%$$

This all means that the system experiences 1 error in every 2.5 million bits and is error-free 85% of the time.

**Section 3.4  Review Questions**

1. What is the bit-error rate for a system that experiences 3 bit errors in a transmission of 512 K bytes of data?
2. What is the error-free second percentage of a system that experiences three seconds with errors in a total transmission time of two hours?

# SUMMARY

Error detection and correction methods are necessary to assure the integrity of the data sent from one location to another. The types of methods used support both asynchronous- and synchronous-type data streams. Asynchronous error detection is facilitated by the use of a parity bit with each character of data sent. Error correction for asynchronous data utilizes the LRC/VRC method, which duplicates the parity process (VRC) and examines each character by bit position (LRC). Synchronous data streams apply CRC or checksum for error detection and the Hamming code for error correction. Table 3-2 summarizes the error methods discussed in this chapter and supplies a quick comparison reference for them.

**TABLE 3-2**

### Error Methods Summary

| Error Method | Data Type | Detection Corrections | Number of Errors Detectable |
|---|---|---|---|
| Parity | Asynchronous | Detection | One per Character |
| LRC/VRC | Asynchronous | Correction | One per Message |
| Checksum | Either | Detection | Unlimited |
| CRC | Synchronous | Detection | Unlimited |
| Hamming Code | Synchronous | Correction | One per Message |

| Error Method | Overhead |
|---|---|
| Parity | One Bit Added per Character |
| LRC/VRC | One Bit per Character Plus LRC Character |
| Checksum | Checksum Character at End of Message |
| CRC | CRC Bytes at End of Message |
| Hamming Code | Hamming Bits Inserted into Data Stream |

## QUESTIONS

**Section 3.2**

1. Which error-detection method is used most frequently with asynchronous data streams?
2. How many errors can parity reliably detect in a single character?
3. Which type of logic gate is used to generate parity bits?
4. List the uses of an exclusive OR gate.
5. What is one drawback when using parity for error detection?
6. Define overhead in terms of serial data messages.
7. How does ARQ facilitate the correction of messages with errors in them?
8. What is the advantage of using ARQ for error correction?
9. What is the prime limitation of the LRC error-correction method?
10. What is meant by forward error correction?
11. How is the LRC character generated?
12. What does a bad VRC match signify?
13. How does the LRC pinpoint the bad bit in a character?
14. What is the hexadecimal value of the LRC character for the following message using regular (7-bit) ASCII:

Our Last Date

15. What is the hexadecimal value of the LRC character for the following message using extended (8-bit) ASCII:

The Yellow Brick Road

**Section 3.3**

16. Using the CRC-4 "divisor" used in this chapter, what is the CRC-4 character for this regular ASCII message:

This 1

17. Using the CRC-4 "divisor" used in this chapter, what is the CRC-4 character for this regular ASCII message:

Lefty

18. Draw the CRC-6 block diagram for the circuit using this "divisor": 1000101
19. Draw the CRC-6 block diagram for the circuit using this "divisor":

$$X^6 + X^5 + X^1 + 1$$

20. What is the checksum character, in hexadecimal, for the following synchronous message using extended ASCII?

Come to our aid now!

21. What is the checksum character, in hexadecimal, for the following synchronous message using extended ASCII?

mmiller@devry-phx.edu

22. How many Hamming bits are used for the message in question 7?
23. How many Hamming bits are used for the message in question 8?
24. What is the value of the Hamming bits, placed starting with the least significant bit position and in every third place after that for the extended ASCII message "See us"?

(Hamming placement example: . . . .H b3 b2 H b1 b0 H)

25. What is the value of the Hamming bits, placed starting with the least significant bit position and in every other place after that for the extended ASCII message "DeVry"?

    (Hamming placement example: . . . .H b3 H b2 H b1 H b0 H)

### Section 3.4

26. Select the error-detection method you would choose for each of the conditions below. Support your selections.
    a) Direct memory transfer from a hard disk to a computer's RAM memory.
    b) Communication link between your home computer and an associate's computer via modems and the telephone lines.
    c) Messages between workstations on a production line.
    d) Communications between bank automatic teller machines and a central computer.
    e) Surface photographs of Venus sent to Earth by an Explorer satellite.
    f) Computerized accounting network.
27. List the two main test functions of bit-error rate testers.
28. Contrast the functions of bit-error rate and error-free second specifications.

## DESIGN PROBLEMS

1. Design, construct, and verify the operation of a parity generator circuit. The circuit allows selection of even or odd parity and indicates when a parity error occurs. Another option for the circuit is to operate with a choice of 7- or 8-bit data inputs.

2. Design, construct, and verify the operation of a CRC-16 generating circuit. The "divisor" to be used is the one discussed in this chapter. The requirements for the circuit are as follows:

    a) While the CRC-16 bytes are being formed, the original bits are shifted through unchanged.
    b) After the last data bit is shifted out, the CRC-16 is shifted out from the circuit.
    c) Before the data are shifted through and after the CRC-16 is shifted out, the data line is to be in an idle line 1 condition.

Essentially, the design of the CRC circuit follows the process shown in the text for CRC-4. However, this design project is extended to include the logic circuitry in conjunction with the CRC-16 circuit to meet the requirements above.

## ANSWERS TO REVIEW QUESTIONS

### Section 3.1

1. Occurrences of bit errors are infrequent, so error detection and correction schemes are designed to handle these infrequent errors.
2. A significant interference or major system failure has occurred.

**Section 3.2**

1. one
2. If an even number of errors occurred within a character, the parity state would be the same as that for a good character.
3. Retransmission of a message when an error is sensed is requested automatically without intervention from the user.
4. Add, subtract, and compare a single bit. Controlled inverter.
5. Even
6. One
7. A bad character
8. The corresponding bad bit in a character.

**Section 3.3**

1. CRC can detect multiple errors dependably, parity cannot.
2. Larger CRC "divisors" reduce the likelihood that two data streams could produce the same CRC character.
3. 

| | | |
|---|---|---|
| T | 0101 | 0100 |
| h | 0110 | 1000 |
| a | 0110 | 0001 |
| t | 0111 | 0100 |
| ´ | 0010 | 1100 |
| s | 0111 | 0011 |
| space | 0010 | 0000 |
| a | 0110 | 0001 |
| space | 0010 | 0000 |
| 1 | 0011 | 0001 |
| 0 | 0011 | 0000 |
| – | 0010 | 1101 |
| 4 | 0011 | 0100 |
| | 0000 | 1111 = checksum |

4. Reduces the chance that two different messages would produce the same checksum.
5. $M = 35 \times 8 = 280$ bits
   $2^H \geq 280 + H + 1$
   $2^9 = 512 \geq 281 + 9 \geq 290$
   9 Hamming bits
6. The receiver could invert a good bit making it bad.
7. Statistically reduces the chance that a Hamming bit is corrupted.

**Section 3.4**

1. .585 m% or $585 \times 10^{-6}$%
2. 99.96%

# CHAPTER 4

# Open Systems Network Models

## OBJECTIVES

After reading this chapter, the student will be able to:
* define the purpose of each of the levels of the OSI model.
* compare the OSI model to the SNA model.
* describe the structure of the SNA model.
* define and compare the bus, star, and ring network topologies.
* differentiate between direct connections, circuit, message, and packet switching.
* describe the benefits and disadvantages of each switching method.

## OUTLINE

4.1   Introduction
4.2   Data Topologies
4.3   Data Switching
4.4   Types of Networks
4.5   The Open Systems Interconnection (051) Model
4.6   Systems Network Architecture
4.7   SNA Operating Sessions

## 4.1  INTRODUCTION

**network**
a system of interconnected communications stations.

**node**
entry or exit point to a network.

**protocol**
set of rules for successful communication between two or more nodes in a network.

A **network** is a system of interconnected communication stations. These stations can be personal computers (PCs), mainframe computers, dumb terminals, or access points to other networks. To simplify matters at the beginning, we shall refer to all connections or access points into a network system as **nodes.** There are a number of ways of interconnecting nodes to form various types of network topologies that can make up a network. We shall also find that many networks are a composite of several topologies and subnetworks that are interfaced to each other and managed by combinations of physical and software constraints. The rules and specifications for these constraints are known as **protocols.**

Interconnecting several nodes into a viable switch network has several advantages, including:

1. Resources, such as bulk storage, communication links, and printers are shared among the nodes in the network.
2. High-speed data rates for the transfer of information are used on these networks.
3. Messages are sent in smaller units or sections, allowing different nodes to communicate with each other in a near real-time environment unaffected by delays caused by nodes waiting for longer messages to be transmitted and received.

These switched networks are local, distant, or combinations of both. In order to have them perform as required, decisions have to be made on how to handle the transmission of data and establish a means for nodes within the network to access the network. In addition, message protocols and node addressing, as well as message integration are required to reassemble messages from these small units into their proper sequence. Lastly, methods of error-detection and/or correction are included in the protocols to insure the integrity of the data sent and received. These are the issues addressed by open-system network standards.

---

**Section 4.1  Review Questions**

1. What is a network?
2. Define protocol. What is it used for?
3. What is a node?

---

# 4.2  DATA TOPOLOGIES

A communications network supporting more than two stations requires some method to establish a unique link between any two stations within the network. The type of linkage and the processes required to establish it are dictated, in part, by the topology of the network. A **topology** is a description of the physical composition of a system or network. Topology may describe the layout or specify an existing system. For instance, the telephone system with its lines and switching stations is considered a topology. A set of interconnected networks is another example of a system-wide topology. On the other end of the topology spectrum, a description of a network that interconnects a number of node points in a specified manner is a topology as well. There are three common types of these topologies, **bus, star** or **hub,** and **ring.**

**topology**
physical composition of a network.

**bus**
network topology where all nodes are connected to a common communication path.

**star** or **hub**
network topology where nodes are connected through a central controlling station or hub.

**ring**
network where nodes are connected in sequential order.

## Bus Topology

The bus topology, shown in Figure 4-1, uses a common transmission line or path between a single primary node and several secondary nodes. Individual secondaries do not communicate directly with each other. All communication is controlled by the primary station

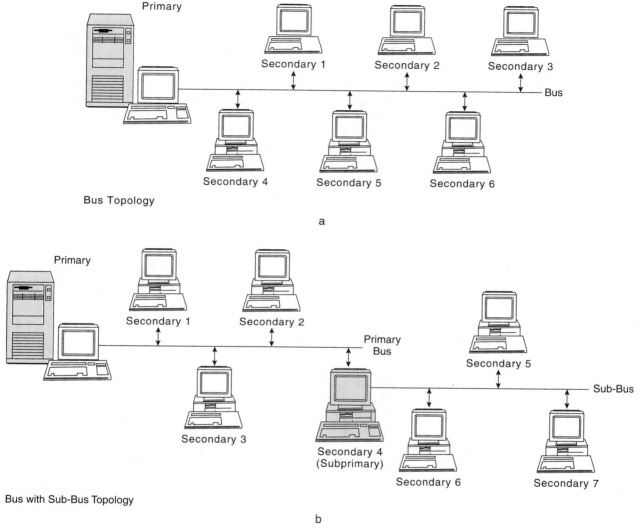

FIGURE 4-1  Bus Topology

and all traffic travels between the primary and one or more secondaries. The primary establishes a connection, or data link, with a secondary by addressing it and waiting for a response from the station addressed.

For instance, if secondary station number 2 in Figure 4-1 had a message to send to station 4, it first had to wait for the primary to access it through a mechanism called a **poll.** When the primary polls a secondary station, it is inquiring if the secondary has any traffic it wishes to send. In the case of our scenario, station 2 has a message to send to station 4 and replies to the primary's poll by sending that message to the primary. The

**poll**
primary message inquiring if a secondary has any traffic to send.

primary will continue its polling routine, checking the remaining stations on the system for messages. After the primary is finished, it notes which secondaries are to receive the messages it has received from the stations it just polled. The primary then starts to query those secondaries to find out if they are ready to receive those messages. This is called a **selection process.**

When the primary selects station 4, it receives an acknowledgment from that station that it is ready to receive its message. The primary then sends the message it originally got from station 2. After station 4 acknowledges that it received the primary's transmission, the communication between station 2 and 4 is completed.

**selection process**
primary message inquiring if a secondary is ready to receive traffic.

---

### EXAMPLE 4-1

Illustrate how a message is sent from secondary station 1 to secondary station 3 in the bus topology system of Figure 4-1.

### SOLUTION

One possible scenario would follow this sequence of events:

a) The primary polls secondary station 1 to see if that station has any messages it wants to send.
b) The secondary responds by sending the message intended for secondary station 3 to the primary.
c) The primary next selects secondary station 3 to find out if that station is ready to receive the message the primary has from secondary station 1.
d) Secondary station 3 replies with a positive acknowledgment indicating it is ready to receive the message.
e) The primary sends secondary station 1's message to secondary station 3.
f) Secondary station 3 responds with another positive acknowledgment indicating it received the message the primary sent.

---

Figure 4-1*b* shows a bus topology where one of the secondary nodes doubles as a primary to an additional network subsystem. Traffic on the subsystem works the same as it does in the main system as long as messages sent and received are destined for stations on the subsystem. The same is true for messages on the main system. Additionally, in this configuration, a station on the subsystem can communicate with a secondary on the main system. The messages are first sent to the subsystem's primary (station 4 in this configuration). This subprimary now must wait for the main primary to select it as a secondary station on the main system. Then the message can be passed onto that primary, which in turn passes it onto the destination station in the main system.

**peer-to-peer bus**
network topology with no controlling primary station.

**contention**
two or more stations attempting to access a network at the same time.

**Peer-to-peer buses** are bus-type networks without a primary station. In effect, all the node points are peer points, meaning that all the stations have equal access to the bus. This has the advantage that all stations can now communicate directly with each other and any station can initiate a data link with another station at any time. The major drawback to this type of system is known as **contention** for the bus.

The lack of a primary or controlling station forces the access to the network to be dependent on a bus arbitration process. Any station wanting to access the bus must first monitor the bus for a set period of time to make sure that there is no activity on the bus. Sensing a clear line, that station can then place its message, destined for any other station on the bus, onto the network. There is, of course, the possibility that two stations might have gotten the same idea at the same time and both of them will try to place a message onto the bus. In that case, a message **collision** occurs. The two stations must cease transmitting their messages and resolve the contention for the bus to allow only one of them at a time to send a message.

**collision**
result of two or more messages sent on the same line at the same time.

## Star or Hub Topology

The star or hub topology used in the system shown in Figure 4-2a places the primary station in the center as a hub with "spokes" connecting it to the other secondary stations. In this system, the primary performs a switching function. It receives messages from one station and routes them to another station on the system. The hub also has a store-and-forward ability to handle several different messages at one time. It stores each of these messages and forwards them when a line is available to the destination station.

---

**EXAMPLE 4-2**

Illustrate how messages from secondaries 1 and 3 are sent to secondary 8 on the star network of Figure 4-2a.

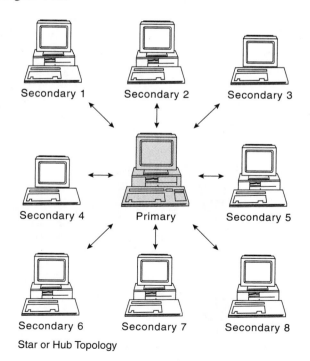

Star or Hub Topology

**FIGURE 4-2a** A Star or Hub Topology

## SOLUTION

For purposes of illustration, we will assume that the message from secondary 1 has a higher priority than that from secondary 3.

1. The primary receives the messages from secondaries 1 and 3 and notes that they are both for secondary 8.
2. The primary stores the lower-priority message from secondary 3 and sends the message from secondary 1 to secondary 8. Actually, the primary switches the incoming lines from secondary 1 to the outgoing lines for secondary 8, allowing the message from secondary 1 to go directly to secondary 8.
3. After the message from secondary 1 is finished, the primary takes the stored message from secondary 3 and sends it to secondary 8.

As was done with the bus topology, secondaries in a star topology can also perform as switching hubs for subsystems as shown in Figure 4-2b. The subsystem in this exam-

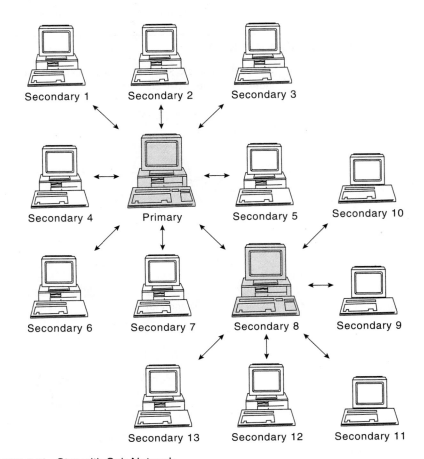

**FIGURE 4-2b** Star with Sub Network

ple uses secondary station 8 as the hub to allow messages to be sent and received among secondary stations 9–13. Any traffic destined for secondary stations 1–8 from 9–13 are first sent to secondary station 8, which recognizes that it cannot switch to another secondary station on its subsystem and must perform as a secondary on the main star and send the message to the main hub to be switched to the appropriate destination.

## Ring Topology

The third common type of topology is the ring topology shown in Figure 4-3a. The ring system was designed to resolve the problem of contention in a peer-to-peer bus system. The stations on the ring are all peer stations with none acting as a primary. A message is sent around the ring in one direction. For example, let's suppose station 2 wants to send a message to station 6. It begins by sending the message to station 3, which determines that the message is not meant for it and sends it on to station 4. Station 4 repeats the process

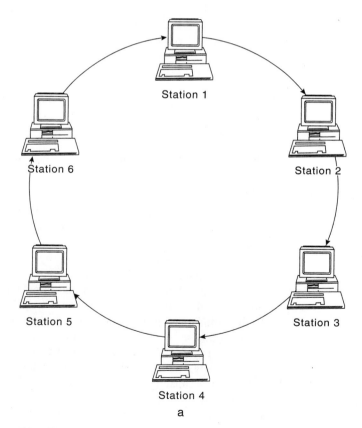

FIGURE 4-3a  Ring Topology

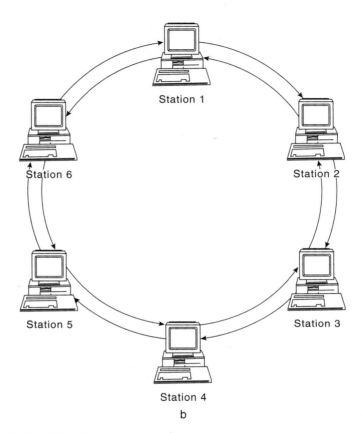

Station 1

Station 6

Station 2

Station 5

Station 3

Station 4

b

**FIGURE 4-3b** Dual Ring Network

as does station 5. When station 6 gets the message, it copies it, sets a status bit to indicate it got the message and sends it along to station 1. Station 1 notes that it has nothing to do with this message and passes it to station 2, the originating station. Station 2 notes that the message was successfully received by station 6 and, therefore, removes the message from the sequence.

Messages sent in a ring can take longer to get to their destination, but the chief problem with a ring system is that when a station on the ring fails, communication stops. While collisions and bus contention can not occur on a ring, the danger of losing the system because of single station failure is a large concern. One solution to that problem, shown in Figure 4-3b, is a dual-ring system. Traffic flows in opposite directions on both rings as long as there is no problem. Under normal operations this effectively doubles the systems capacity. The real benefit is what happens when a station fails. The bad station is shut off from the ring and the ends of the two rings are closed as shown in Figure 4-3c. Messages now flow as they would normally until they reach the bad station.

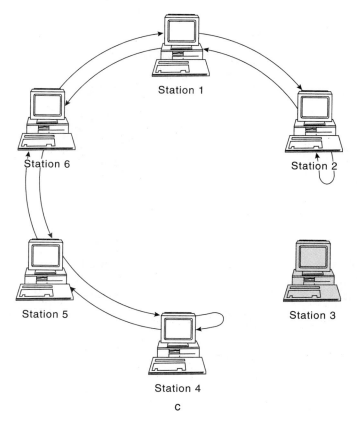

Station 1

Station 6

Station 2

Station 5

Station 3

Station 4

c

**FIGURE 4-3c** Dual Ring with Bad Station 3

Instead of stopping, they are routed back, via the secondary ring, in the direction they came from. As messages reach the bad station from the opposite side, they are looped back again onto the primary ring, thereby retaining the ring integrity until the bad station can be repaired.

Ring topologies can be expanded, like bus and stars, by connecting a station to both ring systems as shown in Figure 4-3d. Station 3 passes messages on each ring normally as long as the messages are destined for stations on their own ring. If a message is to cross from one ring to the other, the connecting station (station 3 in the figure) is responsible for passing them between rings. Keep in mind that these messages have to be passed twice, first when they are sent and then when they are copied and acknowledged.

Another way to overcome the problem of losing a communications system if a ring station goes bad is by use of a hub network operated as a virtual ring. In the context here and throughout the book, the term virtual will always refer to something that is not physically present but operates as if it were. Another term for this type of system is a logical system as opposed to a physical system. Let's use the hub to illustrate the idea.

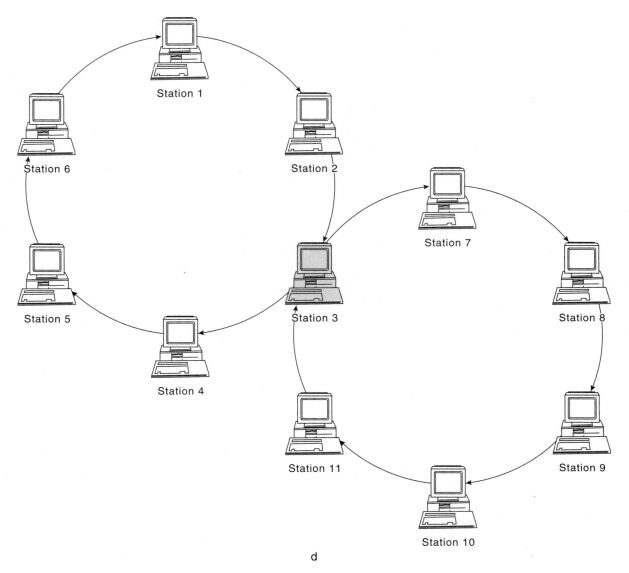

d

**FIGURE 4-3d** Ring with Sub-Ring

In Figure 4-4 we have a physical hub network that looks very much like the one in Figure 4-2. The difference here is that the connections between stations are made first with station 1 connected to 2, 2 to 3, 3 to 4, 4 to 5, etc., until we end up with station 7 connected to station 1. This is illustrated by the arrow connections in Figure 4-4. The arrows indicate the flow of traffic in a ring format from station 1 to 2 to 3 . . . to 7 back to 1. The message handling is done in the same way as it would be in a physical ring. The major difference here is that if a station fails, the hub will reconfigure the system to by-

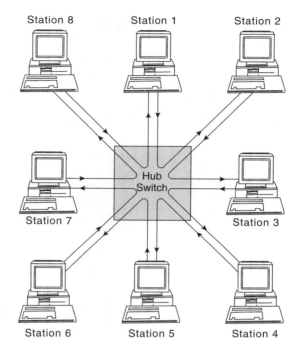

**FIGURE 4-4** Logical Ring

pass that station. For instance, if station 3 fails, the hub would break the connections between stations 2 and 3 and between 3 and 4. It would then connect station 2 to 4, bypassing station 3 entirely. The ring communications continue without including station 3 in the sequence shown by the flow in Figure 4-5.

## Mixed Topologies

Topologies may be mixed in an overall system as shown in Figure 4-6. The key point is that all of the subsystems act independently as long as messages on their systems are destined for other stations on the same system. This allows for a lot of simultaneous traffic to go on. The benefit of mixing systems is to provide a way for diverse networks to communicate with each other and take advantage of the individual benefits of each system.

In Figure 4-6, station A appears to the bus primary as another secondary node on the bus system. All traffic destined for stations other than secondaries 1, 2, or 3 are routed through station A, which acts as a hub for a star system that is comprised of secondaries 4, 5, 6, and the bus system. The hub also provides an access into the ring system, which is connected through another of the hub's spokes. The ring, in turn, includes station C, which provides primary service for the bus topology of secondaries 9, 10, and 11. It is necessary to note that a lot more is involved in interconnecting different networks than is illustrated at this point. Our current purpose is to show you that system topologies can be versatile and interconnected into a cohesive system.

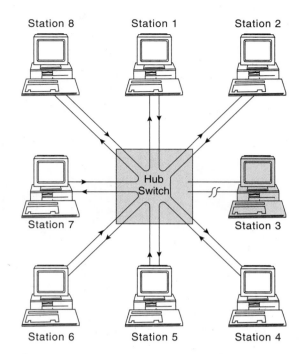

**FIGURE 4-5** Recovering from Station 3 Breakdown

The systems described thus far are fundamental and represent the basic forms that larger, more sophisticated networks take. Local area networks, for instance, utilize all three topology classes, many times intermixing topologies within a well-defined network. Specialized interfacing devices and software manage appropriate protocol and message-format conversions between various subsystems within the entire network. These systems are discussed in succeeding chapters, beginning with a basic two-point communications system and concluding with the numerous forms of networking in place today.

**Section 4.2 Review Questions**

1. List the three network topology types.
2. Which topology can experience contention and collision?
3. What term refers to systems where all the nodes are at an equal level?
4. What is the main disadvantage of a ring system?
5. What would be the advantage of a peer-bus system compared to a primary controlled one?
6. What type of problem does a ring system alleviate compared to a peer-bus system?
7. Name two methods to overcome a ring's main drawback.
8. Define the term *virtual*.

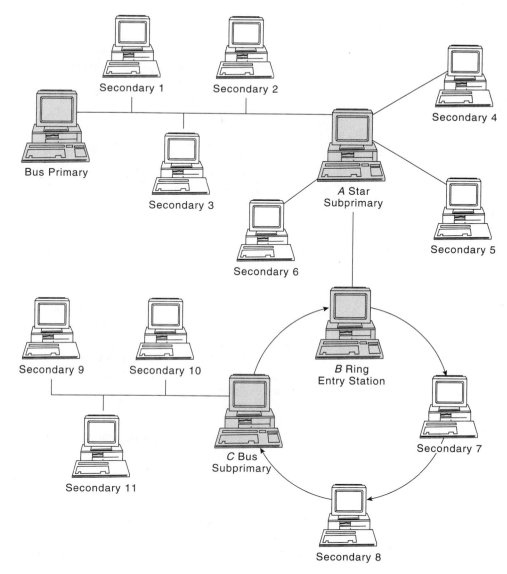

**FIGURE 4-6** Mixed Topology System

## 4.3 DATA SWITCHING

When dealing with multiple nodes or stations on a network, each wanting to send a number of messages to other nodes, there are different ways to manage the actual interconnections between those nodes. One of the most straightforward methods of connecting nodes is illustrated in Figure 4-7. Here the nodes are each directly connected to each

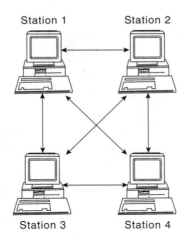

**FIGURE 4-7** Four Stations Directly Connected

other. Each connection is permanent, meaning that a communications link always exists between each node. Each station can communicate with any other station at any given time. This type of data transfer is known as real-time, since messages are sent and received with very little delay. Also, there is no concern of contention for access since each link is separate from all other links.

The main disadvantage of a direct-connected system is in the sheer number of connections required to interconnect all stations. Note in Figure 4-7 that there are only four nodes that are interconnected, but there are six required lines to do the interconnection. That does not seem like a lot, but consider what happens when you begin to increase the number of nodes to be interconnected. To make the process a bit easier to visualize, there is a formula for calculating the required number of interconnections ($N$) for a given number of nodes ($S$). That formula is:

$$N = \frac{(S)(S-1)}{2}$$

(4-1)

For the system in Figure 4-7, substituting 4 for $S$ produces the 6 required connections:

$$N = \frac{(4)(4-1)}{2} = \frac{12}{2} = 6$$

---

**EXAMPLE 4-1**

How many lines are required to directly connect 256 nodes in a single system?

**SOLUTION**

Substituting 256 for $S$ in Equation 4-1 produces our sum:

$$N = \frac{(256)(255)}{2} = 32,640 \text{ lines!}$$

## Circuit Switch

**circuit switch**
A network switching method that physically connects two nodes to facilitate communications.

One alternative to directly connecting nodes in a communications system or network is to use a **circuit switch** in which the nodes are interconnected through a central switch station or hub in a star or hub topology like that in Figure 4-6. For this type of system, the number of connections is equal to the number of stations. For the system in Figure 4-1, 256 lines are needed to interconnect 256 nodes instead of the 32,640 required for direct connections.

Circuit switches (Figure 4-8) utilize a method called store-and-forward to handle the traffic between nodes. Messages are received by the switch from the nodes and stored until the destination node becomes free to accept its messages. The switch then forwards those messages to the destination node. For instance, if station 2 in Figure 4-8 wants to send a message to station 6, it sends it to the switch. If station 6 is currently sending data to another station, then station 2's traffic is stored until station 6 is finished. At that point the message from station 2 is forwarded to station 6. This is definitely not a real-time transfer because frequent delays are experienced on the system.

## Message Switch

Direct connections and circuit switching systems perform their interconnections through a hardware configuration. Two alternate methods involve software solutions to the

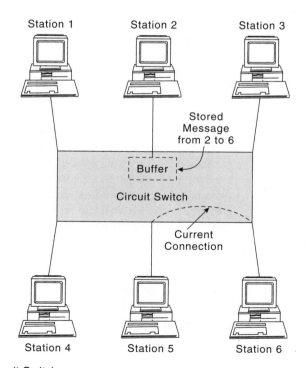

**FIGURE 4-8** Circuit Switch

**message switching**
network switching achieved by storing incoming messages and sending them out as lines become available.

switching problem. **Message switching** utilizes the store-and-forward method to its fullest capacity. In a circuit switch, if the traffic is light, connections may be made immediately without any need to delay the transfer of data. With message switching, all messages are received and stored by the hub and then dispensed later to each destination station. By doing this, the sending stations are immediately freed up once they have sent their messages. They can also send messages destined for several different stations. All of the messages are initially stored. The hub then takes all the messages destined for a given station and sends them one after the other to that station. It does this whenever that station is not in a sending mode and is ready to accept the data.

## Packet Switch

The problems with message switching and the main difficulty with implementing systems using message switching, are the need to allocate a sufficiently large data buffer to hold the incoming messages and the amount of delay it can take for these messages to reach their destinations. Fairly involved software programs are required to manage the routing and storage of these messages. Additionally, if the system is particularly busy, a message could be delayed a long time before it finally is routed to its destination.

**packet**
a small message unit that is part of a larger message.

What is needed is a method of using message switching that allows shorter message "portions," or **packets,** to be sent to each station in a continuous sequence. Each packet contains a header, which includes source and destination identifiers, or addresses, and a packet number. Packets from several sources are routed to their various destinations where they are reassembled into their original messages. The packets are sent continuously as long as there are messages to be sent.

Packets are sent to the various destination nodes in a mixed sequence, which results in stations continually receiving data until all packets are sent and received. For instance, in Figure 4-9, for a four-node system, all four stations could send messages—station 1 to 2, 2 to 4, 3 to 1, and 4 to 2. The messages are divided into packets before they are sent and the packets are routed to their destinations. As such, station 2 could receive a packet from station 1, next station 4 gets one from station 2, then 3 from 1 (notice that station 1 is sending packets to station 2 and receiving packets from station 3).

Station 2 is receiving packets from both stations 1 and 4. Since the packets are short in length, no one station's message is delayed very long—pieces of each message are continually being sent and received. This type of delivery is known as near real-time because of the minimal delay in receiving traffic. The switching method that facilitates this type of delivery is known as **packet switching.**

**packet switching**
a network switching technique that interleaves smaller message units, called packets, and sends them to their destinations.

### Section 4.3 Review Questions

1. How many connections are required to interconnect 100 nodes directly?
2. How many connections are required to circuit switch 100 nodes?
3. Which type of switching is known as real-time and which as near real-time?

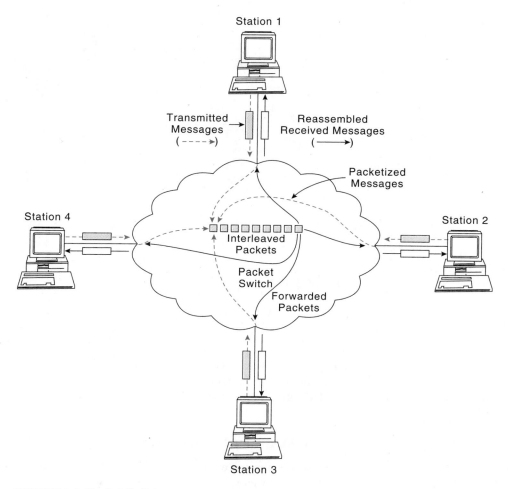

**FIGURE 4-9** Packet Switch

## 4.4  TYPES OF NETWORKS

We have seen what different types of topologies and switching techniques are used by networks. Now we shall take a brief look at how different networks are classified. The basis for classifying networks lies in the physical area that they serve.

### LAN

The very first data networks were developed to interconnect terminals to a mainframe computer located all within the confines of a single building or site as shown in

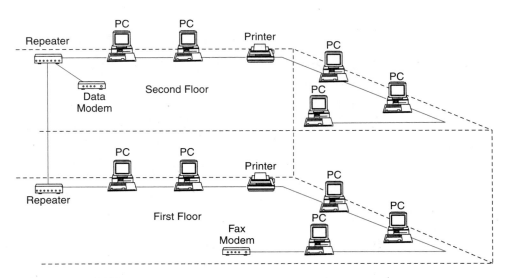

**FIGURE 4-10** LAN

Figure 4-10. These networks were classified as being local to the site, meaning that they did not extend beyond the physical confines of the site. Their designation became known as **local area networks (LAN).**

**local area networks (LAN)**
single building network.

## GAN, WAN, and MAN

The first expansion of networking beyond the confines of a single site extended the network to cover large sections of a country. Something like interconnecting LANs in New York City, Chicago, and Phoenix as in Figure 4-11. Since they covered a wide section of territory, they became known as **wide area networks (WAN).**

**wide area network (WAN)**
large section of a country network.

When banks adopted the use of automatic teller machines, they were connected within the confines of a city and the nearby countryside as illustrated in Figure 4-12. Thus, they became known as **metropolitan area networks (MAN).** When they expanded to interconnect bank branches in several cities throughout the country, they changed from a MAN to a WAN. When they went international, a new designation, the **global area network (GAN),** was adopted. Probably, the best-known GAN is the Internet.

**metropolitan area network (MAN)**
regional network.

**global area network (GAN)**
world wide network.

## CAN

Networks that fell between a LAN and a MAN took on their own designation, which grew out of the most common application for these networks—to tie together many personal computers and other nodes in a university or college campus. By now you have figured this one out—the **campus area network (CAN).**

**campus area network (CAN)**
interconnects several buildings in a restricted area.

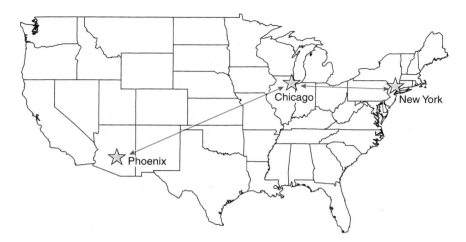

**FIGURE 4-11** WAN

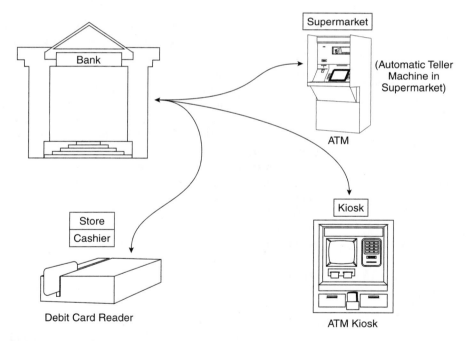

**FIGURE 4-12** MAN

**open systems interconnection (OSI) architecture**
open systems network model standard designed by the ISO organization.

**international standards organization (ISO)**
data networking standardizing committee.

These designations are not just a frivolous exercise. Each type of network has their specific requirements, needs, specifications, and applications. However, they all share one common foundation in that they are all based on a set of standards designed to address their specific needs and goals. A prominent standard that network communications are dependent upon is the **open systems interconnection (OSI) architecture** created and maintained by the **international standards organization (ISO)** consortium.

**Section 4.4 Review Question**

1. Compare the physical coverage of all the network types.

## 4.5 THE OPEN SYSTEMS INTERCONNECTION (OSI) ARCHITECTURE

The flood of companies entering the data communications field throughout the world created a serious problem. Network systems were being created and installed without any concern to the possibility that they may have to interact with other networks. This would only lead to isolation if allowed to continue. Fortunately, a number of organizations evolved with the sole purpose of standardizing all elements of network communications. A number of those organizations have been recognized and accepted by the data communications industry. Among those found in the United States is the **American national standards institute (ANSI), the electronic industries association (EIA),** and the **institute of electrical and electronic engineers (IEEE).** International standards consortium are represented by the **international telecommunications union (ITU)** [formerly the **international consultative committee for telegraphy and telephony (CCITT)**] and by the **international standards organization (ISO).** All of these organizations plus many others are responsible for creating a common set of standards by which multilevel networking can successfully operate. Possibly one of the most significant result of these organization's efforts is the **open systems interconnection (OSI)** communications model developed by ISO.

Open systems architectures are flexible structures set into fixed frameworks. Fundamental activities and requirements are detailed at each level, or layer, of an open-systems network. How these activities and requirements are achieved is left open to individual applications, many of which we shall explore in succeeding chapters.

The concept of an open-systems approach to networking allows any device or system operating with any protocol to communicate with another device or system using its own protocol. Open systems removes the restriction of being forced to operate with a specific set of hardware or software. What makes the open-systems approach work is the ability for communications at a given level to be able to interface with lower and higher communication levels.

The OSI model defines seven distinct levels in its communication model illustrated in Figure 4-13. Each level has a set of specifications and functions that it performs. Any number of communications protocols can operate within a specified level. Messages are formatted according to the rules of that protocol, which might include the addition of header and trailer information, error detection capabilities, and other overhead-type functions. The entire message with its overhead denoted as a **payload** is then encapsulated into the data portion of the next layer's message format and transported using that level's protocol rules. We will see how this operates in future chapters, when we explore the various protocols and systems that operate at each of the OSI model's seven layers.

---

**American national standards institute (ANSI)**

**electronic industries association (EIA)**

**institute of electrical and electronics engineers (IEEE)**

**international telecommunications union (ITU)**

**consultative committee for international telegraphy and telephony (CCITT)**

**open systems interconnection (OSI)**
method by which systems are designed with general specifications to be applicable to any specific protocols and/or physical systems.

**payload**
data unit of a message transmission.

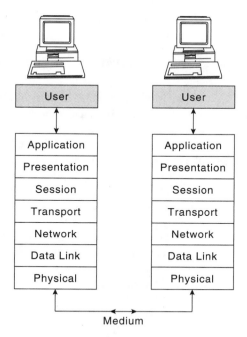

**FIGURE 4-13** OSI Model

## Physical Layer

The physical layer is the lowest layer of the OSI model. It defines the mechanical, electrical, functional, and procedural aspects of the physical link between networks. Mechanical specifications will define items like the connectors and cable types used as the medium of the communications link. Electrical specifications cover aspects that affect the impedance, operating power, attenuation, frequency and bandwidth limitations, and any other electrical quantities that affect the transfer of data.

Functional aspects define the use of any interconnecting lines such as data and control functions. Procedural concerns define the lowest type of protocol rules that are required to assure a good communications link. In Chapter 5 we will look at specific physical-layer protocols and see how each of these specifications apply.

## Data-Link Control Layer

The next layer above the physical layer is the data-link layer, which is the first software protocol layer of the OSI model. This layer assumes that the physical link is in place. The physical link performs its job totally in disregard of the actual meaning of any of the data it transports. To the physical plant, all information it carries is *raw data*—that is, simply ones and zeros without any regard to what the ones and zeros represent or are to be used for. The data-link layer takes on the first responsibility of interpreting some of the binary information being transferred.

The data-link layer is responsible for providing a reliable communications link between two nodes across the physical link. It specifies the data format, sequence, acknowledgment process, and error-detection methods used at the lowest software level. It interprets groups of characters or bits to indicate when a message begins, where it came from, where it goes, whether it is a poll or a selection, and if any errors occurred during the transmission of the data. It does this on a node-to-node basis with disregard to what that node is. That is, the node could be the entry to another network, an end station on a current network, or put to any other use. The data-link process only assures that the message data (payload) it carries is delivered from one node to another. The receiving node will take the payload and pass it along to the next step in the communication system.

The addresses used to determine source and destination nodes at the data-link level are usually the only time a physical address for these entities is used. Data-link addresses will specify a physical node point.

The remaining upper level protocols will use virtual addresses—mainly because they do not know the physical addresses. It is up to the data-link layer to make the final translation between a virtual address and the physical address. The advantage of using virtual addresses in the upper levels is to free those levels from determining the specific physical path their messages must take to be deliverable. Using virtual addresses lets an open system find the most efficient available physical path to send the message through.

**media access control (MAC)**
address management protocol.

**logical link control (LLC)**
link access control protocol.

Two sublayers defined in the data link are the **media access control (MAC)** layer and the **logical link control (LLC)** layer. The MAC layer covers the address management functions and is responsible for network access and control. LLC manages flow and error control, automatic requests for retransmission (ARQ) methods, and message acknowledgments and handshake processes. Chapter 7 describes the most common data-link protocols in use on networks.

## Network Layer

How data are routed through a network is defined in the network layer. Flow control of packet size information as well as congestion avoidance are a concern of the protocols in this layer. Virtual circuits, which are logical routes messages can take through a network, are established and maintained at this level. A single virtual circuit may find any number of physical routes to complete the connection. Packets are formed into frames to include virtual circuit addresses and other pertinent data as well as upper-level payloads. Protocols at the network level are then responsible for the delivery of these frames in the same sequence as they were sent.

There are two types of virtual circuits used at the network layer, connectionless and connection-oriented. These are similar to the same type of telephone line link connections discussed in Chapter 2. Connection-oriented circuits are established before any messages are sent. They are permanent and guarantee a network-level link will always be there when needed. They do have the drawback that they continually use system resources even when there are no messages to be sent. Connectionless circuits are also known as "bandwidth on demand" circuits, which establish a connection when they are needed. The address field of the outgoing message is examined and a connection is created "on the fly" as it is needed to complete the communication. When the message is finished, and if there are no more messages sent between the same two nodes, then the connection is dropped. The advantage is a conservation of system resources that are not maintained during quiet, no-message periods. The downside is that a connection is not guaranteed. If there is a high traffic time, then a node may experience delays trying to get on and establish a link on the network.

Chapter 9 covers communication protocols that address the network layer. These include Integrated Digital Services Network (ISDN), Asynchronous Transfer Mode (ATM), and Frame Relay, which address other layers as well. Chapter 10, covering the Internet, contains possibly the most often-used network-layer protocol, the IP portion of TCP/IP, the Internet Protocol.

## Transport Layer

The interfacing between the applications software and the available hardware is covered in the transport layer. Network connections are created and maintained with consideration toward cost, quality of service, addresses, and error recovery. Essentially, this layer is responsible for the reliable data transfer between two end nodes and is sometimes referred to as the host-to-host layer. Processes at the transport level include:

1. The mapping of transport addresses onto the network.
2. Error detection and recovery to minimize data loss and time lost due to retransmission of frames.
3. Segmentation or fragmentation of messages to maximize transmission efficiency.
4. Flow control between the network layer below and the session layer above the transport layer.
5. Sets quality of service (QoS) for network layer packets to assure end-to-end message integrity and sequence traffic flow control.

**blocking**
keep a node from receiving
messages not addressed to it.

In addition, a process called **blocking** is employed when requested by sending stations. Blocking is used to prevent a particular node from receiving specific frames until the sending node is ready for that receiving node to get them.

There are three types of transport services based on the quality of service each provides. Type A service establishes network connections with an acceptable residual error rate and acceptable rates of signal failures. This type of service is designed to provide methods of acknowledgment, which could signal a retransmission of frames when corrupted data is detected. Type B service maintains acceptable residual error rates but does not tolerate signal failures. In this case, a detected signal failure will cause the communication link to be disrupted until the cause of the signal failure is corrected. Type C service will not pass frames to the sessions layer if errors or signal loss are detected, and will require network layer protocols to correct the problems first. Thus, the transport layer protocol will guarantee the sessions layer a higher quality of service in the delivery of its traffic.

In addition to the three types of service provided by transport layer protocols, the protocols at this level are further classified into one of five classes. Class 0 with the simplest type A quality of service, includes minimal error recovery and is used primarily for straight text transmissions. This class assumes that the network-layer protocols will provide flow control. Class 1, known as the basic error recovery class, operates as a type B protocol, providing increased error recovery capabilities. This class accepts packet type message format at the network level and still relies on that layer to handle flow control.

Class 2 is an enhanced class 0, type A protocol that adds the ability to multiplex communication channels onto a single communication link. This class takes over the responsibility of sequencing packets within the multiplexed environment. Class 3 combines class 1 and 2 services, extending type B error recovery capabilities to the multiplexed communication links. The last of the service classes, class 4, is a type C class, which includes full error detection and recovery capabilities. This class of service assumes that the network layer is inherently unreliable and takes over the job of maintaining the highest quality of service.

In Chapter 10, the Internet, the transport layer is well represented by the transport control protocol, which is the TCP portion of the TCP/IP suite of Internet protocols.

## Session Layer

The remaining layers of the OSI model reach into the higher end of a communication process structure. The session layer concerns file management and overall networking functions. Access availability and system time allocations are included at this layer. Session-layer protocols provide a method by which presentation stations can organize, synchronize, and manage the transfers of information between themselves. Included in these processes is the ability to hold messages until the originating node is ready to release them. In this process, the session-layer protocol is taking advantage of the transport layer's blocking scheme. Further, the session-layer protocol can be directed to discard all data destined for a node without the node being aware that the data was ever sent, if the initiating node requests it through a RESET function. Traffic at the session level can be prioritized if desired by the network manager. Processes at the session level also include:

1. Connection and disconnection of any node from the network.
2. Authentication of user access.
3. The binding of process names to network addresses.
4. Permitting multiple applications to share a virtual circuit.
5. Provide a user interface to a network or a connection between a user and a central host node.
6. Fault recovery if a break in service occurs.

## Presentation Layer

Higher level interfacing requirements, including data compression, format conversion, and encryption, are detailed in presentation-layer protocols. These include character set and code translations, format and syntax resolution, and/or data transformation. At this level, devices are treated as virtual devices or nodes, being defined as logical rather than physical entities. The three basic forms of protocols used in the presentation layer are:

1. Virtual terminal protocol, which is used to allow different types of terminals to support different applications.
2. Virtual file protocol, which handles code conversions within files, file communication, and file formatting.
3. Job transfer and manipulating protocol, which controls the structure of jobs and records.

**abstract syntax notation (ASN)**
encryption method used at the presentation level.

A separate protocol function known as **abstract syntax notation (ASN)** specifies file data structure and the use of encryption to provide a measure of security to communications at the presentation level.

## Application Layer

We finally reached the human-interface level known as the application layer. This layer defines the user interface to lower-level layers and various application processes. The applications layer is separated into several service elements, each addressing distinct application usage. These are:

**common application service element (CASE)**

**specific application service element (SASE)**

**specific user service element (SUSE)**

**file transfer and management (FTAM)**

**file transfer protocol (FTP)**
Internet protocol used to download and upload files.

**simple mail transfer protocol (SMTP)**
e-mail protocol.

1. The **common application service element (CASE)** takes care of services from the lower-layer levels independent of the nature of the user application. This element sets guidelines for the application's required quality of service.
2. The **specific application service element (SASE)** deals with handling large amounts of data, including database access, bulk data transfers, and remote job entry.
3. The **specific user service element (SUSE)** deals with actual user requirements for access and use of the network.
4. **File transfer and management (FTAM)** supports the transfer of files containing data or application programs.

Once again, the Internet provides us with a number of more well-known protocols addressing the applications layer. Among them are these two that are covered in Chapter 10: **file transfer protocol (FTP)** and **simple mail transfer protocol (SMTP).**

**GOSIP, government open systems interconnection profile**

**directory user agent (DUA)**
interface user to system.

**directory system agent (DSA)**
system-to-system interface.

## The Government OSI Profile (GOSIP)

**GOSIP, government open systems interconnection profile,** was adopted as a mandatory requirement for government networks in August 1990. It uses the seven-layer model approach and assigns specific protocol usage to each layer in an attempt to fully standardize government networks. At the applications layer, access is based on the X.500 global directory service specification, which details a hierarchy that a user passes through to gain access to the network. It starts with each user assigned to a **directory user agent (DUA),** who then communicates through a **directory system agent (DSA)** moving deeper into the system. The DSA can communicate to another DUA or pass the access onto another DSA, depending on the type of access the user requires. Eventually, a DSA will connect to the user's requested application allowing the user access to that application and the session can begin.

A major specification at the presentation layer of the GOSIP system is the X.400 e-mail standard originally listed in 1984. X.400 details how e-mail messages are formatted, initiated, and received. The presentation-layer application of GOSIP offers this service in a connection-oriented format.

The session and transport layers are also connection-oriented, possibly because the assumption is that the communications systems would be heavily used and there would be many critical communication networks requiring guaranteed connection services.

Network layer is connectionless, packet-switched service that handles the transport-layer traffic on an as-needed basis, providing a connection path as messages are entered into the network at this level.

The data-link layer uses the high-level data-link protocol that we will study in detail in Chapter 7.

Finally the physical layer has a number of different specifications in use depending on the type of network in use. Specific areas at this layer are defined for fiber optic, coaxial, and copper wire media along with the type of topology—bus, ring, or hub to be used.

While the OSI model is the one that has dominated the network industry, it certainly is not the only model used. **International business machines (IBM)** always attempted to maintain a proprietary system for its network system, yet finally yielded to pressures to make it adaptable to the OSI model. However, for quite a number of years, all IBM network communications systems was based on their **systems network architecture (SNA)** model.

**international business machines (IBM)**

**systems network architecture (SNA)**
IBM's open systems architecture model for network communications.

---

**Section 4.5  Review Questions**

1. What is the advantage of an open-systems type communication model?
2. List the seven layers of the OSI model.
3. Which standards organization developed the OSI model?
4. What is the name of the government open systems model based on the OSI model?

---

## 4.6  SYSTEMS NETWORK ARCHITECTURE

In contrast to the open-systems concept employed by the OSI model, IBM originally developed a closed, proprietary system based on its own standards used solely for IBM computers interconnected through a local type of dedicated network. However, in September 1973, IBM introduced a set of network standards to allow various IBM and non-IBM systems to be interconnected into a common LAN network. This specification, called the systems network architecture (SNA), is a multilayer model similar to the OSI model discussed in Section 4.5.

To some degree the concepts of the SNA model provided the source work for the development of the OSI model. Protocols and specifications for each SNA layer provide a means for data generated at one location to be successfully transmitted to any other location on the network, regardless of the originating and terminating data formats. The actual transformations and physical routing to complete the communications are entirely transparent to the users involved.

By 1980, over 2,500 SNA networks were in use and the number has increased steadily despite the advent of the OSI model. This is directly due to the fact that IBM PCs, as well as other devices and mainframe computers, can be interfaced into an SNA network.

## SNA Layers

Figure 4-14 is a general diagram illustrating the similarities, in a broad sense, between the SNA and OSI communications models. It is important to state at this point that even though similarities between the OSI and SNA functions are shown, there are many detailed differences between the specifications of each model. Drawing parallels between the two is done mainly as a point of reference and should not be taken literally.

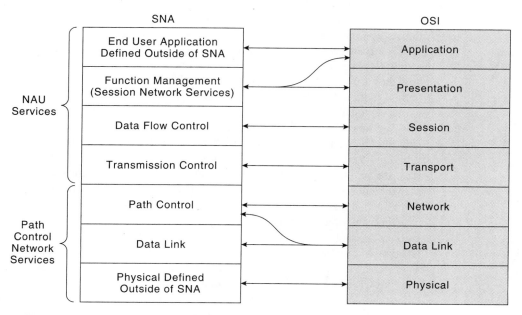

**FIGURE 4-14** Comparison of SNA and OSI Layers

On the lowest scale is the physical layer. The designers behind SNA choose to leave physical specifications to existing standards such as RS232C, discussed in Chapter 5, and to concentrate instead on making physical connections to the SNA network compatible to the network. Physical access to the network is provided through specialized nodes that we will explore later on in this chapter.

Functions included in OSI's data-link layer finds equivalence in portions of SNA data-link and path-control layers. SNA utilizes its own data-link protocol, Synchronous Data Link Control (SDLC), which is thoroughly examined in Chapter 7, to manage data-link access control.

The path-control layer specifies the virtual circuit that a message takes through the system to reach its destination. The use of virtual circuits means that messages will find the most direct route through the network. No physical route is established beforehand, avoiding contention, and unnecessary delays in message routing. However, the SNA standard does allow use of permanent virtual circuits, which are dedicated paths through the system for frequently used data links within the network. In a sense, they are similar to the leased lines used by the telephone company in that the same path is used by two end points each time the data link is established between them. Actual physical paths taken to complete virtual links, even permanent virtual circuits, may differ each time the same link is connected. Message segmenting (similar to the formation of packets) and sequencing are defined at the path-control layer. The setting of transmission priority levels and blocking functions are also found at the path-control layer.

Other parts of the path-control, as well as the transmission-control, layer have equivalent functions in the transport layer of the OSI model. The transmission-control layer is responsible for keeping track of the status of the sessions between users, assuring well-paced and sequenced data flow and possible coding and decoding processes. Message headers, which define request-message and response-message type data are defined in this layer as well.

The fourth SNA layer, the data-flow control layer, deals with responsibilities similar to those of the OSI sessions layer. Identifying the type and mode of data messages and how they are grouped by frame number and address, as well as handling data-flow distribution, is included in this layer. An additional function called bracketing groups related messages together.

The function-management layer, like its OSI counterparts, the applications and presentation layers, supplies the user interface to the network. Details on configuration services, analyzing failures, operator services, interfacing to lower layers, and session services are found in this layer. The conversion of network names supplied by users to network addresses used throughout the system is performed at this level.

Function management services are divided into two sublayers, the function-management data services, which coordinate the interface between the user and the network, and the network addressable unit (NAU) services manager, which is responsible for data-flow control and transmission-control of the lower levels. The services of this level are further categorized as end-user and session-network services. End-user services are further subdivided into session presentation services and applications-to-applications services. The former include establishing a common format between end users and service control between dissimilar devices. Applications-to-applications services include specialized applications functions such as database accesses and protocol updating.

Session network services encompass three subgroups, network operator services, configuration services, and session services. Network operator services facilitate communication among network operators and system service control points (SSCP). System service control points are programs that manage the network and establish and control interconnections that allow users to communicate within the network.

Configuration services are responsible for establishing and disconnecting data links, delivering application programs to SNA nodes, and maintaining network names, addresses, and status. Lastly, the session services perform tasks similar to OSI transport layer services. Capital among these is the translation of logical names into network addresses, which is performed by a sessions-layer control point called a network directory service.

A sixth layer called the application layer is not actually specified, but is, instead, left to external standards and protocols to perform the final user interface into the system. These could amount to the formats and protocols used by the user before they are translated into the formats required by the SNA network. Access to the network is facilitated through the use of a general node point called a **logical unit (LU),** which assures that the incoming data to the network are presented in the correct format.

**logical unit (LU)**
virtual note that services user terminals.

## Logical Units

Logical units are programs that assist in providing translations between the data used by the peripheral devices and the message segment formats required for data handling on the SNA network. There are seven types of logical units (LU) specified by IBM in the SNA model. Some of the more commonly used LUs are:

LU 1—For units that interface directly with SNA character streams. Very little translation is required.

LU 2 and 3—Used to handle IBM 3270 mainframe data streams.

LU 6.2—Used for interfacing units using a general data stream. This data stream does not include headers or trailers, which allows more data to be handled during the same time period as other logical units. The software programs for the LU 6.2 describe the formats and protocols for communication among distributed transaction programs in the SNA network, providing cooperative processing among distributed dissimilar systems connected to the network.

## Physical Units

**physical unit (PU)**
actual hardware that interfaces to user.

**cluster controller**
the hardware and software that manages several stations or nodes.

The physical counterpart to the logical unit is the **physical unit (PU).** Physical units, in their simplest form, include the hardware to interconnect a single peripheral or terminal to the network. Other physical units form the interface for a host computer or front-end processor that could be used to control the network. A more complex physical unit manages the interfacing of many peripheral devices to the network through a device known as a **cluster controller.** Logical units and physical units gain access to the network as software and hardware nodes.

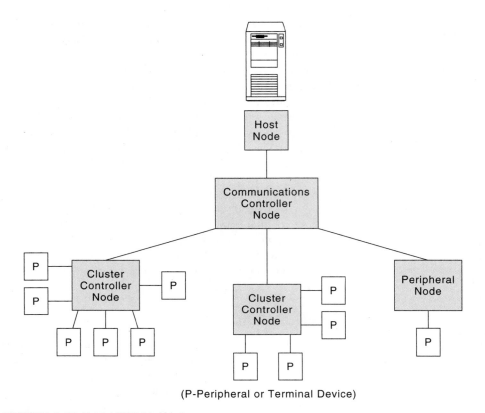

(P-Peripheral or Terminal Device)

**FIGURE 4-15** Basic SNA Network

## SNA Nodes

There are three basic forms of nodes—Host, Communications Control, and Peripheral nodes. Their relationship to the network is illustrated in Figure 4-15 which is a block diagram of a basic SNA system, subarea. The host, or type 5, node provides control over the other nodes within its subarea through system service control-point software. Often the host node contains a peripheral device directly connected to it to allow human interface to the node.

Communications controllers (type 4 nodes) are a subarea node that does contain a system service control point. They maintain control over the peripheral nodes connected to them. Generally, they are interfaced to a host node, providing indirect connections between the host and user peripherals.

At the lowest level of node control are the physical controllers. These fall into two main categories, the peripheral node (type 1 node), which interfaces to a single peripheral device, and the cluster controller (type 2 node), which manages connections to many peripherals. An additional type 3 node is set aside but not yet defined by the SNA standard.

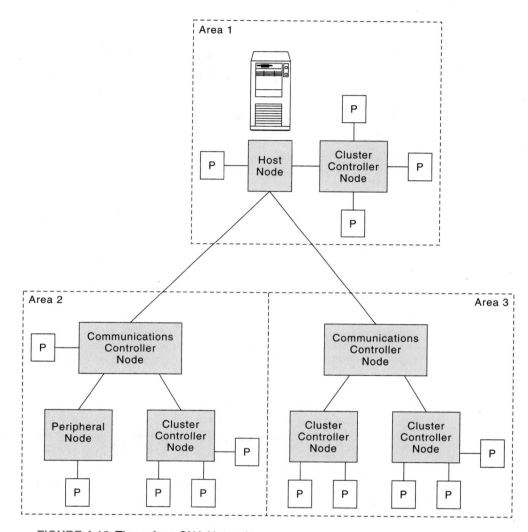

**FIGURE 4-16** Three-Area SNA Network

The system may be expanded into several subareas as shown in Figure 4-16. Here the host node has direct control of a cluster controller within its subarea and other communication controllers within two other subareas. Additional communications controllers within the subareas are again interfaced to peripheral control nodes of varying sizes. It is the host node that has direct access to the network path.

An additional unit classification detailed in the SNA standard is the **network addressable unit (NAU).** NAUs are functional areas that encompass a designated set of logical units, physical units, and system service control points to provide services necessary to

**network addressable unit (NAU)**
SNA subunit responsible for moving data through the network.

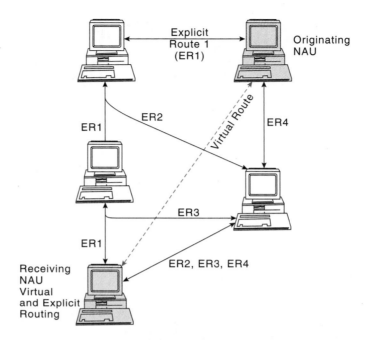

**FIGURE 4-17** Virtual and Explicit Routing

move information through the network. Each NAU area has an address that is used to identify itself to other NAUs.

Lastly, a path control network (PCN) is a low-level component that controls the actual physical routing and flow of data through the system. Here explicit routes are assigned to virtual routes as the need arises (Figure 4-17) to assure rapid and accurate data transmission between nodes. Note that there are several possible physical or explicit routes for a given virtual route.

In Figure 4-17, the virtual route, indicated by the dotted line, connects an originating NAU in the upper right corner with a receiving NAU in the lower left. The explicit routes that could be taken to complete this virtual route start from the originating NAU along line ER1 or ER4. You can follow ER1 through the intermediary node in the upper left corner down to the receiving station. A second route (ER2) leaves the upper left node and travels through the node in the lower right corner before heading to the receiving station. Now follow ER1 to the middle left node and there is another possible routing (ER3) directed through the lower left node. Path ER4 is shown going through the lower right node and on to the receiving station, but there could be additional explicit routes, which are not shown, that can complete the connection back along ER2 or ER3 from the lower right node. Each of these explicit routes would all achieve the same result, to connect a virtual communications path between the originating and receiving nodes.

**Section 4.6  Review Questions**

1. Why do you think IBM opened its network model to a more open systems type of network?
2. Which OSI layers correspond to these SNA layers?
   a) Path-Control Layer.
   b) Transmission-Control Layer.
   c) Data-Flow Layer.
   d) Function-Management Layer
   e) Application Layer
3. What are the main functions of a logical unit (LU)?
4. What is the function of a cluster controller?
5. What is the difference between a virtual route and an explicit route?

## 4.7  SNA OPERATING SESSIONS

Each action throughout the system, from the time data entry is made via a physical unit until it leaves the network through another physical unit, is defined as a SNA operating session. A session is a logical condition connecting two entities through a communication link to facilitate a succession of data transmissions between them. For instance, logical unit to logical unit sessions provide the actual means for end users to communicate with one another. These sessions are established on a "need-to-use" basis to minimize unnecessary activity on the network. The circuit established by this type of session is referred to as a **switched virtual circuit (SVC).**

**switched virtual circuit**
virtual path through a network that is created when it is needed as a message is being transmitted.

System service control point to system service control point sessions are established on a multiple domain (subarea) network at the time the network is engaged and they remain active until the system is shut down. They are used to allow the necessary control service information to be passed between system service control point units, providing cross information transfers between subarea domains. A domain defines a set of SNA resources that are known to, and managed by, a systems-service control point.

**permanent virtual circuit (PVC)**
logical path through a network that is there once the system or network is engaged and remains there until it is shut down.

System-service control point session circuits are often called a **permanent virtual circuit (PVC)** since they remain in place from the time the network is up and running until it is shut down. The SVC and PVC concept will reappear in applications for many other protocols.

Within a given domain, a session between the system-services control point and its logical units is required before the logical unit (and hence the end user) can access the network. Keeping in mind that system-service control points establish the services they manage and control the interfaces to the network, the need for a session between it and the logical units within its domain becomes apparent.

There are also sessions between the system-services control point and any physical units directly connected to the controlling unit. These physical units provide direct human interface for the controlling processes through peripheral devices connected to the physical unit. There are no specific sessions between physical units. However, communications between adjacent physical units in any subarea may be required to transfer applications programs or to provide initialization or system shutdown commands.

Once the system is active and the necessary sessions are established, data in any format and from any source can be placed on the network through a physical unit directed and interpreted by the logical unit's software. Within the logical unit, system-service control point programs manage the translation of the entered data to the format usable by the SNA network. Once the translation is completed, the data are dynamically routed to their destination via the most direct route available. This route exhibits the least propagation delay between communicating users.

At the receiving logical unit, the data format is translated, if necessary, into the form usable at the receiving end-user station. The transformed data are sent to the user through a physical unit. Because of the separate translations at the originating and receiving ends, the protocols and formats at each end can be unique and different. Devices regular serviced by the SNA network include IBM personal computers and mainframes, workstations, line printers, display terminals, and network servers (which are discussed in more detail in a later chapter).

### Section 4.7 Practice Questions

1. Define *session.*
2. List the types of SNA sessions.

# SUMMARY

The design, creation, and operation of various networks have been facilitated by adherence to the concepts and standards outlined in an open-systems architecture. Open systems refers to the ability to use a communications framework without concern about specific operating platforms or protocols. Any appropriate protocol or system designed to fit into an open-systems model will function as expected. The two prime standards for open-systems architecture models are International Standards Organization's Open Systems Interconnection (OSI) model and IBM's Systems Network Architecture (SNA) model. Both list a set of standards and requirements for protocols that operate in one or more of each model's layers. The layered approach sets a hierarchy of communications protocols from the user entry point down to the actual physical medium that carries packets of information. Open-systems modeling allows for considerable flexibility in actual equipment and protocol usage at each layer within a complete system.

# QUESTIONS

## Section 4.2

1. What is a topology?
2. List the basic network topologies.
3. Define bus topology. What is its main drawback?
4. Compare a primary-controlled bus network to a peer bus network.
5. What is meant by a poll on a bus network?
6. What is meant by a selection on a bus network?
7. What is the benefit of a peer-bus network compared to a primary-controlled bus network?
8. What is the main difficulty with a peer-bus network?
9. What must each station on a peer-bus network do before it can attempt to gain access to the bus?
10. What is meant by a collision on a peer-bus network?
11. How does a ring network overcome access contention?
12. What is a ring network's main drawback?
13. Describe how a hub-based virtual ring maintains the ring when a node fails.
14. How are dual rings used to maintain ring integrity when a node fails?
15. What is the name of the circulating message on a ring when none of the nodes has any traffic to send?
16. When is a message removed from a ring?
17. In a virtual ring, each node is connected to the hub in a star configuration. Why is this network called a virtual ring if it really is a hub or star network?
18. Does a hub network always have to be a virtual ring?

## Section 4.3

19. Direct connections between nodes have the advantage that there is always a link between each station. What is the main drawback of a directly-connected system?
20. How many connections are required to interconnect 1,024 nodes into a network using direct connections?
21. Why are direct connections referred to as real-time links?
22. What type of switched connection can also operate in a real-time mode?
23. Contrast message and packet switching.
24. Which switching technique is known as near real-time? Why is it referred to as near real-time?
25. Which switching technique is also known as store-and-forward?
26. How does a store-and-forward system affect the delivery of data traffic?

## Section 4.4

27. The original three network types were LAN, MAN, and WAN. Describe how they differ from one another.
28. CAN and GAN network designations have been added to describe networks that are different from the original three. How do they differ?

**Section 4.5**

29. What is meant by an open system?
30. Why are establishing standards important in network communications?
31. What is a payload?
32. What is raw data?
33. What are the responsibilities outlined in the physical layer of the OSI model?
34. What is the significance of the data-link layer of the OSI model?
35. What is differences between addresses at the physical and data-link layers of the OSI model?
36. What is the advantage of using virtual addresses?
37. What does the MAC layer define?
38. What does the logical link control layer define?
39. What is the difference between connectionless and connection-oriented circuits?
40. Given an advantage for a connectionless circuit and one for a connection-oriented circuit.
41. Which OSI layer defines how data is routed through a network using virtual addresses?
42. Name two routing protocols.
43. Which OSI layer first addresses concerns over quality of service and error recovery?
44. Define what is meant by blocking.
45. How do type A and type B transport services differ?
46. What is the significant difference in type C transport service that the other two do not do?
47. List three of the processes attributed to the sessions layer of the OSI model.
48. List three things that are addressed in the presentation layer of the OSI model.
49. What is the main purpose of the applications layer of the OSI model?
50. What is the purpose of the abstract syntax notation protocol?
51. What is the main standard specification used at the applications level of GOSIP?
52. What is the main standard specification used at the presentation level of GOSIP?

**Section 4.6**

53. What physical specifications are detailed in SNA's lowest layer?
54. Which SNA layers deal with establishing the data link between nodes?
55. Which SNA layer specifies the virtual circuit used to deliver messages to a node point?
56. List two functions of the path-control layer besides handling virtual addresses and circuits.
57. What is the term that describes grouping related messages together at SNA's data-flow control layer.
58. What three subgroups are SNA session layer services divided into?
59. What is a system-service control point?
60. Which sessions control point is responsible for translating logical addresses to network addresses?
61. Which SNA applications level unit is responsible for translations between user protocol formats and SNA formats?

62. Which SNA applications unit is responsible for the hardware to interconnect peripherals to the SNA network?
63. Describe the purpose of a cluster controller.
64. What is the function of a host node in a SNA subarea?
65. What is the function of a communications controller node in a SNA subarea?
66. What is the function of a physical controller node in a SNA subarea?

**Section 4.7**

67. Define an explicit route.
68. How are explicit routes used to implement a virtual route?
69. What is another name for the type of service provided by a switched virtual circuit?
70. What is another name for the type of service provided by a permanent virtual circuit?

# DESIGN PROBLEMS

1. You want to connect six personal computers in the quality assurance department into a local area network. Determine what type of switching you would use for the network and support your choice.

2. At a later date, you are directed to redesign the network in problem 1 to add ten workstations located on the production line. These workstations are to be on a separate network from the quality assurance network. However, both networks are to be interconnected. Which type of network would be considered? What type of switching would you use for the production line? What type of switch would you use to interconnect the two networks? Support all of your choices.

# ANSWERS TO REVIEW QUESTIONS

**Section 4.1**

1. A network is a system of interconnected communication stations.
2. A protocol is a set of rules and specifications for the transfer of data over a communication link.
3. A node is an access point in a network.

**Section 4.2**

1. bus, ring, and star (or hub)
2. bus
3. peer-to-peer
4. one station is down, the ring is down
5. Any node can communicate with any other node, which speeds up communication since they do not have to go through the primary to communicate.
6. Alleviates access contention.

7. Dual rings and hub-based virtual rings are used to overcome a ring's basic downside.
8. Virtual means that the entity is not physically present but the system operates as if it is.

### Section 4.3

1. 4950 connections
2. 100 connections
3. real-time: direct connections and some circuit switched connections near real time: packet switching

### Section 4.4

LAN—single building
MAN— "city" wide area
CAN—campus area covering several sites
WAN—covers major portion of a country
GAN—worldwide

### Section 4.5

1. Communications can be established using any protocol or hardware.
2. physical, data-link, network, transport, presentation, session, and applications
3. International Standards Organization (ISO)
4. GOSIP

### Section 4.6

1. To allow interaction and interfacing to other network systems.
2. a) network and data-link
   b) transport
   c) session
   d) presentation and applications
   e) applications (or none)
3. Logical Unit (LU) is a general node point that translates between user data format and SNA format.
4. A cluster controller is a device that manages a number of peripheral devices and provides access for them to the SNA network.
5. A virtual route indicates which two nodes are linked together. An explicit route is the actually physical route that makes the connection between those end points.

### Section 4.7

1. A session is a logical condition connecting two entities together through a communications link.
2. LU to LU, SSCP to SSCP, SSCP to LU, and SSCP to PU

# CHAPTER 5

# OSI Physical Layer Components

## OBJECTIVES

After reading this chapter, the student will be able to:
- convert between parallel and serial digital data using a UART.
- determine how a modem converts between digital data and analog symbols.
- define the difference between FSK, PSK and QAM modems.
- understand how physical interfaces allow computer terminals and modems to function together.
- track the communication flow of data between two end points using the physical components defined in this chapter.

## OUTLINE

5.1   Introduction
5.2   Units of a Communication Link
5.3   RS232C Interface
5.4   RS449 Interface Standard
5.5   RS422 and RS423 Inteface Standards
5.6   FSK MODEMS
5.7   Additional Types of MODEMS
5.8   V.34 and V.90 MODEMS

## 5.1 INTRODUCTION

**frequency shift keying (FSK)**
modulation technique in which the frequency of a tone is altered by the state of binary bit data.

Early home data communications applications revolved around the use of a **frequency shift key (FSK) Modem** connected to a **personal computer (PC).** Today, operating with FSK links is considered low-speed data communications, utilizing data rates between 300 and 2400 bps. Low- and medium-speed (2400–9600 bps) communication systems

**universal asynchronous receiver/transmitter (UART)**

**modulator/demodulator (MODEM)**
used for converting between digital data and analog data forms.

consist of the data terminal, such as a PC, with an internal **universal asynchronous receiver transmitter (UART),** and a **modulator/demodulator (Modem)** at each communication site.

## 5.2  UNITS OF A COMMUNICATION LINK

The basic units of a low-speed data communications system are illustrated in Figure 5-1. Regardless of the original source of information, it is eventually presented to the communications network as a parallel binary word, usually of either 8 bits (byte), 16 bits (word), or 32 bits (long word or double word). The deciding factor on the size of the parallel data (and, hence, the actual devices used in the data communications system) is the size of the data bus employed by the terminal or personal computer that supplies the data. The functional block that receives the parallel data from the terminal and converts it to a serial stream of data to be sent to a receiving station is the UART.

### UART

The UART accepts the parallel data information and converts it into an asynchronous serial data stream. Asynchronous data streams require that each character in the data stream

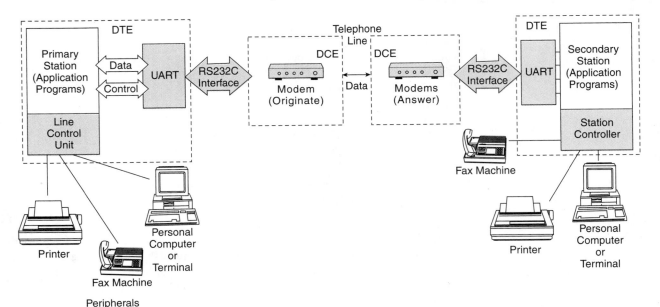

**FIGURE 5-1** Basic Units of a Low-Speed Data Link

begins with a start and ends with a stop-framing bit. Many include a parity bit for error detection. These bits are added to the character data by the UART. On the receiving side, the UART detects the presence of a start bit at the beginning of a data character and reads in the serial stream. After removing the framing bits (often referred to as "stripping them") and checking for parity errors, the data is converted back to parallel form and supplied to the terminal via a data bus interface.

Once a communication link has been physically established, the transmitting side of each station places a mark, or logic 1 level, on the data lines. As long as the data link remains active without data present on the line, the line remains in an **idle line one,** or **marking,** state. The beginning of a digital character is detected by the reception of a start bit, which is a change from the mark condition to logic zero, or **space** state. Character information bits follow the start bit, then the parity bit and finally the stop bit. The number of stop bits actually determine the minimum time a line must remain at the idle mark condition between characters. Each character is synchronized with the UART's receive circuits by the detection of the start bit. Figure 5-2 shows a complete asynchronous serial data stream for an ASCII character *Q*.

There are quite a number of UART devices on the market today. Most of them are manufactured on a single IC chip, which is incorporated onto a circuit board installed in the computer terminal itself. One such IC, the **asynchronous communications interface adapter (ACIA)** produced by Motorola is used in this text to illustrate the application of UART functions. The ACIA is selected since it has the basic elements of most UARTs. Figure 5-3 is a block diagram and pin assignment for the MC6850 ACIA.

The ACIA is comprised of three main functional units: transmit, receive, and control units. It was designed to be used with Motorola's 6800 microprocessor-based computing systems. However, it is sufficiently universal for application with any **central processing unit (CPU).** A CPU is the "brains" of a computer system such as an IBM PC. It directs the operation of all the units in the computer by fetching and executing instructions. Control signals from the CPU direct the movement of data and the execution of operations within the computer, based on the decoding of the instructions.

Eight bidirectional data lines (D0–D7) interface between the ACIA and a computer's data bus to allow parallel byte size data to be transferred between the CPU and ACIA. Three chip selects (CS0, CS1 and $\overline{CS2}$) along with a register select (RS) input lead allow the CPU to directly access the registers within the ACIA. The chip selects are asserted as a result of address-decoding circuitry within the computer. The address is a part of the instruction that identifies where data is to be fetched or sent by the CPU. The selection is made by decoding the address placed on the address bus by the CPU. The decoded signals are used to turn on a specific chip's chip selects (CS).

**idle line one**
state of a communication line when active and not handling messages.

**marker or marking state**
logic 1 state.

**space**
logic zero state.

**asynchronous communications interface adapter (ACIA)**
Motorola's version of a UART.

**central processing unit (CPU)**
controlling unit of a computer.

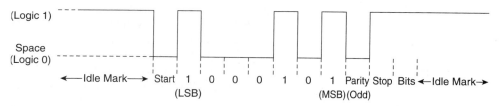

**FIGURE 5-2** Asynchronous ASCII Q

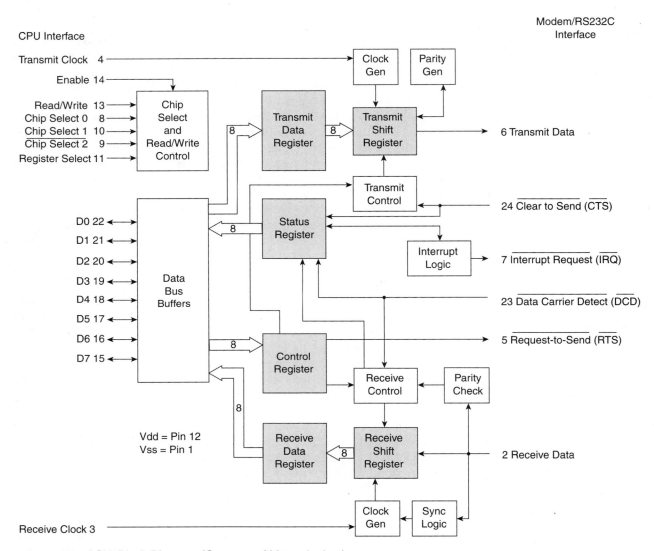

**FIGURE 5-3** ACIA Block Diagram *(Courtesy of Motorola, Inc.)*

The register select (RS) is usually driven by the lowest address line and is used, along with the R/W line, to select which register in the ACIA is being accessed by the CPU's address in accordance to Table 5-1.

The enable (E) input is connected to the computer's clock signal to provide timing and synchronization of ACIA register access to the CPU. The remaining CPU interface signal is the R/W (Read/Write) input line that determines if data is being read from or written into the ACIA.

**TABLE 5-1**

**ACIA Addressing**

| RS | R/W | Register |
|----|-----|----------|
| 0 | 0 | Control Register |
| 0 | 1 | Status Register |
| 1 | 0 | Transmit Data |
| 1 | 1 | Receive Data |

On the modem interface side, transmit (TX) and receive (RX) data lines allow serial data transfers between the ACIA and the modem. The transmit and receive clocks (TX CLK and RX CLK) are used to establish the transmission data rate and the rate at which the receiver samples and shifts in data. $\overline{CTS}$, $\overline{RTS}$ and $\overline{DCD}$ are the clear-to-send, request-to-send, and data-carrier-detect handshake signals discussed in the next section on the RS232C.

Before the ACIA can be used to send and receive data, it needs to be programmed by the computer's operating software to configure it to transfer data in a given format and at a specific data rate. This programming is accomplished by writing a word of control data to the ACIA's control register.

## Data Character Options Found in UARTs

To be universal, a UART—and the ACIA is no exception—has to be able to recognize different data character formats and operate at different data rates. Since system data bus sizes are usually fixed, a byte-size data bus forces 8 bits of digital data to be sent to the ACIA by the central processing unit. However, not all 8 bits are necessarily used because some character codes use less then 8 bits to define a character. The original ASCII code is an example of such a code since it only requires 7 bits to represent a single character. Other codes like EBCDIC and extended ASCII use 8 bits per character. The ACIA is flexible enough to manage selection between 7 or 8 bit data words.

The number of stop bits, which determine the minimum intercharacter idle time, is selectable between 1 or 2 bit periods as required by the hardware or software protocol being used.

Error checking for asynchronous data is commonly performed by the use of parity as explained in Chapter 3. UARTs, including the ACIA, provide selection options of using an even or odd parity system or not to use parity at all. In the first two cases, the appropriate parity bit state for each character is generated and added to the character's data stream. When selecting no parity, some UARTs add an extra bit that is always in a low state. Others do not add a bit if no parity is selected. The ACIA falls into the latter category.

The last operation performed by a UART is a combination of hardware signaling and software programming. Parallel data is converted to serial data and shifted out at a specific rate. This data rate is determined by the frequency of the TX CLK input and an internal divide ratio set in the ACIA. The clock frequency is divided by a ratio factor that generates the shift timing for the serial data stream. For the ACIA, the ratio options available are 1, 16, and 64, with 16 being the most commonly used.

Examples of de facto standard data rates used by frequency shift key (FSK) systems are 100, 300, 1200, and 2400 bps. The 100 and 300 bps rates have now all but disappeared from use while the 2400 bps rate has emerged as the highest standard rate used for FSK communications on telephone lines.

---

### EXAMPLE 5-1

A serial data stream is to be transmitted at a rate of 1200 bps. What is the frequency of the transmit clock (TX CLK) required to generate that data rate if a ratio of 16 is used?

### SOLUTION

The internal circuitry of the UART divides the TX CLK by the ratio factor to produce a shift pulse. This shift pulse determines the rate at which the serial data bits are transmitted. For this example, a TX CLK of 19.2 KHz is required to generate a data stream at 1200 bps using a ratio of 16.

---

The options discussed in this section are selected by programming the appropriate state of the bits in the ACIA control register (Figure 5-4), which is accessible by applying

| 7 | 6 | 5 | 4 | 3 | 2 | 1 | 0 |
|---|---|---|---|---|---|---|---|
| Receive Interrupt | RTS | Transmit Interrupt | Word Size | Stop and Parity | | Counter Divide | |
| 0 = Dis | 0 = On | 0 = Dis<br>1 = En | 0 = 7 bit | 0 = 2 Stop<br>1 = 1 Stop | 0 = Even<br>1 = Odd | 0   0 = ÷ 1 | |
| | | | | | | 0   1 = ÷ 16 | |
| 1 = En | 1 = Xmt (Transmit Interrupt Enable) | 0 = RTS Off<br><br>1 = RTS on Transmit Break | 1 = 8 bit | 0 = No parity | 0 = 2 Stop<br><br>1 = 1 Stop | 1   0 = ÷ 64<br><br>1   1 = Reset | |
| | | | | 1 = 1 Stop | 0 = Even<br>1 = Odd | | |

FIGURE 5-4 ACIA Control Register

an address that turns on the ACIA's chip selects with address line A0 low. This sets the register select (RS), connected to A0 low. A write operation using this address, sends an 8-bit data word to the ACIA control register.

The two least significant bits, b0 and b1, of the control register serve two purposes. When they are both high, they cause the ACIA to reset, which clears any pending interrupt requests and sets status register flags to their inactive states, except the clear to send (CTS) and data-carrier-detect (DCD) signals, which always reflect the state of those inputs to the ACIA.

The other combinations select the divide ratio between the transmit clock and data rate on the transmit data line. This same ratio is used to set the sampling rate for the ACIA receive side. Control register b1 is set low and b0 high to select the divide by 16 ratio. When this is done, the data rate is one sixteenth of the transmit clock (TX CLK) frequency. The opposite states of b1 and b0 select a divide ratio of 64, which would require a clock frequency of 19.2 KHz to produce the 300 bps data rate of Example 5-1. If b0 and b1 are both set low, the divide ratio selected is 1, which means the data rate and clock frequency would have the same value.

Control register bits b2, b3, and b4 are used to select the character format to be used. b4 allows a choice between 8 or 7 bit character length. When b3 is high, the number of stop bits is set to one. When this bit is low (most of the time), the number of stop bits is two. The one exception occurs when b4 is high and b3 is low. For this case, b2 is used to select between 1 or 2 stop bits as shown in Figure 5-4. Additionally, for this set of conditions, parity is not being used. For other combinations of b4 and b3, b2 is used to select between even (b2 = 0) and odd (b2 = 1) parity. Possibly, a more clear look at these two bits can be facilitated with Table 5-2.

Control register bits b6 and b5 determine whether transmit interrupt requests are enabled or disabled. These bits also effect the state of the request to send ($\overline{\text{RTS}}$) handshake output of the ACIA. Details on the RS232C handshake signals are in Section 5.3. Lastly, control register bit b7 enables or disables receive interrupt requests.

## TABLE 5-2

### ACIA Character Format

| b4 | b3 | b2 | Data Size (Bits) | Stop | Parity |
|----|----|----|------------------|------|--------|
| 0 | 0 | 0 | 7 | 2 | Even |
| 0 | 0 | 1 | 7 | 2 | Odd |
| 0 | 1 | 0 | 7 | 1 | Even |
| 0 | 1 | 1 | 7 | 1 | Odd |
| 1 | 0 | 0 | 8 | 2 | None |
| 1 | 0 | 1 | 8 | 1 | None |
| 1 | 1 | 0 | 8 | 1 | Even |
| 1 | 1 | 1 | 8 | 1 | Odd |

**EXAMPLE 5-2**

What hexadecimal data is sent to the ACIA control register to configure it to handle ASCII asynchronous data with odd parity and two stop bits. Disable both interrupts and assert $\overline{RTS}$. Use a divide ratio of 16.

**SOLUTION**

ASCII uses a 7-bit character code, which requires b4 to be low. b3 and b2 are set to 0 and 1, respectively, to pick odd parity and two stop bits. b5 and b6 are low to disable transmit interrupts and assert $\overline{RTS}$. Receive interrupts are disabled by making b7 low. Lastly, b1 is set low and 0 high to select a divide ratio of 16. In binary, the control register looks like:

| b7 | b6 | b5 | b4 | b3 | b2 | b1 | b0 |
|----|----|----|----|----|----|----|----|
| 0  | 0  | 0  | 0  | 0  | 1  | 0  | 1  |

which is 05 as a hexadecimal value.

## UART Receiver

The reason for using a divide ratio between the input clock frequency and the data rate is to assist the receiver in assuring a successful data reception by the UART. The input-receive data line is constantly monitored for a character's start bit, which is sensed as a change from an idle line one to a zero logic state indicated as point A in Figure 5-5. An ASCII $Q$ discussed earlier is used for this illustration. Once a low-level state is detected, it is necessary to verify that this condition is an actual start bit (as opposed to a noise pulse or drop out condition). To do this, the receiver monitors the receive data line for half a bit time period, which is point B in Figure 5-5. For a divide ratio of 16, there are 16 clock cycles per data bit time, so it takes 8 clock cycles to occur before the center of the start bit can be verified. If the level is still low at this time, the low condition is accepted as a start bit and the character can be shifted in.

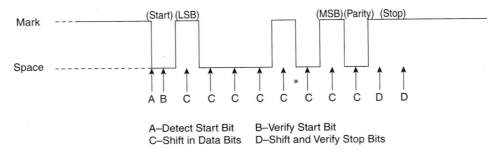

A–Detect Start Bit     B–Verify Start Bit
C–Shift in Data Bits   D–Shift and Verify Stop Bits

**FIGURE 5-5** Receive Timing/Data Relationship for ASCII Q

Shift pulses, points C in Figure 5-5, are generated every 16 clock cycles following the verify of the start bit. This shifts the data bits into the UART receiver at the center of each bit period given ideal circumstances (the transmit clock at one end is exactly the same frequency as the receive clock at the other end of the data link). The shifting continues until the last stop bit, point D, is pulled into the receive shift register. The UART circuits then resume monitoring the receive data line for the next start bit, repeating the process for each character received. Because of this, the synchronization of each character occurs with each start bit.

Because each data bit is shifted in at its center, the transmit clock of the transmitter and receive clock of the receiver do not have to be *exactly* the same. It would be fairly difficult for these clocks, one at the originating station and one at the answering station to be exactly the same frequency. They are as close as possible, but rarely exactly the same frequency. This difference is called *clock skew* and allows the receive clock to be significantly different from the transmit clock without loss of any data information. The reason for this is that the shift process is reinitialized with every start bit. As long as the receive clock frequency allows the sampling and correct shifting in of every bit of a single character word, successful data reception occurs.

Dividing the clock frequencies by 64 instead of 16 divides the sampling rate into smaller increments. This increases the accuracy in the detection and shifting in of the characters. It also provides for larger variations between the receive and transmit clock frequencies.

## Checking UART Performance

The ACIA, like most UARTs, contains a status register to facilitate the transfer of data between it and a terminal or CPU and to check the progress of the UART's operation. Figure 5-6 shows the ACIA's status register. Only one bit in that register is associated with the UART's transmit function. This bit is the transmit-data-register-empty (TDRE) bit. When a word is sent to the ACIA to be converted to serial and shifted out, it is first

| b7 | b6 | b5 | b4 | b3 | b2 | b1 | b0 |
|-----|-----|------|-----|-----|-----|------|------|
| $\overline{IRQ}$ | PE | OVRN | FE | $\overline{CTS}$ | $\overline{DCD}$ | TDRE | RDRF |

RDRF — Receive data register full
TDRE — Transmit data register empty
DCD  — Data carrier detect
CTS  — Clear to send
FE   — Framing error
OVRN — Overrun
PE   — Parity error
IRQ  — Interrupt request

**FIGURE 5-6** ACIA Status Register

loaded into a data buffer register and then transferred to a shift register. Since the data is now in the shift register, the buffer register is empty and available for the next data word. This is indicated by a logic 1 state in the TDRE status bit. The CPU can then transfer another word to the ACIA. It is loaded into the transmit buffer register and remains there until the other word is fully shifted out.

The TDRE bit remains low as long as data still occupies the transmit buffer register. After the previous data word is fully shifted out, the current contents of the buffer register is transferred to the shift register and the buffer register, again, becomes available (empty) for another data word indicated by a logic 1 in TDRE once more.

A similar indication is provided for the receive side by the *receive-data-register-full (RDRF)* bit. This bit is high when a received data word is fully shifted into the UART and transferred from the receive shift register to the receive buffer register. The CPU is obligated to read the received data word from that buffer before the next data word is fully shifted in. If the application program fails to perform this read, the next word is transferred to the receive data buffer and the **overrun (OVRN)** error status bit is set.

Other error indications in the status register are parity error (PE) and framing error (FE). Parity error indicates corrupted data and framing error detects the absence of a stop bit after the data word has been fully received.

An **interrupt request (IRQ)** indication causes an interrupt request signal to be sent to the CPU under any one of the following conditions, if the interrupt request function has been enabled in the control register:

1. TDRE or RDRF is high.
2. a receive error has occurred.

The last two status bits monitor two handshake signals from the modem, clear-to-send (CTS) and data-carrier-detect (DCD). The functions of these signals are discussed in Section 5.3 on the RS232C interface.

To verify that transfer of serial data is correct, a regular oscilloscope can be used, but it is difficult to keep track of the data as it is sent one bit at a time. A special type of test equipment called a **datascope** is available to facilitate the reading of serial data. It displays and holds a serial stream of data after the data has been applied to the datascope's vertical input. The first bit sent to it would be the start bit. This is followed, in time, by the least significant bit of the character word, the remaining character bits, parity, and stop bits. Figure 5-7 illustrates how a transmitted character would appear on the datascope.

**overrun (OVRN)**
error indication of when a computer fails to read incoming data in time.

**interrupt request (IRQ)**
interrupt request signal informing the CPU to divert from its current program processing to execute a new program.

**datascope**
test equipment used to monitor a serial data stream.

---

### Section 5.2 Review Questions

1. Which ACIA control pins are used to allow selection of the ACIA control register to be accessed by the CPU? What state must they be in?
2. Which handshake signals are provided for control signals between the ACIA and modem?

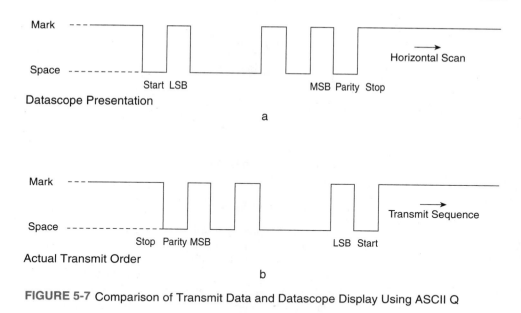

**Datascope Presentation**

a

**Actual Transmit Order**

b

**FIGURE 5-7** Comparison of Transmit Data and Datascope Display Using ASCII Q

3. What clock frequency is required to achieve a 1200 bps data rate with a divide by 64 ratio?
4. What are the character data sizes selectable using the ACIA control register?
5. In order to use parity with an 8-bit data size, how many stop bits can you have using the ACIA?
6. What is the significance of verifying the start bit at its center?
7. What are the three receive errors that can be checked by reading the status register?
8. What do TDRE and RDRF indicate?
9. How does a datascope view differ from the a regular picture of data transmission?

# 5.3 RS232C INTERFACE

**data communications equipment (DCE)**
data communications equipment of data circuit terminating equipment.

**data terminal equipment (DTE)**

Several interface standards have been formulated to interconnect any one of the many modems also known as data sets or other **data communications equipment (DCE)** on the market to any of the many **data terminal equipment (DTE)** systems made today. One of the more popular and widely used standard is EIA's RS232C. This standard designates the physical connector type used, the electrical characteristics, and functions of the signals applied to the lines of the interconnect cabling between the terminal and modem. The RS232C standard has been incorporated by numerous standards agencies into their suites of protocols and specifications.

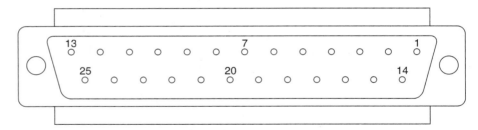

**FIGURE 5-8**  DB 25 Connector for RS232C

## RS232C Functions

Figure 5-8 is a picture of a typical 25-pin connector defined as part of the RS232C standard. Table 5-3 lists the functions associated with each pin of the connector. Refer to this table during the following discussion.

Serial data is sent out on the transmit data (TX data) line, BA. The RS232C standard specifications require that the output voltage level of a driving amplifier for mark (logic 1) data lies between –5 and –15 volts and for space (logic 0) data lies between +5 and +15 volts. Terminating circuits receive serial data on the receive data (RX data) line, BB, which recognizes a mark (logic 1) level as a voltage of a least –3v and a space (logic 0) level of at least +3 volts providing for a 2-volt noise immunity between sending and receiving data. Thus the data lines operate using negative logic since a logic 1 is more negative than a logic 0.

For all control lines, the voltage levels have the same values as the data lines, but the logic type is positive logic. An "ON" condition uses positive voltages while an "OFF" condition is indicated by negative voltage values. The main control signals used to establish a data link are:

1. RTS (Request-to-Send)—sent by the UART to the modem, this signal begins the process by which a data link is established. The modem, after a short delay, responds by asserting its **CTS (clear-to-send).** The purpose of these local handshake signals is to establish a communications link between the modem and the UART. The delay between RTS and CTS allows the modem circuits to come on and settle before data can be sent to it by the UART.

**clear-to-send signal (CTS)**
modem response to a request to send.

2. **CD (carrier detect), DCD (data carrier detect)** or **RLSD (receive line signal detect)**—when a modem detects a RTS input from the UART, one of its responses is to start sending a mark logic 1 tone onto the telephone line. On the receive side, when a modem detects the presence of the mark tone on the telephone line, it responds by asserting the CD signal as a remote handshake signal. The UART detects an active CD and interprets it to mean that a remote station is connected online and the data link is completed.

**carrier detect (CD), data carrier detect (DCD), receive line signal detect (RLSD)**
detect received signal after link is established.

3. **DTR (data terminal ready)** from DTE and **DSR (data set ready)** from the modem—these lines are used to inform the two main units (DTE and modem) in the data link that they are electrically ready to establish a data link. In some systems, these signals are connected directly to positive voltage sources. In this

**data terminal ready (DTR)**

**data set ready (DSR)**

**TABLE 5-3**

**RS232C Pin Functions**

| Pin # | EIA Symbol | Common Abbreviation | Function | Originate From |
|-------|------------|---------------------|----------|----------------|
| 1 | AA | GND | Ground | DTE/DCE |
| 2 | BA | TxData | Transmit Data | DTE |
| 3 | BB | RxData | Receive Data | DCE |
| 4 | CA | RTS | Request-To-Send | DTE |
| 5 | CB | CTS | Clear-To-Send | DCE |
| 6 | CC | DSR | Data Set Ready | DCE |
| 7 | AB | COM | Signal Return | DTE/DCE |
| 8 | CF | RLSD | Rcv Line Signal Detect | DCE |
| 9–11 | | | Not Used | |
| 12 | SCF | SRLSD | Secondary RLSD | DCE |
| 13 | SCB | SCTS | Secondary CTS | DCE |
| 14 | SBA | STxData | Secondary TxData | DTE |
| 15 | DB | TxSigC | Transmit Signal Timing | DCE |
| 16 | SBB | SRxData | Secondary RxData | DCE |
| 17 | DD | RxSig | Receive Signal Timing | DCE |
| 18 | | | Not Used | |
| 19 | SCA | SRTS | Secondary RTS | DTE |
| 20 | CD | DTR | Data Terminal Ready | DTE |
| 21 | CG | SQ | Signal Quality | DCE |
| 22 | CE | RI | Ring Indicator | DCE |
| 23 | CH/CI | DSRS | Data Signal Rate Select | DTE/DCE |
| 24 | DA | TxSigT | Transmit Signal Timing | DTE |
| 25 | | | Not Used | |

case, they only indicate that the RS232C cable is firmly connected between the two units. Other systems generate an active DTR when the DTE is up and running (actually in a ready state). Modems in this type of system assert DSR shortly after the modem is powered and operating, again, essentially saying that it is "ready."

4. Common Signal Return—this line establishes a common signal reference point between the modem and DTE.

5. RI (ring indicator)—This line is asserted by modem capable of detecting an incoming telephone ring signal. These modems are equipped to automatically answer the call. A DTE connected to an auto-answering modem responds to the RI signal by asserting its RTS line beginning the process required to establish the data link.

The remaining lines are used for secondary channel signaling and for timing control for modems requiring data stream synchronization. A signal quality detector line CG is available for DCEs that have the electrical means for error detection incorporated into them.

Secondary channel data rates are much lower than the data channel. Their primary use is for control signal information. Frequently, they are configured to send acknowledgment or control data in the opposite direction from the data channel. When they are configured this way, the secondary channel is referred to as a **back channel.**

An international standards organization based in France, formerly known as the **international consultative committee for telegraphy and telephony (CCITT),** and currently as the **international telecommunications union (ITU),** has issued a specification very similar to RS232C. This specification, V.24 and its relationship to RS232C is shown in Table 5-4.

**back channel**
channel that is used to convey control data between two stations in a data link.

**consultative committee for international telephony and telegraphy (CCITT)**

**international telecommunications union (ITU)**

---

**TABLE 5-4**

### Comparison of RS232C and V.24

| RS232C | Pin Number | CCITT V.24 Designation | Signal Name (V.24) |
|--------|-----------|------------------------|--------------------|
| GND | 1 | 101 | Shield (Ground) |
| TxData | 2 | 103 | Transmit Data |
| RxData | 3 | 104 | Receive Data |
| RTS | 4 | 105 | Request to Send |
| CTS | 5 | 106 | Clear to Send |
| DSR | 6 | 107 | Data Set Ready |
| COM | 7 | 102 | Signal Common Return |
| RLSD | 8 | 109 | Receive Line Signal Detect |
| None* | 9 | None | +12V V.24 Only |
| None* | 10 | None | −12V V.24 Only |
| None* | 11 | 110 | Signal Quality Indicator |
| Pins | 12, 13, 14 | None | Not Used by V.24 |
| TxSig | 15 | 141 | Transmit Clock (DCE) |
| SRxData | 16 | None | Not Used by V.24 |
| RxSigC | 17 | 115 | Receive Clock |
| None* | 18 | 142 | Local Loopback |
| SRTS | 19 | None | Not Used by V.24 |
| DTR | 20 | 108.2 | Data Terminal Ready |
| SQ | 21 | 140 | Remote Digital |
| RI | 22 | 125 | Ring Indicator |
| DSRS | 23 | 111 | Data Rate Select |
| TxSigT | 24 | 113 | External Transmit Clock |
| None* | 25 | 142 | Test Indicator |

*RS232C pins designated with "None" are not used.

## Electrical and Physical Specifications

A maximum of +/–25V can be applied to any RS232C line without damage to any DTE or modem circuits. Other limitations and specifications imposed by the RS232C standard include:

1. Cable impedance (330 Ω minimum for driver lines and 7,000 Ω maximum for terminating load circuits).
2. A maximum line capacitance of 2,500 pf on any terminating line.
3. Capacitance per foot of RS232C cable is not to exceed 50 pf/foot.
4. A maximum cable length between DTE and modem of 50 feet results from meeting the two capacitance specifications. At 50 pf/foot, a grand total of 2,500 pf would result from a 50-foot cable. Longer cables would violate the 2,500-pf limit.

Numerous integrated circuit manufacturers produce buffer interfaces to facilitate the translation of common digital logic levels (0 = 0V and 1 = 5V for TTL as an example) into RS232C levels. One such manufacturer, National Semiconductor, produces an RS232C line receiver (Figure 5-9) and line driver (Figure 5-10. The driver, a DS1488,

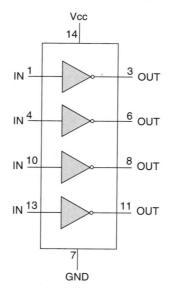

| Vin (min) | Vcc | Vout |
|-----------|-----|------|
| Logic 0 3.0V | 5.0V | 0.4V max |
| Logic 1 -3.0V | 5.0V | 2.8V min |

**FIGURE 5-9** RS232C Line Receiver

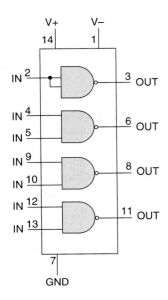

**FIGURE 5-10** RS232C Line Driver

| Both Inputs | V+/V- | Vout | Iout |
|---|---|---|---|
| Logic 0 = 0V | 9.0V | 7.0V | 10 ma |
| Logic 1 = 5V | 9.0V | -6.8V | 10 ma |
| Logic 0 = 0V | 13.2V | 10.5V | 10 ma |
| Logic 1 = 5V | 13.2V | -10.5V | 10 ma |

converts TTL levels to one of two pairs of RS232C levels (+/–9V or +/–13.2V). The actual output levels are dependent on the supply voltage applied to the chip, allowing this chip to be incorporated in a wide variety of RS232C interface applications. The level selections are shown on the chart in Figure 5-10.

The line receiver, a DS14C89A, accepts voltages as low as +/–3V from the RS232C lines and converts them into TTL logic levels. Note, in the charts for both interface chips, that there is a logic change as well as a voltage change. That is, TTL logic 1 (5V) is changed into a negative voltage and logic 0, into a positive voltage. This is in keeping with the mark and space RS232C logic [negative voltage for a mark (logic 1) and a positive voltage used for a space level (logic 0)]. To use these interface chips to translate control signals (like RTS and CTS), an inverter is required on the TTL side so that the logic levels of on and off will correspond to the correct RS232C levels (on—a positive voltage and off—a negative voltage).

**FIGURE 5-11** RS232C Breakout Box—Datacheck #7 *(Photo Courtesy of GN Nettest, Inc.)*

## RS232C Breakout Box

A device called a breakout box (Figure 5-11) is used to monitor and test devices that use the RS232C interface. Light emitting diodes (LEDs) are used to constantly monitor the lines. Some boxes monitor all of the RS232C lines, while others just monitor the initial set comprised of RTS, CTS, CD, RxData, TxData, DSR, and DTR. Some boxes are equipped with switches and test access points that allow users to make or break each line and/or inject test signals onto each line. The access points also allow other monitoring equipment, such as oscilloscopes, datascopes, and power meters to be connected to the RS232C data and control lines. These breakout boxes are of great assistance in determining what is occurring electrically between terminal equipment and modems.

## Null Modem

There are times that it is desirable to connect two terminals over a short distance without the need for a modem. Essentially, since these terminals are in close proximity of each other, there is no need to use telephone lines to interconnect them. In this case, a "null modem" (Figure 5-12) can be used to do the connection. A null modem is actually an RS232C cable, which has a few wires crossed. Transmit and receive data between both ends are criss-crossed so that the transmit data of one terminal is fed into the other on the receive data line. Data terminal ready (DTR) and data set ready (DSR) lines are also cross-coupled so that one terminal's DTR output becomes an indication on the other's DSR input. Lastly, the request-to-send (RTS) output of each terminal is fed back to its own clear-to-send (CTS) and to the other terminal's carrier-detect (CD) input. A common line between both terminals supplies a common ground reference for electrical voltages and signals. By using the null modem, two terminals can establish a data link without the need for a modem. An active RTS from one terminal signals its own CTS and the other

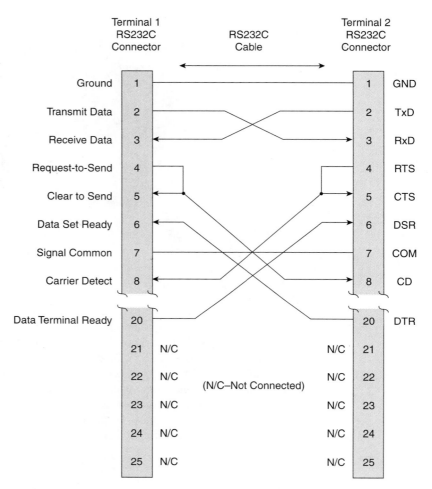

FIGURE 5-12 Null Modem

terminal's CD. Similar action is established at the other terminal and the data link is set. Data can now flow in serial digital form between the two terminals.

RS232C has experienced wide acceptance and usage for systems with maximum data rates up to 20 Kbps, the maximum rate specified for this standard over relatively short distances. Requirements for an interface for systems using higher data rates and/or desiring more distant interconnections led Electronics Industries Association (EIA) authors to develop the RS449 standard. This standard specifies signal and parameters similar to RS232C, but with different cabling specifications that can handle faster data rates over longer distances.

**Section 5.3 Review Questions**

1. What is the significance of the RS designation for RS232C?
2. What are the range of voltages for data logic 1?
3. How is the maximum cable length for a RS232C interface determined to be 50 feet?
4. What is the ITU specification for RS232C?
5. What is the purpose of a breakout box?
6. What is a null modem used for?

# 5.4 RS449 INTERFACE STANDARD

The RS449 standard contains most of the signal lines and electrical specifications outlined in the RS232C. Table 5-5 lists the RS449 functions by pin number for the two connectors, a 37-pin main channel connector and a 9-pin secondary channel connector specified by the RS449 standard. Included in the table are notations for equivalent RS232C signal lines. Note that EIA selected more appropriate circuit designations for these control signal lines. Abbreviations closer to the line functions are used.

RS449 specifies two connectors (Figure 5-13), one with 37 pins for the main data channel and one containing 9 pins for secondary channel signaling. The increased number of pins are used to accommodate the added functions detailed in the standard. A number of pins share the same function to assist in reducing line capacitance. This allows for data transfer rates up to 2 Mbps over cables that can reach a maximum of 200 feet in length.

**loop back**
test condition in which the transmitted data is looped back into its own receiver.

New functions include control indications for local and remote **loop-back** test functions. Local loop-back (LL) indicates the system is in the local loop-back test mode. Data placed on the transmit data line is looped back through the modem onto the receive data line to the DTE. This mode is used to test the modem functions. Since data is not placed onto the medium, the test is restricted to the local equipment at a station. In contrast, the remote loop-back test mode (indicated by an active RL line), checks the telephone system connections and lines as well as the modem at both stations. The data sent by a station is looped back through the remote modem onto the telephone lines back to the sending station. The remote DTE is not included in this test mode. A third control line, test mode (TM) signals the DTE that the modem is in one or the other test mode. A similar type of control signal, the data mode (DM) indicates that the modem is in a normal data transfer mode. This last control line replaces the DSR function of the RS232C standard.

The modem can be transferred to a standby channel by using the select standby (SS) signal. The modem returns a status indication that it is connected to the standby channel by asserting the standby indicator (SB) line. If the DTE desires to prepare the DCE to receive a new line signal, it asserts the new signal line (NS). This causes the receive-data line (RD) to a mark state and disables the carrier-detect (CD) line. Once the modem receives

## TABLE 5-5

### RS449 Pin Assignment

*37-Pin Connector (Main Channel):*

| PIN NO. | RS232 | CIRCUIT | NAME | DIRECTION |
|---------|-------|---------|------|-----------|
| 1 | | | Shield | Common |
| 2 | CI | IS | Signal Rate Indicator | From DCE |
| 3 | | | Not Used | |
| 4, 22 | TD | SD | Send Data | From DTE |
| 5, 23 | DB | ST | Send Timing | From DCE |
| 6, 24 | RD | RD | Receive Data | From DCE |
| 7, 25 | RTS | RS | Request-to-Send | From DTE |
| 8, 26 | DD | RT | Receive Timing | From DCE |
| 9, 27 | CTS | CS | Clear-to-Send | From DCE |
| 10 | | LL | Local Loop-Back | From DTE |
| 11, 29 | DSR | DM | Data Mode | From DCE |
| 12, 30 | DTR | TR | Terminal Ready | From DTE |
| 13, 31 | RLSD | RR | Receiver Ready | From DCE |
| 14 | | RL | Remote Loop-Back | From DTE |
| 15 | RI | IC | Incoming Call | From DCE |
| 16 | CH | SF | Select Frequency | From DTE |
| | | SR | Signaling Rate Selector | |
| 17, 35 | DA | TT | Terminal Timing | From DTE |
| 19 | SG | SG | Signal Ground | Common |
| 20 | | RC | Receive Common | From DTE |
| 23 | CG | SQ | Signal Quality | From DCE |
| 28 | | IS | Terminal in Service | From DTE |
| 32 | | SS | Select Standby | From DTE |
| 34 | | NS | New Signal | From DTE |
| 36 | | SB | Standby Indicator | From DCE |
| 37 | | SC | Send Common | From DCE |

*9-Pin Connector (S—Secondary Channel):*

| | | | | |
|---------|-------|---------|------|-----------|
| 1 | | | Shield | Common |
| 2 | SCF | SRR | Receiver Ready | From DCE |
| 3 | SBA | SSD | Send Data | From DTE |
| 4 | SBB | SRD | Receive Data | From DCE |
| 5 | SG | SG | Signal Ground | Common |
| 6 | | RC | Receive Common | From DCE |
| 7 | SCA | SRS | Request-to-Send | From DTE |
| 8 | SCB | SCS | Clear-to-Send | From DCE |
| 9 | | SC | Send Common | From DCE |

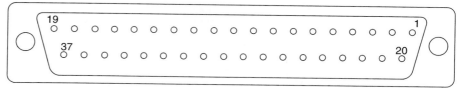

37 Pin Connector

a

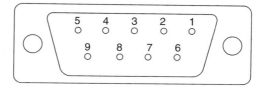

9 Pin Connector

b

**FIGURE 5-13** RS449 Connectors

the new signal, it places it on the receive-data line and again asserts the carrier-detect line. The send common (SC) and receive common (RC) lines indicate which direction data is flowing (out on transmit-data line or in on receive-data line).

The secondary channel lines are included in the 9-pin connector/cable along with a few duplicated functions found in the 37-pin connector. These leads are summarized in Table 5-6 with their equivalent RS232C functions.

---

**Section 5.4  Review Questions**

1. Compare the data rate and cable distance specifications for RS232C and RS449.
2. What is the loop-back process used for?

---

## 5.5  RS422 AND RS423 INTERFACE STANDARDS

The RS232C and RS449 specifications detail basic requirements for interfacing DTE and DCE equipment. Two additional standards specify electrical and mechanical requirements for interconnecting any two individual circuits used for transferring serial data. They are the RS422 for balanced circuits and RS423 for unbalanced circuits. Balanced networks, as shown in Figure 5-14a, generate two outputs that are the complements of one another. The voltage (or power) of each output (V1 and V2) can be measured in relation to a common reference point (signal ground, for example). However, the full output is generally measured differentially between both outputs (V3).

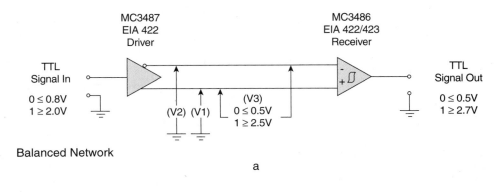

Balanced Network

a

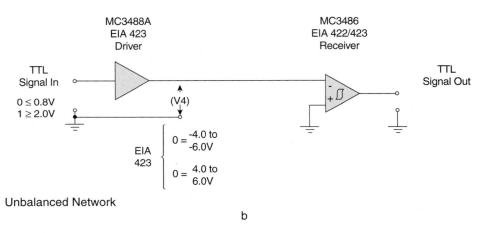

Unbalanced Network

b

**FIGURE 5-14** Balanced and Unbalanced Interfaces

The output of an unbalanced circuit is taken from a single output with reference to a common return such as ground (V4 in Figure 5-14*b*). Balanced networks are used to assist in the elimination of noise experienced on the lines between the generator and receiver units. Unlike the desired outputs, noise on one output is approximately identical to the noise from the other output. Since they nearly agree, the potential difference between the noise levels is negligible. This reduction in noise power is called common-mode rejection and results in less noise presented to the receiver from the generator.

The RS422 and RS423 specifications set requirements for balanced and unbalanced digital interface circuits. The details for each are similar enough that this section will only illustrate the type of information referred to in the RS422 specification. Exact information for the 423 standard is left to the individual requiring its use.

Figure 5-15 shows the output-loading circuits for an RS422 balanced generator. It shall be referenced in discussion involving RS422 electrical specifications. RS422 specifies that a generator is to be able to produce a voltage between 2 and 6 volts for each logic level, while sourcing a 100-$\Omega$ differential impedance. In Figure 5-15 this is represented by 50-$\Omega$ load resistors between each output and signal ground. The differential output (Vo) is sensed from output A to output B across the combined 100-$\Omega$ resistance.

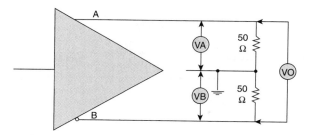

**FIGURE 5-15**  RS422 Generator Output

The generator is limited to 6 volts from either output to ground as well as from output to output. Essentially, this says to the designer of the receiving end that, at no time, are the inputs to the receiver to expect more than a 6-volt potential. The polarity of the voltage generated determines the logic level. RS422 does not specify which logic level matches which polarity. It does, however, state that the magnitude of the voltages for each polarity is not to differ by more than 0.4 volts. That is, if a logic 1 is +5 volts, a logic 0 must fall into the range of –4.6 volts to –5.4 volts. While the specification does not assign polarity to logic level, it does state that a logic 1 is a mark and is to be the same level as a control line off condition. Logic 0, a space, is to be the same polarity as a control line on signal. This conforms to the RS232C specifications for data and control line signals.

The input impedance of the balanced receiver is greater than or equal to 4k ohms. Internally, the receiver is to be able to respond to a change of +/–0.2 volts as a logic switch. That is, the receiver must be able to detect a 200 millivolt signal and determine the correct logic level. Once the logic level is detected, the receiver must indicate that level as long as the common mode input voltage remains between +/–7 volts. Receivers are built to withstand a maximum differential input of 12 volts.

Another area specified by RS422 is the length of the cable between the generator and receiver. As with the RS232C standard, RS422 specifies the capacitance per foot as the limiting factor to a cable's length rather than attempting to specify actual lengths. The capacitance of a cable is specified as 52.5 pf/meter (or 16 pf/foot) terminated into 100 $\Omega$s. In addition, for data rates under 90 Kbps, the maximum length of a cable is 1,200 meters (approximately 4,000 feet). Figure 5-16 is a graph of the usable data rates and maximum cable lengths expected when using the RS422 standard.

The rational for presenting information for the EIA standards here is to provide a feel for the type of information contained within them. Realize that there are a lot of additional details within each standard and that the specifications discussed here are done so that the student can perceive how their use can facilitate the standardizing of the interconnection of various devices.

**Section 5.5 Review Questions**

1. How do balanced circuits reject noise?
2. What is the maximum input voltage for a balanced RS422 interface?
3. What is the range of output voltages from an RS423 unbalanced driver?

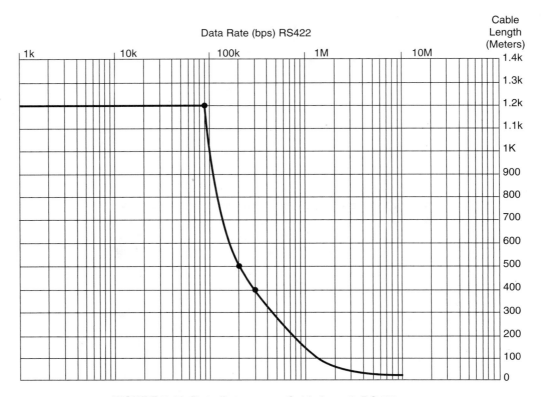

**FIGURE 5-16** Data Rate versus Cable Length RS422

## 5.6 FSK MODEMS

The last functional block in a basic low-to-medium speed data link is the modulator/demodulator or modem, data set, or DCE. It is the job of the modem to convert the serial digital data stream from a UART to a series of audio sinewaves or tones. For a frequency shift key (FSK) modem, one tone represents a logic one (mark) and another is used for a logic zero (space). On the receive end, these tones are converted back to digital levels by the receiver's modem. Figure 5-17 illustrates the functional blocks of a FSK modem.

On the transmit side, serial data is supplied by the UART through a RS232C interface, on the TX data input and fed to a **voltage-controlled oscillator (VCO).** The VCO acts as a frequency modulator by converting DC voltages into different analog sinewaves. These sinewaves have a different frequency for each level of DC voltage applied. Hence, a tone at one frequency is produced for a logic one's DC voltage and a second tone at a different frequency is produced for the space logic DC voltage level.

**voltage-controlled oscillator (VCO)**
A circuit that produces an AC signal whose frequency is determined by the DC voltage at its input.

The request-to-send (RTS) input from the terminal (or UART) initiates the modem's RTS/CTS delay, which varies with each modem type and requirements. The general pur-

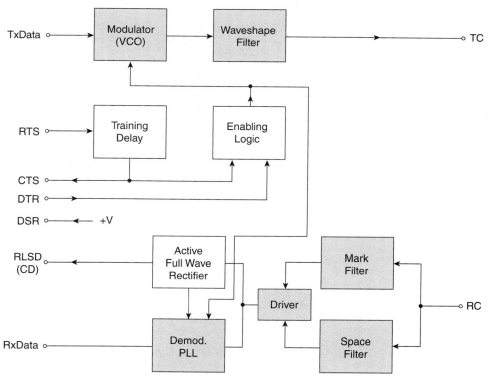

**FIGURE 5-17** Modem Functional Blocks

pose of the delay is to give modem circuits time to turn on and stabilize before handling data from the UART. Once the delay is completed, the modem generates a clear-to-send (CTS) signal back to the terminal completing the local data link handshaking. An active CTS enables the voltage controlled oscillator, which, sensing idle line ones from the UART, immediately outputs a mark tone on the telephone lines.

Data-terminal ready (DTR) and data-set ready (DSR) inputs to the modem and DTE, respectively, can be as simple as a steady logic level one. They are used for signaling the terminal and modem that they are physically interconnected (usually through a RS232C cable). These logic inputs must be sensed before the modem or terminal will react to the RTS/CTS sequence.

The mark signal received on the input to the receive side of the modem passes through the mark filter to the carrier, data carrier, or receive line signal detect (CD, DCD or RLSD) and demodulator units. An active full-wave rectifier, in the carrier-detect section, converts the tone into a DC voltage that is sent out of the modem on the CD line to the DTE to complete the remote handshake necessary to establish the data link. The remote station may already be online, having gone through the same RTS/CTS sequence performed by the initiating station. It is also possible that the remote site will establish its local link through its RTS/CTS sequence in response to an active CD. The choice is selected by the application software that controls the system and the type of modem used. In either case, once the CD lines at both stations are active, a data link is established and actual data can now be sent back and forth through the modems.

The mark signals from the mark filter and space signals (which are passed through a space filter) are detected by a **phase lock loop (PLL)** circuit in the demodulator section. Three operational blocks form a PLL. They are a phase detector, a buffering amplifier and a VCO (Figure 5-18). For a FSK demodulator, the VCO is designed to produce a squarewave signal at a frequency (fc), which lies between the mark and space frequencies. This frequency is never sent between modems since it is neither a mark or space logic level. The purpose of the phase detector is to produce a DC voltage (the PLL's error voltage) whenever there is a difference between the phases of a reference waveform (fc) and input sinewave (fin).

Different systems require different modes of data communication operation. Systems operating in full duplex have different needs then those operating in half-duplex or simplex. A primary controlling station that initiates the call in a full-duplex data link uses an *originate station,* while the secondary remote station requires an *answer station.* The main difference in these modem types is the mark and space frequencies used to represent data logic ones and zeros.

The Bell Telephone 103 modem is a full-duplex, 300-bps modem that can be manufactured as an originate or an answer modem. As an originate modem, the transmit mark and space frequencies are 1,270 and 1,070 Hz, respectively. The receive frequencies are 2,225 Hz for a mark and 2,025 Hz for a space. To operate with a 103 originate modem, a compatible answer modem must transmit mark and space tones at the originate's receive frequencies and be able to demodulate the originate's transmit mark and space tones. The use of different frequency pairs for each transmission direction (to the originate and to the answer station) allows the data transmissions to occur simultaneously. Secondary tones produced from any mixing of these signals on the transmission medium are rejected by the receive filters in each modem.

Some modems operating in half-duplex echo back the transmitted data to the sending station's video screen or printer. Called *echo-plex,* this allows the sender to check the in-

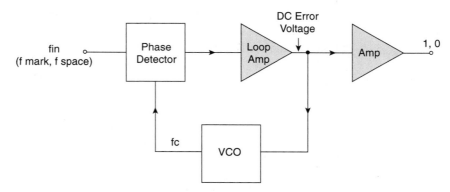

**FIGURE 5-18** Phase Lock Loop

formation being sent at the time it is being transmitted. This echo is permissible because, in half-duplex, actual data transmission direction is one way at a time. A primary can send to a secondary, but the secondary must wait until the primary is finished before it can send data to the primary. Since the receive side is not being used by a station while it is transmitting, the data sent out is routed through the station's receiver and echoed back.

### Section 5.6  Review Questions

1. Which section of a FSK modem is used to convert digital data to analog data?
2. Which section of the FSK modem is used to detect incoming analog signals?
3. Why are originate and answer modems required for full-duplex operation?
4. What is the purpose of the echo-plex function?

## 5.7  ADDITIONAL TYPES OF MODEMS

**symbol**
A single electrically measurable signal, which represents groups of one or more data bits.

**symbol rate**
Rate at which symbols are transmitted in symbols per second (sps).

A symbol element is any electrically measurable characteristic of a waveform that can be used to hold information. For instance, in the previous section, frequency shift keying (FSK) modulation was used for the transfer of data between two stations. Each binary bit was represented by a number of cycles of a sinewave signal on the telephone lines. One frequency represented a one logic bit and another the zero logic level. In this case, the frequency of the signals is the **symbol** used to represent the data.

The data rates used with FSK modems are limited by the bandwidth of the telephone lines (300 Hz–3 KHz). Each signal symbol transmitted by an FSK modem represented a single data bit so that the **symbol rate** (symbols per second or sps) for data was identical to the data rate and could not exceed 3000 sps. Ideally the symbol rate should be at a rate closer to the center of the telephone line bandwidth (approximately 1,500 Hz). A symbol rate of 2400 sps, while still within the telephone bandwidth, approaches the high side of that bandwidth. If the same data rate of 2400 bps could be transferred at one-half of the symbol rate (1200 sps), its transmission would fall nearer the middle of the bandwidth. The problem is to squeeze the information at 2400 bps into a switching rate of 1200 sps.

Amplitude can also be used to represent digital data. A digital data stream can be fed to a digital-to-analog converter (DAC) to produce an output voltage that varies with the input digital information. A group of bits, (for instance, four bits per group) from the data stream are "collected" and presented to the DAC. A different analog voltage level out results from each of the sixteen combinations of ones and zeros of the four bits. In this manner, each voltage level is a symbol representing four data bits. This form of digital coding is called *amplitude shift keying (ASK)* modulation.

Another electrical aspect of an analog signal that can be used as a symbol is the phase of a sinewave voltage. Using a group of four bits again, a separate phase can be generated for each of the sixteen bit combinations. One advantage of *phase shift keying (PSK)* is that the frequency and amplitude of the signal used remain the same for each symbol. The frequency can be selected to fit comfortably in the middle of the telephone line bandwidth. Further, translating either group's three bits at 2400 bps or four bits at a data rate of 4800 bps into a symbol rate of 1200 symbols per second keeps the switching rate close to the center of the bandwidth as well.

Amplitude and phase aspects of a sinewave can be simultaneously altered to form symbols. This type of modulation is called *quadrature amplitude modulation (QAM)* and is used for data rates of 9600 bps and higher. Use of PSK at these rates cause the difference between the phases to become too small to meet phase hit limits specified for the telephone lines. Recall from Chapter 2 that a phase hit is a sudden change in phase that lasts for a short period of time. The telephone company specifies phase hits of +/−10° as acceptable on the telephone lines. To avoid this problem, the combination of amplitude and phase modulation increases the phase differences while introducing amplitude variations.

The modems used to develop these forms of modulation are made up of several of these functional blocks: balanced modulators, bit splitters, digital-to-analog converters (DACs), linear summers and various filters.

### Section 5.7  Review Questions

1. List three aspects of an analog signal that can be used as a symbol.
2. What is the limiting factor for FSK modems?
3. What is the limitation imposed on PSK modems?

## The Balanced Modulator

The heart of most PSK modulators is the balanced modulator. This circuit allows the phase of a carrier sinewave (fc) to be altered by a modulating digital signal. Figure 5-19 is an example of a transformer-balanced modulator. It is used to illustrate the concept behind balanced modulators. By and large, most balanced modulators today are made from semiconductor devices to avoid the expense, bandwidth limitations, and associated magnetic distortions and interference inherent in transformers.

The reference frequency (fc) is applied to T1 and is coupled through the secondary winding to the diodes D1 and D2 on the high side and D3 and D4 on the return side. The in-phase relationship between the input (primary winding) side of the transformer and the output side (secondary winding) is indicated by the dots at the top side of both transformers. On the return side of the secondary, the signal is 180° out of phase with dotted reference point.

The digital data stream is applied to the center taps of T1's secondary and T2's primary. The current level supplied by the digital circuits is enough to cause the diodes to turn on when the correct polarity is applied. As a point of reference, a logic one is selected to be positive at input A and negative at input B. This forward biases diodes 1 and 3 and turns off diodes 2 and 4 as shown in Figure 5-20. The signal coupled from T1's secondary is not large enough by itself to turn the diodes on, but once the diodes are on

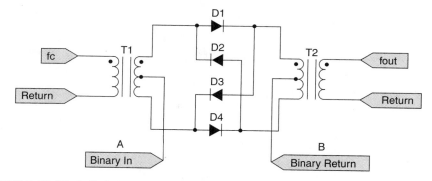

**FIGURE 5-19** Diode Balanced Modulator

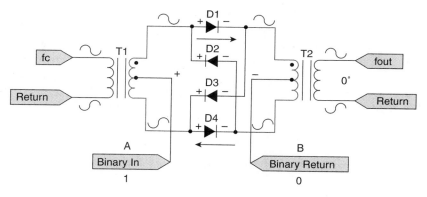

**FIGURE 5-20** Logic 1 to 0° Phase

(from the digital input), this signal easily passes through to the primary of T2. The logic one, in this case, causes fc to be passed to T2 and coupled so that the phase of the output signal is the same as the input signal. Since transformers only pass AC signals, the DC logic voltage does not appear at the secondary of T2.

Reversing the polarity at inputs A and B to represent a zero, turns on D2 and D4 while back-biasing D1 and D3 as seen in Figure 5-21. This time fc is directed to the opposite (non-dotted) end of T2. The output signal coupled to T2's secondary is 180° out of phase with fc at the input. A vector diagram (Figure 5-22) can be drawn to illustrate the phases representing a logic one and a logic zero. Zero degrees of phase lies on the positive side of the *x*-axis and is used as a reference for any phases generated by the modulator. It is important to note that the zero reference is arbitrary and that the interpreting of data is done by relative rather than absolute means. The sinewave form of an AC signal is usually designated as the zero degree reference lying on the *x*-axis. When using this designation, the *y*-axis, which is 90° from the *x*-axis becomes the cosine axis. 180° or negative *x*-axis is denoted –sine and 270° or –90° is the –cosine.

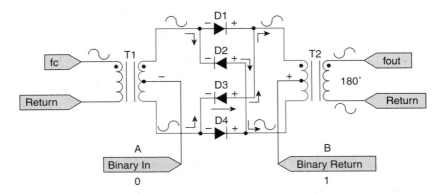

**FIGURE 5-21** Logic 0 to 180° Phase

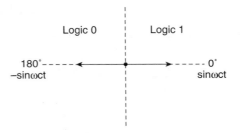

**FIGURE 5-22** Balanced Modulator Phasor Diagram

Logic one in our example generated a signal that was the same phase as the input. This is represented on Figure 5-22 as a vector at zero degrees. This does not imply that the output signal is at zero degrees. What it does imply is that the difference between the reference (fc at the input) and the output is zero degrees. Similarly, the phase of the signal for a logic zero (180° out of phase with reference, fc) is shown lying on the negative *x*-axis. The magnitude of the vectors used on this and subsequent phase diagrams (except QAM) is of unit length. This is easily acceptable since these devices are modulating the phase while keeping amplitude and frequency constant.

## Differential Phase Shift Keying

One aspect of frequency shift keying (FSK) modems is that in order to use them, especially full-duplex modems, the lines they are connected to are required to possess sufficient bandwidth to handle the mark and space signal pairs from the originate and answer modems. To reduce the bandwidth requirements for the connecting medium, it becomes desirable to modulate a characteristic of a signal other than its frequency. Holding the

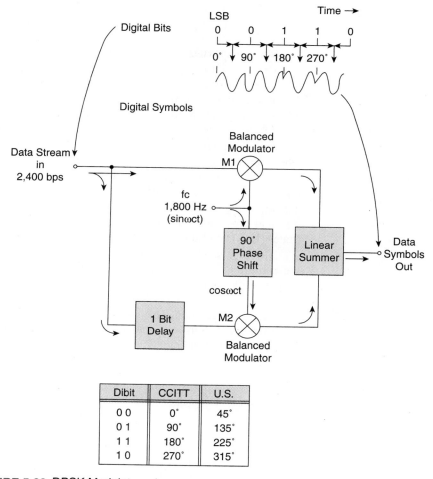

**FIGURE 5-23** DPSK Modulator

| Dibit | CCITT | U.S. |
|-------|-------|------|
| 0 0 | 0° | 45° |
| 0 1 | 90° | 135° |
| 1 1 | 180° | 225° |
| 1 0 | 270° | 315° |

frequency of the signals delivered by a modem constant reduces the amount of bandwidth needed for that signal. One option is to modulate a signal's phase instead of the frequency.

A *differential phase shift keying (DPSK)* modem accepts a serial data stream and produces a single-carrier signal whose phase changes according the digital values of the incoming data stream. As the data stream is applied to the modem (Figure 5-23), it is fed to a balanced modulator (M1) and a one-bit time delay circuit. The output of the delay circuit is fed to another balanced modulator (M2). A carrier frequency (fc) is applied to the first balanced modulator (M1) with a starting reference of 0°(sinωct). This carrier is also shifted 90° (cosωct), which is then applied to the second balanced modulator (M2). Each binary data bit causes its respective modulator to output a signal that has the same phase (due to a logic 0 applied) as the fc reference input to the modulator or a signal that is 180° out of phase with that input (for a logic 1 applied). In the diagrammed example of Figure 5-23, a 0 bit into M1 causes that modulator to produce a $\sin\omega ct$ (0°) signal. A 0 bit into M2 produces a $\cos\omega ct$ (90°) out.

The two signals from the balanced modulators are summed linearly (that is, they are not mixed) and the resulting data symbol is sent to the connecting medium. Effectively, the linear sum of the two signals produces a vector sum of those signals. In the example of 0 logic bits applied to M1 and M2, the $\sin\omega ct$ (0°) and $\cos\omega ct$ (90°) signals have a vector sum of 45° measured with respect to the x-axis fc reference of $\sin\omega ct$ or 0°. Therefore, the symbol representing a 00 combination to the modulators is a 1,800-Hz carrier signal whose phase is 45° from the $\sin\omega ct$ fc reference.

In Figure 5-23, a truth table for the bit combinations applied to the modulators and resulting phase of the symbols are shown. The differential bits applied to the modulators are called dibits since two (*di*) bits are grouped for each symbol. The two bits are the current and preceding bit. The phases specified by standards organizations in the United States (U.S.) and Europe (ITU) is also illustrated.

The key behind differential phase shift keying is that the inputs to M1 and M2 are the current data bit (M1) and the preceding data bit (M2). The symbols produced at the output result from comparing the current data bit with the previous data bit. At the receiver, the reference clock (fc) has to be reproduced. It is recovered from the data stream using a PLL to reproduce the clock. This is frequently called *clock recovery*. The receiver will then establish its own phase for a reference based on a digital *training sequence* it receives from the sending station. This sequence is an established bit pattern that the receiver can decipher to establish a correct reference for its recovered fc clock.

## The QPSK Modem

To encode a number of bits from a serial data stream into a single symbol begins by presenting the bits as a group simultaneously to the circuits that are doing the transformation. This job is done in PSK modems using a circuit called a *bit splitter*. Bit splitters are chiefly shift-registers with controlled outputs. The data is shifted in at the same rate as the data rate. Output registers are loaded at the symbol rate. This causes the bits to be presented to the translating circuits at the symbol rate. Phases of the output symbols will change at the same rate as data changes are presented to the translating circuits.

Figure 5-24 is that of a *quandrature phase shift key (QPSK)* modulator. The serial data input is sent to a bit splitter to produce dibit outputs. The dibit outputs of the bit splitter are fed to digital logic inputs of two balanced modulators. The first bit into the bit splitter is outputted as dibit *a* to the upper-balanced modulator and the second bit (dibit *b*) to the lower one. The reference sinewave (fc) as $\sin\omega ct$ is applied to the input of the top modulator and to a 90° phase shift circuit. The output of the phase shift circuit is applied as the reference frequency to the lower balanced modulator. Since it is shifted 90° from the $\sin\omega ct$ input signal, it is denoted as being a $\cos\omega ct$ signal.

The outputs from the balanced modulators, which will be in phase with their respective fc signals for a logic one binary input and 180° out of phase for a logic zero, are fed to a linear summer. Linear summing avoids developing sum and difference frequencies

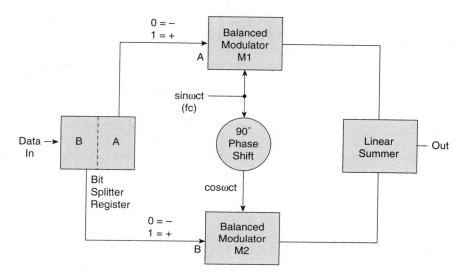

**FIGURE 5-24** QPSK Modulator

normally generated by a nonlinear circuit used for standard radio modulators. A bandpass filter rejects unwanted signals produced by the summer. The result, at the output, is a vector sum of the outputs of the two balanced modulators.

Recall that the output amplitude and frequency of the balanced modulators are identical. The bandwidth requirements for the QPSK modulator are similar to those discussed in the section on balanced modulators. The linear sum of the two modulator signals produces a signal whose frequency is that of the carrier frequency and whose phase depends on the states of pairs of the input serial data. The symbol rate at which the phase of that signal changes is one-half that of the data rate applied to the input of the modulator.

As an example, a carrier frequency of 1,650 Hz is used with a 2400 bps QPSK modulator. The dibits cause the phase of the carrier to switch at a rate of 1,200 Hz. In effect, the rate at the input of 2400 bps is causing the QPSK modulator to send data at a rate of 1200 sps. Both that rate and the frequency of the carrier signal, itself, lie comfortably in the center region of the telephone system bandwidth of 300–3,000 Hz.

For the remaining discussion on medium-speed modems, the following conventions are used:

1. Logic one inputs to the balanced modulator are positive since they produce outputs that are in phase with the fc input.
2. Logic zero inputs are negative since they produce a signal 180° out of phase with the input.
3. The phase shift will be +90° to produce a cosωct reference to the lower modulator.

Keep in mind that these are arbitrary selections. The logic influences could be reversed if we so desired. The phase shifter could also, as easily been designed to yield a positive or negative 90° shift.

Figure 5-25 is the vector diagram for all the possible symbol (phase) results at the output of the QPSK modulator. Two binary bits can form four different combinations of one and zero states. Each combination generates a separate phase of the signal at the output represented by a vector symbol on the diagram. As an example, if both bits are low, the outputs of the balanced modulators will be 180° out of phase with their respective fc inputs. This results in a negative sinωct (180°) signal from the upper modulator output summed with a negative cosωct (270°) signal output from the lower balanced modulator. Since the amplitude of these signals are always the same, we can consider them to be of unit length so the vector sum becomes a signal with a phrase that lies halfway between both signals (at 225°) at a length of 1.414 times the unit length. The vector representing this symbol is shown on Figure 5-25 and is marked 00. The process is repeated for each combination of one and zero states of the dibits from the bit splitter. Note that there are four distinct phase symbols, one for each combination and that they are separated from each other by 90°.

Specifications for a PSK modulator can be drawn from observations made about the QPSK modem, but would not be conclusive without testing them on other PSK circuits. That job having been done (and will be verified when we look into other PSK circuits), the following mathematical relationships are derived to describe a PSK's parameters:

1. The symbol rate (S) in symbols per second (sps) equals the data rate (D) in bits per second (bps) divided by the number of bits per symbol (n):

$$S = \frac{D}{n} \tag{5-1}$$

2. The number of different symbols (M) possible at the output is equal to the radix 2 raised to the nth power:

$$M = 2^n \tag{5-2}$$

3. The size of the bit splitter register and the ratio between bps and sps is n.
4. The amount of phase difference between symbols (P) is 360° divided by M:

$$P = \frac{360°}{M} \tag{5-3}$$

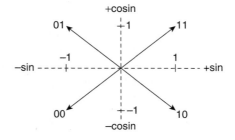

**FIGURE 5-25** QPSK Phasor Diagram

**EXAMPLE 5-6**

For a QPSK modulator, what is the number of symbols possible at its output? What is the phase difference between each symbol and for a data rate of 3600 bps, what is the symbol rate at the output?

**SOLUTION**

Applying equations 5-1 and 5-2, the number of symbols possible for a QPSK is $2^2 = 4$. The phase difference between each symbol is $360°/4 = 90°$. Finally, the symbol rate is the bit rate divided by the number of bits per symbol or $3600/2 = 1800$ sps. This last piece of data says that the phases of the symbols will change at a rate of 1800 symbols per second.

## QPSK Demodulator

In a QPSK demodulator, the reference frequency (fc) is recovered from the incoming bit stream using a clock recovery circuit such as the circuit shown in Figure 5-26. Before actual data is sent to a PSK receiver, it is preceded by a string of an alternating pattern of 1s and 0s called a preamble. The purpose of this preamble is to allow the clock recovery circuit to "lock in" on the reference or clock frequency. The PSK demodulator crudely demodulates the incoming preamble using the free-running clock of a phase lock loop's (PLL) voltage-controlled oscillator (VCO). This crudely demodulated digital type data is then fed back into the clock recovery through a 1/2-bit delay and an exclusive OR as shown in Figure 5-26. The timing diagram, in that figure, illustrates how the preamble of alternating 1s and 0s creates a "clock pulse," which is sent to the PLL. The preamble is

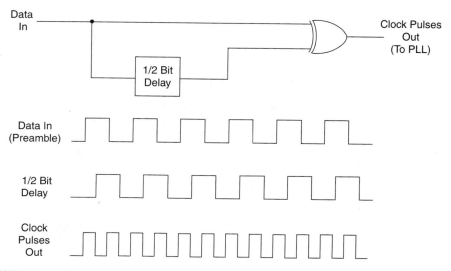

**FIGURE 5-26** Clock Recovery

long enough to allow the PLL to capture and hold lock between the data steam pulses and the VCO of the PLL. In this lock condition, the frequency of the VCO output is the same as the pulse rate generated by the exclusive OR. Since the PLL is locked at this point, the loss of pulse information for a limited time period does not cause the PLL to unlock, thus maintaining correct clock frequency. That is, once actual data replaces the preamble, clocking is not lost for short consecutive strings of 1s or 0s.

One type of QPSK demodulator, shown in Figure 5-27, literally reverses the process used in the QPSK modulator. The recovered clock supplies the reference clock to a balanced modulator and 90° phase shifter. The output of the phase shifter is fed to a second balanced modulator. The clock inputs and fin to the balanced modulators cause these signals to be "multiplied," producing second harmonic signals of the original clock input. These second harmonics are filtered at the output of the balanced modulators leaving DC voltage levels equivalent to amplitude results of the "multiplication." These DC levels are fed to ADCs and then into output shift registers.

## N-Nary PSK Modems

One major limiting factor to the data rate that can be used with any modem on any system is the bandwidth of that system. For binary, essentially noiseless communications, a form of the *shannon channel capacity* (*C*) limitation is used to determine the relationship between the number of different symbols (*n*) a modem can produce and the maximum data rate that modem can be operated at when connected to a specific system. This relationship is summarized, mathematically as:

$$C = BW \, (\log_2 n) = BW \times m \tag{5-4}$$

where *C* is the channel capacity in bits per second;

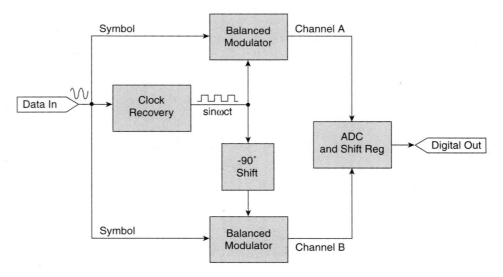

**FIGURE 5-27** QPSK Demodulator

*BW* is the bandwidth of the channel (or medium);
*n* is the number of different symbols produced by the modem;
and *m* is the number of bits per symbol.

For an FSK modem, where $n = 2$ (mark and space tones), $m = 1$ bit per symbol, and the $BW = 3,000 - 300$ or $2,700$ Hz, the theoretical maximum data rate is:

$$C = 2700 \ (\log_2 2) \ = \ 2700 \times 1 = 2700 \ bps$$

Above 2700 bps, it is necessary to transmit more bits per symbol in order to use the same system bandwidth of 2,700 Hz. QPSK modems use two bits per symbol, which increases the maximum data rate, according to the Shannon capacity equation, to $2 \times 2700 = 5400$ bps. To operate at higher data rates requires still more bits for each symbol transmitted from the modem.

Figure 5-28 is an N-nary PSK modem, specifically an 8-PSK modem. N-nary specifies the total number of different symbols producible by a particular modem. The 8-PSK modulator produces 8 different symbols, one for each group of three data bits presented at its input. Each symbol is separated in phase by 360°/8 or 45° degrees. The bit splitter shift register is increased in size to generate a single symbol for each group of three bits from the data stream. Bit A, the first bit into the bit splitter and bit B, the second bit are sent to 2-bit digital-to-analog converters (DACs). The third bit (bit C) is sent to one DAC directly, inverted, and sent to the other DAC. Bit C or $\overline{C}$ (depending on which DAC is being referenced) controls the DAC amplitude voltage required to eventually produce the desired symbol vectors. These values are 0.5V when the state of the C or $\overline{C}$ input is low and 1.207V when it is high. The states of bits A and B determine the polarity of the voltage from the DAC. A one on either bit causes its corresponding DAC to output a positive voltage. A zero produces a negative voltage value. The polarity of this voltage determines

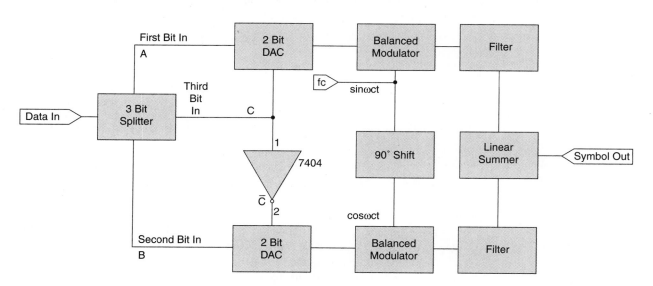

**FIGURE 5-28** 8PSK Modulator

whether the output of the balanced modulators are in or out of phase with the reference (fc) inputs.

The amplitude of the voltage determines the amplitude of the balanced modulator signals. While actual voltages used may not be the same used here, the ratio between the two levels remains the same. This ratio is required to develop the correct phase angle relationships between symbols.

Figure 5-29a is the vector diagram for an 8-PSK modulator. A similar type of diagram (Figure 5-29b) showing just the end points of the vectors is called a *constellation* diagram, and is used for purposes of clarity, particularly as the number of symbols produced by the modulator increases resulting in many more phase vectors in a single quadrant.

Using the "triangle" in quadrant I, consider the symbol vector as the hypotenuse and the individual channel vectors representing the transmitter-balanced modulator outputs as the adjacent and opposite sides of the triangle. The angle between the symbol vector and the axis (which is the phase of the symbol) is found by calculating the inverse tangent using the values of the triangle's "sides."

$$\tan^{-1}(.5/1.207) = 22.5°$$

This makes the phase angle between each vector twice 22.5°, or 45°, which agrees with the earlier computation of 360°/8, where 8 is the total number of symbols produced by an 8-PSK modulator. The magnitude of each symbol is calculated using Pythagorean theorem:

$$A^2 + B^2 = C^2$$

$$\text{solved for } C = \sqrt{A^2 + B^2}$$

Where A and B are the adjacent and opposite sides and C is the hypotenuse representing the symbol vector. For the modem represented by the vector and constellation diagrams of Figure 5-29, the voltage amplitude of the output signals is:

$$C = \sqrt{(.5)(.5) + (1.207)(1.207)} = 1.306\,V$$

---

### EXAMPLE 5-8

Show the steps needed to derive the vector symbol for a binary data stream of 000.

### SOLUTION

The three bits enter the bit splitter and are directed to the 2-bit DACs. The output of channel A's DAC (A=0 and C=0) is −.5V and the lower DAC (B=0 and $\bar{C}$=1) is −1.207V.

Note: since bit C controls the amplitude of the DAC circuit, the amplitude of the outputs of these two DACs are always different.

The DAC outputs are fed to the balanced modulators producing −.5sinωct on channel A and −1.207cosωct on channel B. These are linearly summed to form the appropriate symbol in the third (III) quadrant on the vector diagram. The remaining vector locations on the diagrams, each representing a different symbol and

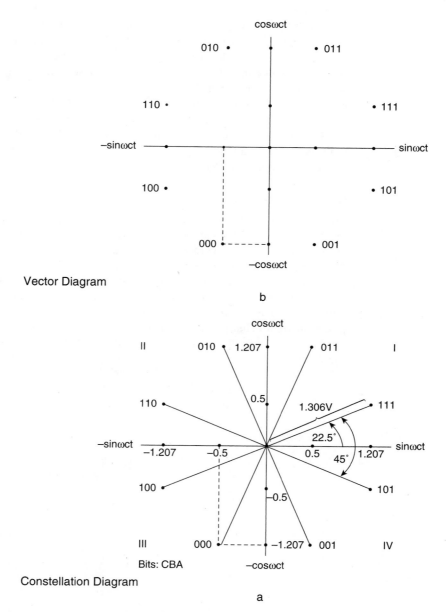

Vector Diagram

b

Constellation Diagram

a

**FIGURE 5-29** 8-PSK Output Symbol Diagrams

combination of bits A, B, and C, are found in a similar manner, to complete the diagrams in Figure 5-29.

Note that the actual voltages and polarities, as well as the lead or lag condition of the phase shifter, are selected by the circuit designer and can be any combination. The benefit of using 8-PSK modems is to increase the data rate beyond the limitation imposed on a FSK or QPSK modem without an increase in bandwidth. For example, a 4800-bps data rate input to an 8-PSK modem using 1,850 Hz for a carrier frequency, causes the phase of the output signal to change at a 1600 sps rate. All of these values fall, comfortably within the 300–3,000 Hz bandwidth of the telephone system. These values have been selected arbitrarily in this case as in earlier examples involving bandwidth considerations. However, the results are illustrative of the types of bandwidths required for circuits driven by these modulators.

For the 8-PSK modem, the Shannon Law ($2,700 \log_2 8$) determines that the upper limit for the data rate using an 8-PSK modem is 8100 bps. Data rates above 8100 bps are transferred using a different form of modulation.

## QAM

To satisfy the Shannon Law, a modem operating at 9600 bps, operating over the telephone system, would require 4 bits per symbol. The circuit produces 16 different symbols at its output. Using standard PSK for modulation creates symbols, which are 360°/16 or 22.5° apart. This phase difference comes close to the +/–10° of phase hit specified as permissible by the telephone company. These phase hits could cause sufficient phase shift in a symbol to produce a data error at the receiver. An alternate method of modulation is required to reduce the phase error risk inherent to PSK modems that begin to approach the phase hit limit. This method, called quadrature amplitude modulation (QAM) incorporates the use of amplitude and phase modulation within a single modulator. Figure 5-30 is the block diagram of a QAM modulator. It uses the circuits we are now familiar with—bit splitters, DACs, balanced modulators and linear summers.

At first glance, the QAM modulator looks very similar to the 8-PSK modulator we just looked at. One immediate difference is that four bits are shifted into the bit splitter

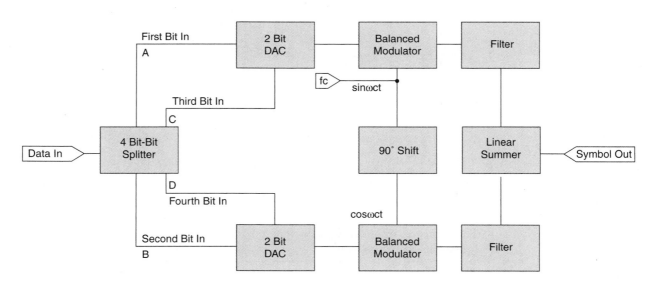

**FIGURE 5-30** 16 QAM Modulator

instead of three. These bits are directed to the DACs with bits A and C sent to the upper-channel DAC and bits B and D to the lower-channel DAC. As with the 8-PSK modulator, these bits create one of four output levels, two of which are positive voltages and two which are negative voltages. The amplitudes are selected to produce phase angle differences between symbol vector phases of 30° as shown on the vector and constellation diagrams of Figure 5-31. Note that there are two vectors in each quadrant that have no phase

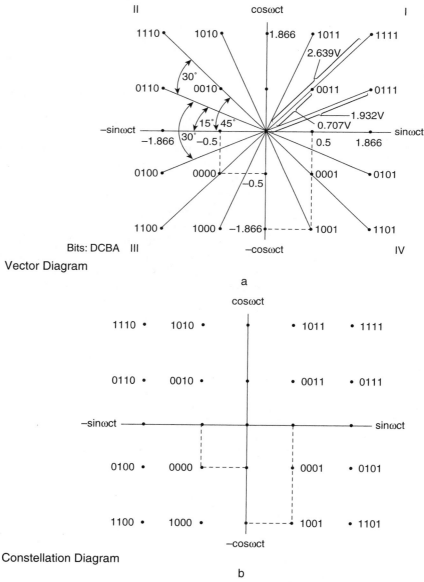

Vector Diagram

Constellation Diagram

**FIGURE 5-31** 16-QAM Output Symbol Diagrams

difference but do exhibit amplitude differences. This occurs because, unlike the 8-PSK modulator, it is now possible for both DACs to produce the same output voltage magnitude. There are twelve different phases and three different amplitudes for the symbols shown on the vector diagram for the QAM modulator.

Depending on the inputs to one of the DACs, the output can be multiples of: 1.34, 5, –1.34 or –5 volts. Combining these with the different phase possibilities from the balanced modulators (SIN ωct, –SIN ωct, COS ωct and –COS ωct) results in sixteen possible vectors at the modem's output as shown in Figure 5-31.

---

### EXAMPLE 5-9

Show how the vectors representing 0000 and 1001 are produced.

### SOLUTION

Assuming the DACs work the same as they did for the 8-PSK modulator, bits A and B determine the polarity (0 = – and 1 = +) and bits $\overline{C}$ and D, the amplitude (0 = .5V and 1 = 1.866V), then the bit combination 0000 puts out –.5V from both DACs. This reverses the phase outputs of the balanced modulators. When summed, the output symbol is –5Vsinωct - 5Vcosωct, which is the middle long vector in quadrant III of the vector diagram, Figure 5-31.

For 1001, bit A = 1, bit B = 0, bit C = 0 and bit D = 1. Channel A's DAC is +0.5V and channel B's is –1.866V. This makes channel A's balanced modulator's output to be 5sinωct and B's –1.866cosωct. These are summed to produce the symbol vector indicated in quadrant IV of Figure 5-31 As with the 8-PSK modem, the remaining vectors can be found in a similar manner.

---

9600-baud modems cause the symbols to change at a rate one-fourth of the data rate or 9600 bps/4 = 2,400 Hz. If the carrier used in the modem has a frequency of 1,870 Hz, then the modem successfully accommodates the 300–3,000 Hz bandwidth of the telephone system. Additional consideration must be given when using a QAM system since amplitude as well as the phase of the symbols is being changed for each quad bit combination. In addition to being susceptible to phase hits, QAM transmissions are also affected by telephone system impairments that affect a signal's amplitude, such as gain hits and interference.

Differences in the modulated signals for each of the primary low- and medium-speed modulators is illustrated in Figure 5-32. The FSK signal (Figure 5-32*a*) shows changes in the signal's frequency as the data bits vary between mark and space states. PSK signals (Figure 5-32*b*) show variations in phase caused by the modulating symbol information. Finally, QAM signals show changes in both phase and amplitude (Figure 5-32*c*). The number of actual amplitude changes have been minimized for this figure to reduce confusion.

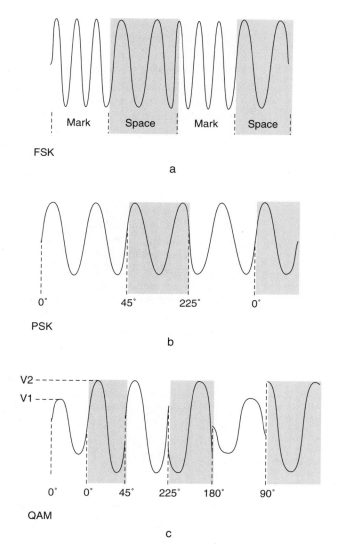

**FIGURE 5-32** Comparison of FSK, PSK, and QAM Signals

**Section 5.7  Review Questions**

1  What is the element of a sinewave that is being used as a symbol at the output of a balanced modulator?
2.  How is a logic 1 different from a logic 0 at the balanced modulator's output?

3. Why are bandpass filters required at a balanced modulator output?
4. How many different symbols are produced by a DPSK modulator?
5. What do DPSK symbols represent?
6. What is the purpose of the bit splitter in a PSK modulator?
7. Why is a linear summer used to generate the output signals?
8. What is the symbol rate at the output of a QPSK modulator with a 3600-bps input?
9. For a PSK modulator with 3 bits per symbol, what is the number of different symbols possible at the output? What is the phase angle between each phase? Given an input data rate of 3600 bps, what is the output symbol rate?
10. What activity must precede the actual demodulation of synchronous data?
11. What is used by the receiver to synchronize between transmitted and received phases of the symbols?
12. Why would a shift register be included with the ADC circuitry at the output of the QPSK demodulator?
13. What is the upper data rate limit for a 16-PSK modem connected to the telephone system, using the Shannon capacity relationship?
14. What is the phase angle between neighboring symbol vectors for a 16-PSK modem?
15. In general, a 16-PSK modem is not used to handle the data rates found in Question 1. Why, do you think, this is so?
16. What is the upper limit of the data rate used by 16-QAM modems connected to telephone system, using the Shannon law? How does this compare to a 16-PSK modem?
17. Why are 16-QAM modems preferred over 16-PSK modems?
18. What additional consideration must be taken into account when using QAM modems?

## 5.8  V.34 AND V.90 MODEMS

The modems examined to this point performed well for low- and medium-speed data rates. Today's modems have long bypassed the 9600-bps barrier to achieve rates up to 56 Kbps with promise of faster ones to come. The ITU standard for two of the more widely popular modems is V.34 for speeds up to 33 Kbps and V.90 for up to 64 Kbps data rates. In actuality, V.34 modems operate at 28.8 Kbps and V.90 is having difficulty pushing 56 Kbps.

V.34 is specified as a two-wire full duplex modem using quadrature amplitude modulation (QAM) techniques to deliver 28.8-Kbps data rates. These modems are suited for both leased or dial-up lines. To establish a link using V.34, these modems use V.8 handshaking techniques. V.8 specifies a 300-bps frequency shift key (FSK) signal using half duplex to exchange linking information. It is used to perform feature and mode negotiations between V.34 modems at both ends of the link. Choices include:

1.  Data mode or text phone operation.
2.  Type of modulation mode available.
3.  Support for V.42 error detection and/or V.42bis compression.
4.  Wireline or cellular operation.
5.  Channel frequency to use for optimum bandwidth.
6.  Desired symbol rate.
7.  Power level to use from 1 to 14 dBm in 1-dB steps.

V.34 modems use adaptive pre-emphasis to assure an even signal level throughout the transmission spectrum. It does this by boosting some signals and attenuating others as needed to precompensate for possible channel distortions it may have uncovered during the handshaking process.

V.90 modems use *pulse-code modulation (PCM)* methods, discussed in Chapter 6, for downstream transmissions, coding 256 different discrete levels into 8-bit codes. These modems operate asymmetrically, with the downstream rate operating at 56 Kbps. The upstream line uses V.34 techniques and only achieves a 28.8-Kbps data rate. If the modem detects that it cannot operate at a full 56 Kbps downstream, it will revert to using V.34 on the downstream link as well.

Ideally, since V.90 modems use PCM to decode 256 levels into 8-bit codes at an 8,000-conversions-per-second rate, they should be capable of sending at 64 Kbps. However, voltage levels close to zero tend to be too noisy to use, V.90 discards codes that replicate this noise. In the final selection, these modems use the most robust 128 codes, which restricts the data rate to 56 Kbps.

**Section 5.8  Review Questions**

1.  What type of modulation is used for V.34 modems?
2.  What process is used by V.90 modems?
3.  What is the purpose of the V.8 process?
4.  What does a V.90 modem do if it discovers that it cannot communicate at 56 Kbps?

# SUMMARY

The move of coded data from the simple dots and dashes of Morse code to using character codes transferred through the telephone system required those character codes to go through many transformations. First they leave the terminal in parallel format and are fed to a UART. In the UART extra bits are added to frame the characters and perform error detection (parity). The data is then shifted out of the UART in serial form to a modem. The modem further translates the binary data into pairs of sinewave signals that can be sent on the telephone lines to a remote site. At the receive end, the signals go through a reverse process to be converted back into parallel character codes for display or use.

Low-speed communications, which perform this type of data link utilize frequency shift keying modulation techniques. Data rates for these systems usually lie between 100 and 1200 bps. Numerous medium-to-large scale integrated circuit devices have been developed to fill the electrical functions of UARTS, modems, and RS232C interface drivers and receivers. These systems can be monitored and tested using specialized equipment like datascopes and breakout boxes.

The process needed to transfer data rates beyond the bandwidth limitations of the telephone lines requires translating the digital data rates into symbol rates that do conform to the bandwidth. To do this, groups of bits were used to create single symbols. These symbols are used to modulate the phase and, in the case of QAM, the amplitude of the signal placed onto the telephone line. These transfers, generally, use synchronous data which requires a synchronizing clock signal to be regenerated, at the receiver, from the data stream itself.

As the data rate is increased, the need to transmit more bits during a single symbol period increases. Low-speed modems operate fine using FSK modulation and one bit per symbol. Moderate speeds are handled using PSK techniques allowing multiple bits to be included in each symbol. To maintain a lower number of bits per symbol but still increase the data rate requires that more than one element of a symbol to be modulated. This is accomplished using QAM modulation. To further increase data rates, more sophisticated technology involving DACs, coding techniques, and bit compression are required. In essence, a form of coding—that is changing the original data to a form using less bits—is utilized to allow higher data rates to be sent on the telephone lines.

The International Telecommunications Union (ITU) has generated a set of standards for modems operating at a number of "standard" data rates. Since these modems are frequently referred to by the ITU specification number, they are listed here in the Table 5-6.

## TABLE 5-6

**Mark and Space Frequencies Specified by Bell 103 and 202 and ITV V.21 and V.23 Specifications.**

| Modem Type | Baud Rate (bps) | Duplex Type | Transmit Space (Hz) | Transmit Mark (Hz) | Receiver Space (Hz) | Receiver Mark (Hz) | Answer Tone (Hz) |
|---|---|---|---|---|---|---|---|
| 103 Originate | 300 | Full | 1,070 | 1,270 | 2,025 | 2,225 | N/A |
| 103 Answer | 300 | Full | 2,025 | 2,225 | 1,070 | 1,270 | 2,225 |
| 202 | 1200 | Half | 2,200 | 1,200 | 2,200 | 1,200 | 2,025 |
| 202 Back | 5 | N/A | 387* | 387* | 387** | 387** | N/A |
| V.21 Originate | 300 | Full | 1,180 | 980 | 1,850 | 1,650 | N/A |
| V.21 Answer | 300 | Full | 1,850 | 1,650 | 1,180 | 980 | 2,100 |
| V.23 Mode 1 | 600 | Half | 1,700 | 1,300 | 1,700 | 1,300 | 2,100 |
| V.23 Mode 2 | 1200 | Half | 2,100 | 1,300 | 2,100 | 1,300 | 2,100 |
| V.23 Back | 75 | N/A | 450 | 390 | 450 | 390 | N/A |

*$\overline{\text{BRTS}}$ = 0; BTD is a mark.
**Makes $\overline{\text{BCD}}$ = 0.
For Bell 202 Back, an inactive $\overline{\text{BRTS}}$ forces BCD low. The absence of one 387-Hz tone on RC forces $\overline{\text{BCD}}$ high (inactive).

# QUESTIONS

### Section 5.2

1. What are the functions of a UART in a low-speed data communications system?
2. Which bits of a serial data stream are the framing bits? What is their function?
3. Which bits in a serially-transmitted character word from a UART are optional? Which are always the same amount regardless of the character code used?
4. What are the errors detected by the ACIA? What causes each to be set?
5. What is the function of the ratio factor between transmit and receive clocks and the data rate?
6. What are the data size, stop bit, and parity options of the ACIA?

### Section 5.3

7. What are the voltage levels for RS232C data on the transmit line?
8. What are the received voltage levels for data on the received line?
9. What are the local handshake signals between a UART and modem? What is their purpose?
10. Which signal is referred to as a remote handshake? What information does it convey?
11. What does DSR and DTR signify? Which units generate each signal?
12. What is the maximum voltage that may be applied to the RS232C transmit or receive data line without causing damage to the equipment?
13. What is the maximum recommended cable length and total capacitance for an RS232C cable?
14. Which RS232C control signal indicates the presence of a tone on the telephone lines?
15. List the minimum control signals used to interface a modem to a UART using an RS232C interface.

### Section 5.4

16. List at least three changes incorporated in the RS449 standard when compared to the RS232C.

### Section 5.5

17. What is the benefit of using RS422 balanced interface circuits?

### Section 5.6

18. What are the main functions of a modem in a low-speed data communications system?
19. Define mark and space tones.
20. Define the purpose of idle line ones.
21. What are the main transmit functional blocks of a FSK modem? Explain the function of each unit.
22. What are the main receive functional blocks of a FSK modem? Explain the function of each unit.
23. Which tone is initially used to cause carrier-detect (CD) line to be asserted by a modem? What logic level does that tone represent?

24. What is the purpose of the DTR and DSR signals between the modem and UART?
25. What is the functional and electrical differences between an originate and answer modem?
26. What type of communication link requires originate and answer modems?
27. Secondary and/or back channels in a modem normally carry what type of information?

**Section 5.7**

28. What three measurable electrical characteristics of a sinewave can be used as a symbol? Which modulation types discussed in the chapter are used for each symbol type?
29. Which circuit is used to phase modulate a sinewave with digital data?
30. How many bits are grouped into a DPSK symbol?
31. How does a DPSK receiver decipher the incoming symbols into binary bits?
32. What is the main advantage of PSK modulation compared to FSK?
33. What is a drawback of PSK versus FSK modulation?
34. How many different symbols are produced at the output of a DPSK modulator? How do they differ?
35. Since a grouping of two bits is called a dibit, what would a group of three bits be called?
36. What type of digital circuit is used as a bit splitter?
37. How many bits are grouped to form a QPSK symbol?
38. Which functional circuit block is used to recover the clock from the input data stream in a PSK demodulator?
39. How many different symbols are possible at the output of a 16-QAM modulator?
40. What is the advantage and disadvantage of a 16-QAM modulator compared to a 16-PSK modulator?
41. What is the symbol rate for a 2100-bps data stream using a QPSK modulator?
42. What is the symbol rate for a 2100-bps data stream using a 8-PSK modulator?
43. What is the phase angle difference between symbols for a QPSK modulator?
44. What is the phase angle difference between symbols for a 8PSK modulator?
45. What is the maximum data rate for a modem connected to the telephone lines, in bps, of a modulator that generates a symbol for every group of 5 bits at the data input?
46. What is the maximum data rate for a modem connected to the telephone lines, in bps, of a modulator that generates a symbol for every group of 12 bits at the data input?
47. A QAM type modulator converts groups of 6 bits into symbols. What are the values of the following parameters for this modulator if the data rate is 12000 bps?
    a) symbol rate
    b) number of different symbols
48. What standard data rate is associated with a QAM modulator?
49. What is the common data rate for V.34 modems?
50. What is the common data rate for V.90 modems?

51. What does a V.34 modem use to negotiate features and operating modes with another V.34 modem?
52. List four options that a V.34 can choose from after it has learned about the remote V.34 modem.
53. What are the upstream and downstream data rates for a V.90 modem?
54. What modulation technique is used with V.34 modems?
55. What is one modulation technique used with V.90 modems?
56. What does a V.90 modem do if it detects that it cannot communicate at 56 Kbps?

# RESEARCH ASSIGNMENTS

1. Many modem manufacturers are producing 2400, 4800, 9600, 28K, and 56K bps modems. Research one of these modems using magazine articles, the Internet, and company data manuals. Your report should include the circuits used in the modem, which specifications the modem adheres to, the standardizing organization whose standards are being used, all of the specifications for the modem, and any other pertinent information about the modem.

2. Research and write a paper (or give a class presentation) on one of the following topics:

a) Two port UART used with 16/32 bit microprocessors (68000 family or 80x86 family).

b) Auto-answer modems.

c) Commercial 28Kbps or 56Kbps modem.

d) Other related up-to-date interface standards, UARTs, and modems.

# ANSWERS TO REVIEW QUESTIONS

**Section 5.2**

1. $RS = 0$, $R/W = 0$, $CS_0 = 1$, $CS_1 = 1$, $CS_2 = 0$
2. $\overline{RTS}$, $\overline{CTS}$, $\overline{DCD}$
3. 76.8 KHz
4. 7 and 8 bits
5. 1 stop bit
6. To make sure that the start bit is "real" and not noise by verifying that it is still low halfway through the bit time.
7. 20.2 KHz to 18.28 KHz, or 1.92 KHz
8. parity (PE), framing (FE), and overrun (OVRN)
9. TDRE: transmit buffer is empty and is available for the next transmitted data word.
   RDRF: received data word is fully shifted in and is ready to be transferred to system.

10. Both stations must have similarly configured UARTS. The mark and space tone pairs are transmitted at different frequencies from the originate station compared to the answer station.

**Section 5.3**

1. RS stands for recommended standard, indicating that there is no enforcing body behind the standard.
2. −5 to −15V from the transmitter, −3 to −15V into the receiver.
3. The maximum distance is derived by dividing the maximum capacitance of 2,500 pf by the maximum capacitance per foot of 50 pf/ft.
4. V.24
5. To allow RS232C lines to be monitored and tested.
6. Allow two computers to connect directly using the serial interfaces without using a modem.

**Section 5.4**

1. RS232C has a maximum cable length of 50 feet, RS449 of 200 feet. RS232C operates at a maximum data rate of 20 Kbps, RS449 at 2 Mbps.
2. Loop-back function is used to test a station by causing its transmitted signal to be looped back directly to its own receiver.

**Section 5.5**

1. Signals are sensed between active and return line inputs to a balanced circuit. Noise appears in phase on both lines and is not detected.
2. 6V
3. 3. ±4 to ±6V

**Section 5.6**

1. VCO
2. PLL
3. They convert transmit tones from each end of the link to different mark and space frequencies.
4. Echo-plex echoes back transmitted data to its own DTE when in half-duplex mode.

**Section 5.7**

1. frequency, phase, and amplitude
2. bandwidth of telephone line (300–3,000 Hz)
3. phase hits of ±10°
4. phase
5. Logic 1 and logic 0 differ by 180° at the output.
6. Filters remove signals generated by mixing action of non-linear diodes.
7. two
8. states of current and preceding bits
9. 8 symbols, 45°, 1200 sps
10. clock recovery
11. preamble
12. to convert symbol rate back to data rate

13. 10.8 Kbps
14. 22.5°
15. The phase difference between phases is too close to phase hit specification of telephone company—possible data errors.
16. 10.8 Kbps, which is the same as the 16-PSK modem.
17. improved phase angle difference (33° compared to 22.5°)
18. QAM modems are also susceptible to amplitude impairments and limitations imposed by the telephone system.

## Section 5.8

1. QAM
2. PCM
3. V.8 negotiates features and operating modes between V.34 modems.
4. V.90 reverts to V.34 (28.8 Kbps) operation if it detects that it cannot operate successfully at 56 Kbps.

CHAPTER **6**

# Higher Capacity Data Communications

## OBJECTIVES

After reading this chapter, the student will be able to:
* determine how to combine several communications channels into one using frequency division multiplexing (FDM).
* determine how to combine several communications channels into one using time division multiplexing (TDM).
* determine the composition of T1 and other digital carriers.
* analyze how data transfers are accomplished using pulse coded modulation PCM).
* apply sampling theorem and quantization to the PCM process.
* extend PCM methods to include delta modulation and companding.
* apply companding concepts to a CODER/DECODER or CODEC data set.

## OUTLINE

6.1    Introduction
6.2    Multiplexing Methods
6.3    Sampling Theorem
6.4    Quantization
6.5    Pulse Code Modulation
6.6    Delta Modulation
6.7    Digital T Carriers
6.8    Companding
6.9    Codecs

## 6.1 INTRODUCTION

The modems and communication systems that have been explored in Chapter 5, for the most part, are utilized for low-to-moderate speed data rates. This data rate is limited by

179

**channel**
Single line of communication.

**frequency division multiplexing (FDM).**
multiplexing techniques where channels share frequency in a band.

**baseband**
single channel communication system in which that channel is the only one occupying the system's bandwidth.

the system's bandwidth (300–3 kHz for the telephone system). The various methods thus far discussed utilize a small portion of this bandwidth to manage a single communication link also known as a **channel** to send data. In these cases, the digital data is converted to an analog signal and sent via modems onto the telephone lines.

Another method of transmitting the digital data in analog form is called **frequency division multiplexing (FDM).** This method allows many voice or **baseband** channels (bandwidth 300–3 kHz) to share space in a much larger total bandwidth. This band of channels is then sent via radio, microwave transmitters, satellites, or along fiber optic cables to a receiver where they are demultiplexed into individual channels destined for designated stations.

## Baseband and Broadband Channels

The term *baseband* has evolved into two distinct definitions depending on whether the area under discussion is general in nature or applies to local area networks presented in Chapter 8. In general terms, baseband refers to a single channel transmission. Thus a single voice channel can be considered a baseband signal whose band is from 300 to 3 kHz. Similarly, if a single channel is used to transport a baseband signal, it is called a baseband channel or system. A point-to-point system as discussed in Chapter 1 is an example of a baseband system. A single channel's signal is the only signal occupying the data link. In this case the signal is analog (output from the modem), not digital.

In the context of local area networks, the term baseband indicates a transmission method, typically used for shorter distances, in which the entire bandwidth of the LAN cable is required to transmit a single digital signal. The limiting aspect of the definition when applied to local area networks is the digital nature of the signal.

**broadband**
multichannel communication system in which the system bandwidth is shared by the channels using it.

In contrast, **broadband** systems are those that allow many channels to share the system's bandwidth. By definition, a broadband channel is a communications link having a bandwidth greater than a voice grade channel and potentially capable of much higher transmission rates. Broadband systems are also called wideband systems. With respect to LAN technology, a broadband channel is a system in which multiple channels access a medium that has a large bandwidth capable of transporting multiple channels of data. Each channel is multiplexed so that it occupies a different frequency slot within the bandwidth of the system and is demultiplexed at the receiving end.

Baseband is a single channel (either analog or digital), which occupies the entire bandwidth (though it may not utilize all of that bandwidth) of a system's medium. Broadband, which is also either analog or digital, refers to multiple channel transmissions in which each channel occupies a portion of the medium's bandwidth.

**Section 6.1 Review Questions**

1. Define a channel.
2. What are the characteristics of a baseband channel?
3. What are the characteristics of a broadband channel?

# 6.2 MULTIPLEXING METHODS

**multiplexer (MUX)**
unit or software that allows
signals to share a single
channel.

One way to increase the capacity of data transmission is to multiplex numerous channels into a single transmission channel, using a **multiplexer (MUX).** This has the advantage of requiring a single transmission medium to carry many channels of information. The downside of multiplexing is that there is a requirement for either additional time or bandwidth to achieve the multiplexing. Two of the more common forms of multiplexing are frequency division multiplexing, requiring a large bandwidth to accommodate all the multiplexed channels; and time division multiplexing, requiring separate time slots for each channel.

## Frequency Division Multiplexing (FDM)

**frequency shift key (FSK)**
modulation technique that uses
changes in frequencies.

In its simplest form, FDM is represented by a **frequency shift key (FSK)** modem. Here two frequencies were used to represent a one and a zero. In a full-duplex system, four sets of frequencies occupied the line with a maximum of two signals on the line at any one time. In effect, the bandwidth of the telephone line was shared by each of the signal frequencies. This example falls short, since it deals with a single channel of communication and does not involve any form of true multiplexing. It does, however, illustrate frequency allocation usage of the telephone system bandwidth. Full-duplex operation is performed by allocating the transmit frequencies of the originate and answer modems into different frequency bands within the bandwidth.

FDM applies itself to the increased usage of a network's capacity by utilizing as much available frequency bandwidth as is allocated to the system. Applications of FDM include fiber optic links, microwave, and satellite communications that have bandwidths in the megahertz and gigahertz ranges.

## Voice Band

A single voice or baseband channel occupies (ideally) a 0–4 kHz bandwidth as allocated for a telephone subscriber circuit. Actual operating bandwidths for the voice channel is 300–3 kHz. The remaining band area (0–300 and 3–4 kHz) is unused and acts as a builtin

**guardband**
a portion of unused bandwidth
between channels.

**guardband.** Simply put, a guardband is an area that acts as a buffer between adjacent channels to avoid one channel's data from crossing over into another channel. It is similar in concept to the "dead" area between broadcast channels on a FM radio.

The first channel in an FDM system occupies the frequency band from 0–4 kHz. A second channel could then begin at 4 kHz and end at 8 kHz and still have a 4-kHz bandwidth and contain the same guardband area. Putting these two channels together produces a total bandwidth of 0–8 kHz. These two channels, of course, are not going to be sent

**plain old telephone system
(POTS)**
telephone company system.

using **plain old telephone system (POTS)** lines, since the upper end of the telephone line bandwidth is only 4 kHz.

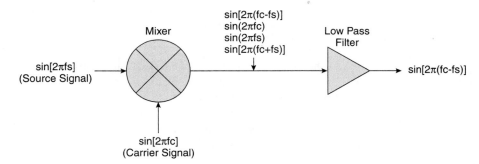

**FIGURE 6-1** Mixer Blocks

All voice grade channels have a bandwidth of 0–4 kHz. In order to be transmitted as part of a broadband signal, the voice channel is mixed with a carrier frequency. A mixer (Figure 6-1) is a nonlinear circuit that produces the original signal and the sum and difference frequencies of two input waveforms. When a balanced modulator is used as the mixer, the voice channel and carrier signals are suppressed at the mixer's output. All frequencies above the difference frequency are filtered out, so that the mixer/filter circuit only produces the difference frequency.

---

**EXAMPLE 6-1**

Illustrate the results from mixing a voice grade channel (0–4 kHz) with a 64-kHz carrier signal.

**SOLUTION**

Refer to Figure 6-2 to observe the frequency spectrum produced by the mixing action. The original voice channel signal and the carrier signal are suppressed and do not appear at the mixer output. The difference of the carrier and the voice channel (60–64 kHz) is the lower sideband signal, while the sum (64–68 kHz) fills in the upper sideband area. The upper band is filtered using a low-pass filter following the mixer so that only the difference frequencies (lower sideband) is produced from the mixer. A second and third channel can be produced at higher frequencies to this example's channel by mixing them with higher carrier frequencies.

---

## FDM Groups

The telephone company has a set of standards which specify the frequency allocations for a broadband FDM system. The standard begins with 12 voice channels, each mixed with carrier signals that are 4 kHz apart from each other. These carrier signals range from 64 kHz to 108 kHz. The resulting difference frequencies are combined into a single

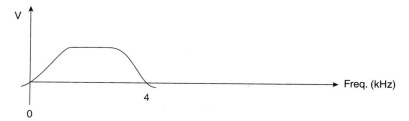

Original Voice Band Channel

a

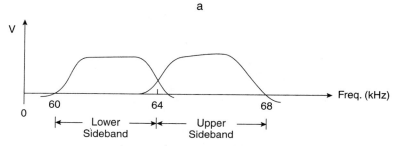

Voice Band Mixed with 64 kHz Carrier Signal

b

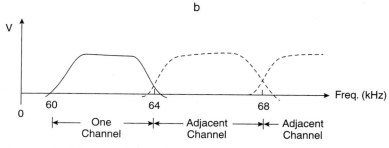

Filtered Channels (3 Adjacent Channels)

c

**FIGURE 6-2** Results of Mixer/Filter Operation

*group* channel by linearly summing the outputs of several mixers as shown in Figure 6-3. Each mixer mixes a voice channel (0–4 kHz) with a carrier signal. The carrier signals are 4 kHz apart, placing each channel adjacent to the next in the overall frequency spectrum of the group (Figure 6-4). The bandwidth of the group ranges from 60 kHz (the carrier of the group 12 minus 4 kHz voice channel) to 108 kHz (the carrier of group 1, 108 kHz, minus 0 kHz, the lower end of the voice channel) for a total bandwidth of 60–108 kHz = 48 kHz. This 48-kHz bandwidth can be verified by multiplying 12 channels times 4 kHz per channel, resulting in a 48-kHz total bandwidth at this point.

Notice that the 60–108 kHz bandwidth is well above the telephone system 300–3 kHz bandwidth. This group channel is meant to be sent by mediums other than voice grade telephone lines. It could be sent on a fiber optic cable, which has a larger bandwidth (discussed later), but it is more frequently used for radio, microwave, and

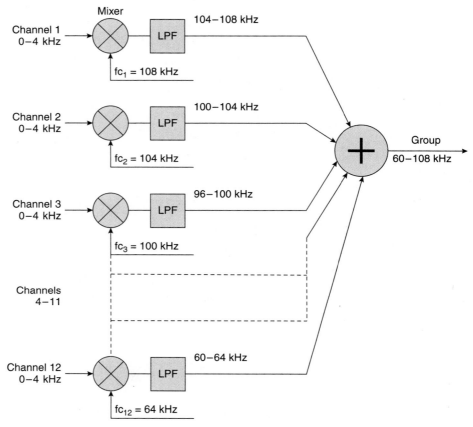

(Note: LPF-Low Pass Filter)

**FIGURE 6-3** Formation of a Twelve-Channel Group

satellite transmissions. Some of these have extremely large bandwidths compared to 48 kHz. As such, larger groupings are created in similar manner to take full advantage of these large bandwidths.

---

**EXAMPLE 6-2**

What is the bandwidth range of channel 7 of a group transmission?

**SOLUTION**

The carrier frequency (fc) for each channel is 4 kHz less than the previous channel. Channel 1's fc is 108 kHz, channel 2's is 104 kHz, etc. This makes channel 7's fc equal to 84 kHz. Mixing 84 kHz with the 0–4 kHz channel range and then filtering produces a bandwidth of 80–84 kHz for channel 7.

---

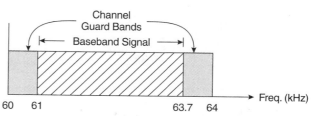

Channel 12 Frequency Spectrum after Mixing with 64 kHz Carrier Signal

a

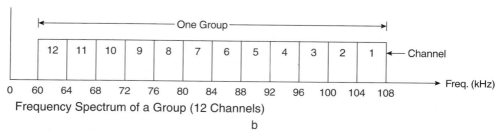

Frequency Spectrum of a Group (12 Channels)

b

**FIGURE 6-4** FDM Group Formation

**carrier pilot**
reference signal that sets the
amplitude of a FDM group.

Each channel within the group may vary in amplitude since they are created from 12 separate sources. To regulate the amplitude of the group, a **carrier pilot** signal at 104.08 kHz is added to the group. Like attenuation/gain distortion described in Chapter 2, this carrier pilot is used to set the amplitude reference for the group. The amplitudes of the separate channels are adjusted to match the carrier pilot so that the group will have a uniform amplitude. The carrier pilot is also used to generate a carrier detect (CD) signal to indicate that a remote link has been established.

**supergroup**
60 channel FDM signal.

By repeating the process described for creating groups, larger groupings are developed as actual transmission frequencies increase. Five groups (60 channels) are assembled to form a **supergroup,** which fills a bandwidth between 312 and 552 kHz. The actual bandwidth (552 – 312 = 240 kHz) equals the 5 groups ( 5 × 48 kHz), which in turn, is equivalent to the 60 channels (60 × 4 kHz) that make up the 5 groups. Group 1's carrier pilot is used as the supergroup's carrier pilot for amplitude regulation. The next level formed from 10 supergroups is the **mastergroup.** An additional carrier pilot at 2840 kHz is inserted at the mastergroup level for amplitude regulation at that level. Table 6-1 summarizes the bandwidth assignment of the telephone company's FDM standard allotment.

**mastergroup**
an FDM signal containing 10
supergroups (600 voice
channels).

**jumbo group**
FDM signal made up of
6 mastergroups.

Various other grouping levels are also categorized: a **jumbo group** is composed of 6 mastergroups; each mastergroup encompasses 10 supergroups; one supergroup contains 5 groups, each with 12 channels. A jumbo mux combines three jumbo groups into a single channel. By performing a little multiplication, it is determined that 3 × 6 × 10 × 5 × 12 = 10,800 channels are carried by a jumbo mux system.

An optional grouping specified is a mastergroup mux, which allows 2 to 6 mastergroups (1,200–3,600 channels) to be combined. Generally, use of a mastergroup mux transmission is terminal. That is, a mastergroup mux is used instead of a jumbo group or jumbo group mux.

**TABLE 6-1**

**Frequency Assignment for Telephone Company FDM**

| FDM Level | Number of Channels | Bandwidth Range | Bandwidth |
|---|---|---|---|
| Voice Channel | 1 | 0–4 kHz | 4 kHz |
| Group | 12 | 60–108 kHz | 48 kHz |
| Supergroup | 60 | 312–552 kHz | 240 kHz |
| Mastergroup | 600 | 564–3,084 kHz | 2,520 kHz |
| Jumbo Group | 3,600 | 564–17,544 kHz | 17 MHz* |
| Jumbo Mux | 10,800 | 3.0–60 MHz* | 57 MHz* |

*Values approximate.

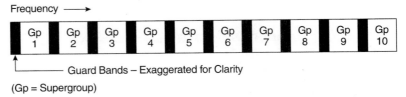

Frequency ⟶

Guard Bands – Exaggerated for Clarity

(Gp = Supergroup)

**FIGURE 6-5** Mastergroup with Guard Bands

Guardbands, as mentioned earlier, are used to prevent interference between adjacent channels. At the group and supergroup levels, the builtin guardband of the voice channels is sufficient to assure this. Starting with mastergroups, additional spacing is inserted into the frequency spectrum. Notice that the 10 supergroups, each with a bandwidth of 240 kHz, should form a 2,400-kHz mastergroup. In reality, the mastergroup is specified to occupy a bandwidth of 2,520 kHz. The additional 120 kHz is assigned as ten additional 12-kHz guardbands between the supergroups, which make up the mastergroup. Guardbands of 12 kHz also precede and follow the mastergroup (Figure 6-5). Equally, 6 mastergroups require 15,120-kHz (15.12 MHz) bandwidth for a jumbo group. Jumbo groups are allotted bandwidths of 16,980 kHz, providing for additional guardbands of 310 kHz each between mastergroups in the jumbo group (Figure 6-6).

Essentially, the microwave and satellite transmitter carriers are modulated by the master or jumbo groups of channels and transmitted to receivers where the signals are demodulated and demultiplexed by mixing the transmitted grouping using the same carrier frequencies that formed the original groups. This process is repeated until each individual channel is recovered and sent on to its final destination.

The advent of FDM systems resulted in a secondary effect that has been capitalized on. The band below 564 kHz is not used by mastergroups or larger groups. That is, the band from 0–564 kHz is unused by these transmissions. It was soon discovered that non-voice communications signals (control, test data, etc.) presented at high-data rates could

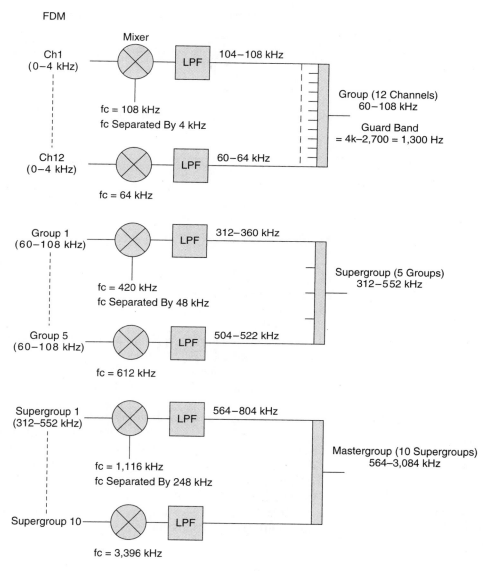

**FIGURE 6-6** Forming of Multichannel Distribution Using FDM

be sent along in this band, with a mastergroup transmission as shown in Figure 6-7. Signals of these type are referred to as **data under voice (DUV).**

**data under voice (DUV)** method that utilizes normally unused band areas for supplemental voice transmissions.

## Time Division Multiplexing

Using the point-to-point conventional means of transferring data on a baseband system previously discussed in the text, sending three messages of varying lengths is accomplished in

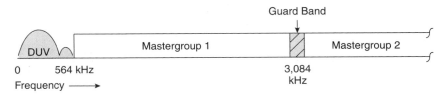

**FIGURE 6-7** Jumbo Group with Data under Voice (DUV)

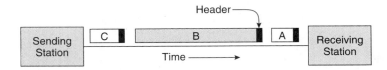

**FIGURE 6-8** Conventional Transmission

the following manner. Message A is transmitted first, followed, in time, by message B, and then message C as shown in Figure 6-8. This is a simple and straightforward process. The main drawback is that message C must wait until messages A and B are sent before it is transmitted.

Expand the problem by applying it to a multipoint system (Figure 6-9), where messages are destined for different secondaries. Secondary station 3 in Figure 6-9 would like to start receiving its message at about the same time as station 1 got message A and station 2 received message B. To resolve the problem, all three messages are reformed into smaller parts called **packets.** These packets are of equal length, resulting in more packets for longer messages (message B in Figure 6-10a). The packets comprising messages A, B, and C are interleaved and assigned time slots as shown in Figure 6-10b. A header (shaded area), containing address and packet sequence number information, precedes each packet. The interleaved packets are transmitted and received by the destination stations. The appropriate packets (determined by a destination address in the header) are extracted by each station as they are received and reassembled (by packet sequence number included in the header) into their original message form. This is the essence of a **time division multiplex (TDM)** system.

## TDM Methods

Two basic forms of TDM are in use today, **synchronous TDM (STDM)** and **asynchronous TDM (ATDM),** also known as **statistical TDM (STATDM)** or **STAT MUX.** STDM systems assign time slots of equal length to all packets regardless if whether or not anything is to be sent by each station with an assigned time slot. As an example, if message A was not included in the previous illustration, its allotted time would still be allocated and used, taking up time and bandwidth on the system. The TDM frame would appear as it does in Figure 6-10, but time slots for the A message would not contain information.

**packets**
a section of a message.

**time division multiplex (TDM)**
multiplexing technique that assigns channel signals to time slots.

**synchronous TDM (STDM)**
synchronous time division multiplex.

**asynchronous TDM (ATDM)**
asynchronous time division multiplex.

**statistical TDM (STATDM)**
statistical time division multiplex.

**STAT MUX**
statistical multiplex.

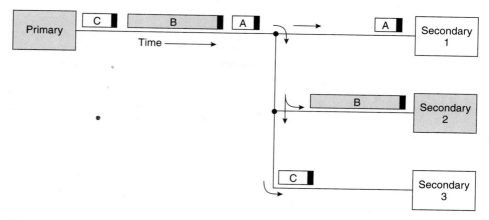

**FIGURE 6-9** Conventional Multipoint Communication

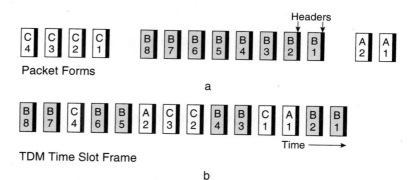

**FIGURE 6-10** Packet Time Allotment

STDM systems are comparatively easy to implement once the applications software allocates the time slots. STAT MUX systems, on the other hand, are more complex, but they allow for a means of reassigning time slots when there are those not in use. Essentially, STAT MUX networks assign time slots only when they are to be used and deletes them when they are idle. The total time used for a STAT MUX frame varies with the amount of traffic currently being handled.

Broadband systems manage many channels of communication on a single line at a given time by dividing the systems bandwidth so that each channel occupies a portion of that channel. These channels are multiplexed onto a single carrier and demultiplexed at the receiving end (Figure 6-11). This requires that the bandwidth of the system be large enough to contain all the channels. Time division multiplexing, on the other hand, uses a baseband system requiring less bandwidth. Only a single channel's signal appears on the system at any given time. Multiple channel use of a TDM system relies on sharing transmission time periods rather than a system's bandwidth.

Messages from channels A, B, and C can be sent one at a time (message A followed by messages B and C) but the same problem exists that was involved with the multipoint

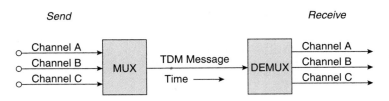

*Send*                                                      *Receive*

**FIGURE 6-11** Multichannel TDM System

**channel**
single line of communication.

system in Figure 6-9. There is a period of time that elapses before message C is received by its destination station. Compound this problem by realizing that each **channel** in this example is a multipoint system. That could delay a message from one station to another, using conventional methods, by a significant amount of time.

STAT MUX systems are most suitable for these high-density, high-traffic applications. A constant flow of messages are assigned time slots and interleaved as each channel on the send side becomes active and requires communications with another channel. If a channel does not have any traffic, its time slots are deleted and reassigned to an active channel. This way, the interconnecting media achieves a higher state of efficiency usage than with STDM systems.

### Section 6.2 Review Questions

1. Why can't POTS be used to transport FDM channels?
2. Which frequency resulting from the mixing action of an FDM modulator is used on the FDM channel?
3. What is the actual process behind FDM?
4. What methods are used to send FDM channels?
5. What process is used to demultiplex FDM signals?
6. What is gained by multiplexing packets into time slots?
7. Compare STDM and STAT MUX methods of time division multiplexing.

## 6.3 SAMPLING THEOREM

**analog to digital converter (ADC)** and **digital to analog conversion (DAC)**
circuits or devices that convert between analog and digital signals.

A technique for sending analog information on digital carriers involves translating voice or other analog signals into a train of pulses, which are, in turn, coded into digital format. The digital data is then transmitted and the process reversed at the receiving end. In effect, this process combines sampling with **analog to digital conversion (ADC)** and **digital to analog conversion (DAC)**. An advantage gained by this method is in the reduction of noise effects usually inherent in analog transmissions.

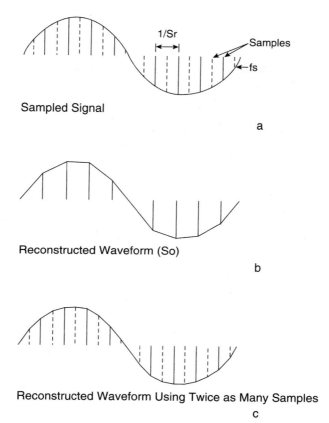

Sampled Signal

a

Reconstructed Waveform (So)

b

Reconstructed Waveform Using Twice as Many Samples

c

**FIGURE 6-12** Sampled Signal

The process of digitizing an analog signal starts by dividing the original signal into uniformly spaced pulses or samples as shown in Figure 6-12. The amplitudes of the sample pulses rise and fall with the amplitude of the original signal. Specifically, this part of the process is a form of *pulse amplitude modulation (PAM)* since the original signal is being separated into individual pulses (or samples), each of which has a different amplitude based on the amplitude of the original signal. At the receive end, these samples are used to reconstruct the original signal. The more frequently the samples are taken, the more accurate is the reconstructed waveform. Notice in Figure 6-12, that the bottom signal is sampled twice as much as the top signal. The recovered signals shown below each example show the difference caused by the differences in the sampling rates.

## Nyquist Sampling Rate

To determine the minimum number of samples to use to replicate the original waveform reliably, a sampling theorem developed by Nyquist is used. This theorem states that for a given signal, f(s), the minimum sampling rate, known as the **Nyquist sampling rate (Sr)** to assure accurate recovery of the signal at the receiving end is twice the frequency of the highest sinewave element sin(2πfst) of the original signal or as a formula:

$$Sr = 2[\sin(2\pi fst)] \tag{6-1}$$

Original signals are sampled at rates at or above the minimum sampling rate to assure that the original signal is accurately replicated. If the sampling rate were less than twice the highest fundamental sinewave frequency, then a distortion called *aliasing,* or *foldover,* occurs. To understand what this means, consider sampling the voice signals on the telephone lines that contain signals from 300 to 3 kHz. Essentially, the sampling process causes mixing, which is similar to that used for regular amplitude modulation (AM) to result. This process creates the sum and difference frequencies as well as the original signals that were mixed. For a sampling circuit, these are Sr, sin(πfst), Sr-sin(2πfst) and Sr+sin(2πfst). Filters are used to remove all but the difference and original f(s) signals. With an Sr = 2sin(2πst), the frequency range of the original signals reside next to the difference frequencies. If the sampling rate is higher than 2sin(2πst), there is a gap between one group and the other. The problem of foldover occurs when Sr is less than 2sin(2πst).

---

### EXAMPLE 6-3

Show the differences between sampling a voice channel (300–3 kHz) using sampling rates at and below 2sin(πfst).

### SOLUTION

The minimum sampling rate is twice the highest frequency component of f(s) or $2 \times 3$ kHz = 6 kHz. Mixing the voice band with 6 kHz and removing the higher-frequency elements, produces the original voice channel (300 to 3 kHz) and the difference band 6 kHz – (300 to 3 kHz) = (3 kHz to 5.7 kHz). These two bands are illustrated in Figure 6-13a. Using 4.5 kHz for Sr as an arbitrary value that is less than 2sin(2πst) results in the original voice band and a difference frequency band of 4.5 kHz – (300 to 3 kHz) = (1.5 to 4.2 kHz). Figure 6-13b illustrates those two bands. Note that there is a foldover of the original band and the difference frequency band from (1.5 to 3 kHz).

---

## Sampling Using TDM

Two or more signals can be sampled at different times. The samples are all sent and extracted at the receiver according to their time relationship. In essence, the samples of one signal will be assigned to specific time slots while samples from other signals occupy different time slots. Figure 6-14 illustrates the concept of varying sampling times for different signals.

For simplicity, signals A and B have the same frequency and are sampled at the same Nyquist rate. Signal A is sampled beginning at time t0. Samples are spaced at 1/Sr time

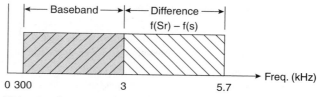

Sampling Rate (Sr) = 2 × Signal Rate (s)

a

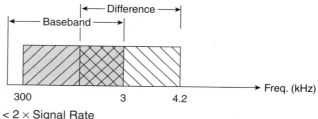

Sampling Rate < 2 × Signal Rate

b

**FIGURE 6-13** Aliasing (Fold-Over) Distortion

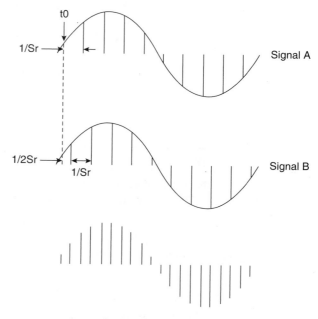

Combined Pulse Train

**FIGURE 6-14** Sampling Two Signals at One Time (TDM)

periods. The first sample for signal B is taken at a time of t0+ 1/2Sr or halfway between the first two samples for signal A. Each succeeding sample for signal B occurs 1/Sr time period from the last sample. This places the samples for signal B in between the samples for signal A. All the samples are sent sequentially. The receiver extracts the sample pulses, starting at t0 and occurring each 1/Sr following t0. From this, the receiver recreates signal A. The receiver also extracts the samples starting at t0 + 1/2Sr and each succeeding sample 1/Sr away. From these samples, signal B is reconstructed.

## Natural Sampling

**natural sampling**
sample peaks follow actual sampled signal.

**flat top**
sample and hold that maintains peak values between samples.

Samples are done by generating a short pulse at a specific time. The amplitude of the pulse is equivalent to the amplitude of the signal at the time of the sample. The width of the pulse is designated tp and the time between pulses (1/Sr) is Tr. The shape of the pulses themselves come in two forms. One is called **natural sampling,** in which the peak of the pulse follows the signal's actual shape. The second pulse form is a **flat-top** shape, in which the peak amplitude is held flat by the sample and hold circuit, rather then being allowed to follow the actual signal amplitude change exactly. These shapes are shown in Figure 6-15. Several different flat-top formats based on the type of pulse sampling performed are illustrated in the figure.

The original signal is shown in Figure 6-15$a$. Sampling pulses set at Sr are drawn in Figure 6-15$b$ and underneath are the natural sampling (Figure 6-15$c$) and flat-top (Figure 6-15$d$ pulse trains. For flat-top sampling, the reconstructed signal (So) shown in dotted lines in Figure 6-15$d$, for a given signal f(s) is represented by the relationship:

$$So = \frac{tp}{Tr} \, [\sin(2\pi fst)] \tag{6-2}$$

where tp is the time period for the sampling pulse and Tr is the reciprocal of sampling rate (Sr). Since tp/Tr is the duty cycle of the sampling signal, the relationship of So to $\sin(2\pi fst)$ is a direct factor of that duty cycle.

---

### EXAMPLE 6-4

What is the value of an output signal derived from an input signal f(s) equal to 4Sin($2\pi$ft) sampled at a rate of 40 kHz, with a sample pulse width of one microsecond?

### SOLUTION

Entering the values into Equation 6-2:

$$So = 1\mu \times 4Sin(2\pi ft) \, / \, (1/40 \text{ kHz}) = 0.16Sin(2\pi ft)$$

Observe that the amplitude of the reconstructed signal is not the same as that of the original signal. However, the signal itself is correct (that is, it is Sin($2\pi$ft)). The amplitude is recoverable by using an amplifier to boost So back to its prior value.

---

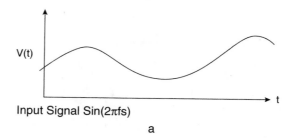

Input Signal Sin(2πfs)

a

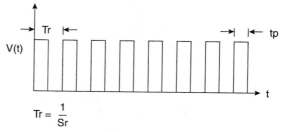

$Tr = \dfrac{1}{Sr}$

Sampling Pulses at Rate Sr. (Pulse Exaggerated for Clarify)

b

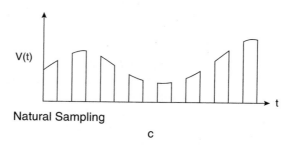

Natural Sampling

c

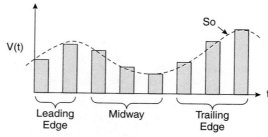

Recovered Signal So = $\dfrac{tp}{Tr} \times \sin(2\pi fs)$

Flat-top Sampling

d

**FIGURE 6-15** Sampling Types

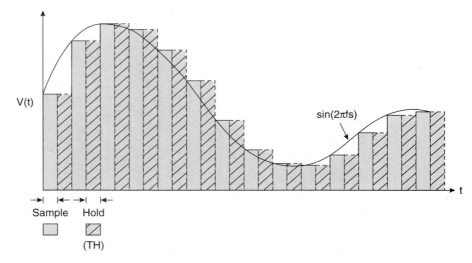

**FIGURE 6-16**  Sample-and-Hold Waveform

## Sample and Hold

Notice that there is a period of inactivity between sample pulses in Figure 6-15. During the reconstruction process, a method to abridge this gap is required to more accurately reproduce the original signal. This method is called sample and hold. A sample pulse's amplitude is detected and that value retained until the occurrence of the next sample pulse (Figure 6-16). For this method to be effective, the hold time between samples (TH) is relatively small compared to the time period of the original signal.

The most common method used for sample-and-hold circuits is to employ a capacitance at the output of a buffer amplifier. The capacitor is charged to the sample pulse value. When the amplitude falls to zero between pulses, the capacitor remains charged to the pulse value. The next sample pulse causes the capacitor to charge or discharge to that value. Again, that value is held until the next pulse arrives. Figure 6-17 is an example of a sample-and-hold circuit.

### Section 6.3  Review Questions

1. What is the significance of the Nyquist Sampling rate?
2. What type of modulation is effectively used by the sampling method discussed in this section?
3. Describe the difference between natural and flat-top sampling.
4. Why is the use of a sample-and-hold circuit desirable?

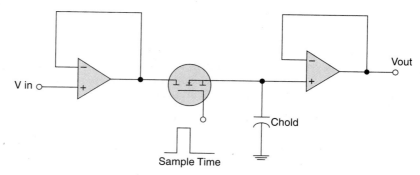

**FIGURE 6-17** Sample and Hold Circuit

## 6.4 QUANTIZATION

**quantization**
process of approximating sample values by using fixed discrete values.

**Quantization** is the process of approximating sample levels into their closest fixed value. The values are preselected and since they are fixed, they are easy to encode. The quantized waveform experiences either quantum changes in amplitude or no change in amplitude. Given a signal, f(s), with peak voltage points of Vh and Vl, the size, S, of a quantum step is determined by this relationship:

$$S = (Vh - Vl)/n \qquad (6\text{-}3)$$

where $n$ is the number of steps or levels between Vh and Vl.

Figure 6-18 illustrates the relationship between f(s) and a quantized example. Note that the quantized waveform appears similar to a sample-and-hold approximation of the same f(s). The difference is that the normal sample-and-hold result has voltage levels equivalent to the voltage level of f(s) at the point the sample was taken. The quantized levels are those fixed levels that are the nearest to f(s) at the point the sample is taken.

**quantization error** or **quantization noise**
difference between quantized signal and sample levels.

The difference between the sample-and-hold and the quantized results is called the **quantization error** or **quantization noise.**

Essentially, the process of sampling and quantizing a signal is a form of pulse amplitude modulation (PAM), where the samples produce pulses of varying amplitudes. When these amplitudes are restricted to discrete quantized values, they can be assigned specific binary codes using a method called pulse coded modulation.

**Section 6.4  Review Questions**

1. Define the process of quantization.
2. What is quantization noise?

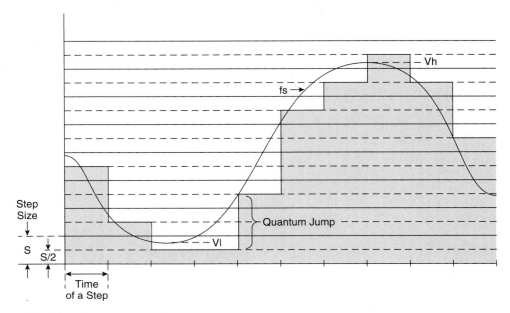

**FIGURE 6-18** Quantized Signal

## 6.5 PULSE CODE MODULATION

Figure 6-19 summarizes the theoretical process of developing a *pulse code modulation (PCM)* effect. The range of voltages for signal f(s) is divided into discrete quantized steps (S). The signal is sampled at each step with the resulting amplitude of the samples coded into binary values. The binary equivalents are actually associated with analog values midway between step amplitudes to minimize errors. These binary codes are shown at the bottom of the figure. The original waveform is transmitted as a serial stream of binary bits representing the quantized levels of each of the samples. At the receive station, the binary bits are decoded into the quantized samples and the original signal is reproduced from the resulting samples. Quantizing errors resulting from original levels that were not exactly the same as the quantized step value appear in the replicated waveform.

PCM systems typically sample analog inputs at 8,000 times a second and converts each sample into an 8-bit binary code. By doing this, the PCM system can support a transmission rate of 8,000 × 8-bits or 64 Kbps.

A typical PCM transmitting system (Figure 6-20) is comprised of a low-pass filter and sampling circuit followed by a quantizing and encoding unit. These last two functional blocks are combined into a single analog-to-digital (ADC) converter. The most

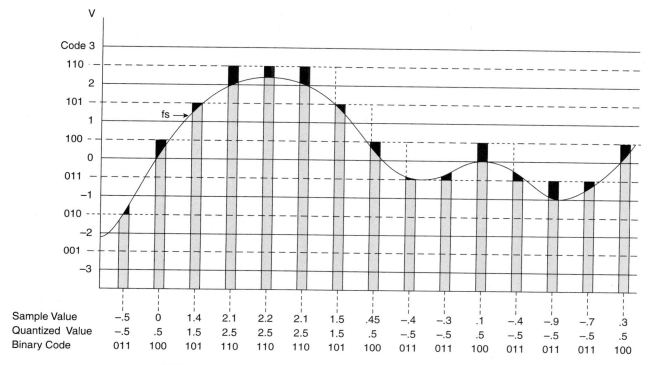

| | | | | | | | | | | | | | | | |
|---|---|---|---|---|---|---|---|---|---|---|---|---|---|---|---|
| Sample Value | −.5 | 0 | 1.4 | 2.1 | 2.2 | 2.1 | 1.5 | .45 | −.4 | −.3 | .1 | −.4 | −.9 | −.7 | .3 |
| Quantized Value | −.5 | .5 | 1.5 | 2.5 | 2.5 | 2.5 | 1.5 | .5 | −.5 | −.5 | .5 | −.5 | −.5 | −.5 | .5 |
| Binary Code | 011 | 100 | 101 | 110 | 110 | 110 | 101 | 100 | 011 | 011 | 100 | 011 | 011 | 011 | 100 |

**FIGURE 6-19** Coding a Quantized Signal

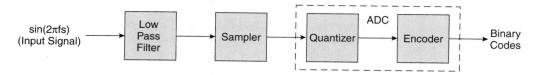

**FIGURE 6-20** PCM Transmit Blocks

commonly used analog-to-digital converter for this purpose is a successive approximations converter. A simplified schematic for such a circuit is shown in Figure 6-21.

Initially, the digital-to-analog converter (DAC) flip-flops are preset by the start of
conversion (SOC) signal so that the flip-flop in the most significant bit position is set to
one and the remaining flips reset to zero. The resulting reference voltage from the R-2R
digital-to-analog converter (DAC), which is midway between 0V and maximum voltage,
is compared to the sample's voltage level. If the level is lower than the digital-to-analog
converter analog voltage, the most significant flip-flop is reset. If the opposite is true, the
flip-flop is left set. The next flip-flop is set and the comparison is made again. This
process is repeated until all of the flip-flops have been tested—signaled by an active end

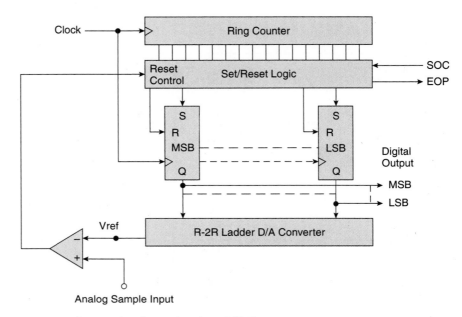

FIGURE 6-21 Successive Approximations A/D Converter

of process (EOP) indication. The final result is the digital value that is the closest approximate value at Vref to the input analog sample. The process can be repeated for the next sample by re-asserting the SOC signal. The rate at which samples are taken is limited by two factors: the clock rate input to the ADC and the rate at which SOC is asserted. SOC can not be asserted until EOP indicates the end of a conversion. This signal is set $n$-clock pulses after SOC is set. Here, $n$ is the number of flip-flops or digital bits of the ADC. In essence, the analog-to-digital converter (ADC) does the encoding and quantizing in the same process. It is rare that the digital-to-analog converter's output exactly matches the sample's level. Instead the actual level is the fixed (quantized) value nearest, but below the sample's actual value.

## PCM Decoder

The PCM decoder (Figure 6-22) reverses the process. The digital data is fed serially into the decoder. Each one of the data bits are reshaped to remove distortions caused by the transfer along the interconnecting medium used. After shaping, they are fed into a digital-to-analog converter to produce the quantized samples they represent. These samples are held and filtered to recreate the original signal, f(s). The differences between the original signal and the re-created one result from quantization error and any possible bit errors that might occur in the transmission. The later errors can be detected and/or corrected before the data stream is decoded.

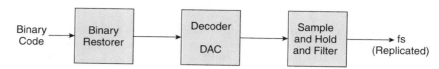

**FIGURE 6-22** PCM Decoder

## Differential PCM

In an attempt to reduce the number of codes sent by a PCM system, a slight variance on sampling method is used by a system known as differential pulse code modulation (DPCM). The binary codes in this technique represent the difference between two samples rather than the samples themselves. Since most differences between neighboring samples are small and happen more frequently than larger changes, a method called **entropy encoding** is used to assign smaller digital codes to these differences. DPCM works well with data that is reasonably continuous and exhibits small gradual changes such as photographs with smooth tone transitions.

**entropy encoding**
assigns smaller digital codes to smaller changes in samples.

Another way to reduce the size of binary codes is to employ **predictive coding.** With this method, the value of the next sample is predicted based on the value of the current sample called the *predictor.* If the prediction is correct, a very small code is used. However, if the prediction is wrong, the code generated will be longer than a normal PCM code for the predicted sample.

**predictive coding**
code developed by predicting a sample's amplitude based on the current sample.

An additional type of PCM, known as **adaptive differential pulse code modulation (ADPCM)** is commonly used to compress audio signals. In this case, quantization step sizes adapt to the current rate of change in the signal being compressed. That is, as the signal rate increases, step sizes decrease to capture a more accurate picture of the signal amplitude changes. As the rate drops, samples are made less frequently and step sizes increase. This method does introduce additional quantization noise at the lower audio frequencies. The actual compression ratio resulting from this method is no more than an order of 4:1.

**adaptive differential pulse code modulation (ADPCM)**

---

### Section 6.5 Review Questions

1. What are the functional sections of a PCM modulator?
2. What PCM functions does an analog-to-digital converter perform?
3. Why is a sample-and-hold circuit required in the PCM decoder?
4. How does DPCM differ from PCM?
5. What is the advantage gained by DPCM?
6. What does the use of predictive coding achieve?
7. What does adaptive DPCM vary to achieve compression?

# 6.6 DELTA MODULATION

**delta modulation**
based on a difference in sample
levels rather than the levels
themselves.

Pulse code modulation codes each quantized sample into a binary code, which is sent and decoded at the receiver. Another form of coded modulation is called **delta modulation.** The purpose behind this form of modulation is to minimize effects of noise without increasing the number of bits being sent. This increases the signal-to-noise ratio, improving system performance. The idea behind delta modulation is to take samples close enough to each other so that each sample's amplitude does not vary by more than a single step size. Then instead of sending a binary code representing the step size, a single bit is sent signifying whether the sample size has increased or decreased by a single step. This process is illustrated in Figure 6-23. The original signal is first sampled and quantized as with pulse code modulation. If the sample currently being coded is above the previous sample, then a binary bit is set to a logic one. If the sample is lower than the previous sample, then the bit is set low.

---

**EXAMPLE 6-5**

What binary code is sent for the original signal in Figure 6-23, using delta modulation?

**SOLUTION**

As shown in Figure 6-23, the original signal is quantized at a given sample rate. A sample whose amplitude is one step size above the previous sample generates a logic 1 bit. One that is lower generates a logic 0 level. The bit pattern generated by the signal in Figure 6-23, starting with the first sample on the left is:

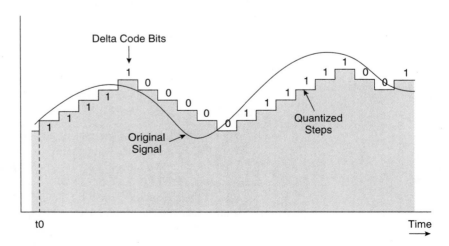

**FIGURE 6-23** Delta Encoding of Quantized Waveform

1111100000111111001

To aid in illustrating this result, the 1 and 0 logic levels are shown on the Figure above each step.

The functional block diagram for a delta modulator is shown in Figure 6-24. Basically, samples from the original signal are compared to the output of a staircase generator. If the results of that comparison show the original signal to be larger than the staircase voltage, the comparator is set high. This is sent out as a logic 1 and causes the staircase generator to increase by a step. If the comparator indicates that the staircase voltage is greater then the original signal, than the comparator goes low and causes the staircase generator to decrease by one step.

## Delta Demodulation

The delta demodulator (Figure 6-25) can be as basic as a staircase generator to replicate the staircase equivalent of the original waveform. As the serial-coded delta data is fed into the demodulator, a staircase voltage is generated out. For each logic 1 received, the staircase output is increased by one step and for each 0 received, the output level is decreased by one step. The staircase voltage is then filtered and smoothed out to more closely represent the original signal.

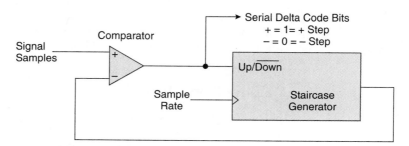

**FIGURE 6-24** Delta Modulator

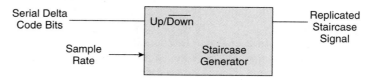

**FIGURE 6-25** Delta Demodulator

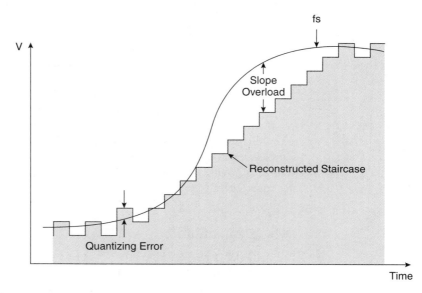

**FIGURE 6-26** Delta Modulation Slope Overload

## Slope Overload

Delta modulation systems are most useful with signals that change at a slow rate. A problem occurs as the rate of the original signal, f(s) increases (Figure 6-26). At the slower slope changes, the replicated staircase tracks fairly close to the original f(s). A rapid increase in slope of f(s) causes a widening gap between the f(s) form and that of the reconstructed waveform at the receiver. This difference is known as **slope overload.**

**slope overload**
large difference between original signal and replicated delta modulated signal.

## Adaptive Delta Modulation

Delta modulation works well as long as the original signal does not have large variations in frequency, which cause slope overload. However, analog signals that are derived from audio or video information vary in frequency. For a delta modulation system using fixed step sizes, this presents a problem. The original signal is quantized and sampled as shown in Figure 6-27 in the solid staircase. Note that signal changes are not accurately represented by the staircase signal, particularly on the rising section of the waveform. The problem can be reduced by increasing the sampling rate, except that would cause more digital bits to be transmitted. Another solution is to vary the step sizes as required to more closely resemble the original signal. This is the concept behind **adaptive delta modulation (ADM).** The dotted staircase of Figure 6-27 shows the objective of adaptive delta modulation. To make the staircase voltage track the original signal more accurately, the step size is increased on the leading edge and decreased on the trailing edge.

**adaptive delta modulation (ADM)**

When using adaptive delta modulation, a long string of 1s or 0s indicates that the staircase generator is continually trying to play catch-up with the original signal. That is,

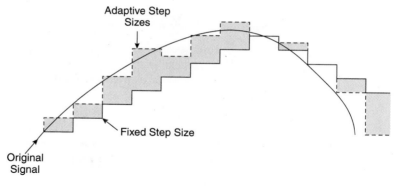

**FIGURE 6-27** Fixed and Adaptive Step Sizing

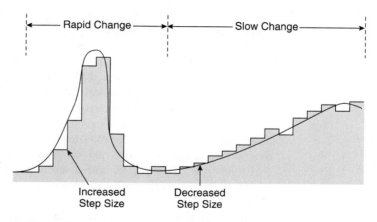

**FIGURE 6-28** Example of Adaptive Delta Modulation

it is trying to continually add or continually subtract step voltages from the output. The more frequently the digital bits change from 0 to 1, the more closely the staircase voltage is tracking the original waveform. An adaptive modulator causes the step size to increase whenever it detects many consecutive 1s or 0s and causes the step size to decrease when it detects alterations in the logic level of the digital data stream. The result of varying the step size is shown in Figure 6-28. The rapidly changing leading edge begins to generate a set of consecutive 1s. Immediately, an adaptive delta modulator increases the step size. As the peak is reached, alternating 1s and 0s are experienced and the modulator reduces the step size. It is increased on the downslope after a string of 0s are detected. The second slope, rising slowly begins to cause alternating 1s and 0s to appear because of the large step size from the previous downslope. The adaptive delta modulator decreases the step size, causing the staircase output to, once again, follow the signal more closely.

Section 6.6  Review Questions
1. How does delta modulation differ from PCM?
2. What is meant by slope overload?
3. How is slope overload avoided?

## 6.7  DIGITAL T CARRIERS

As technology improved the speed at which electronic circuits could dependably operate, the need to increase the rate at which data is transferred also increased. To respond to the need for interconnecting computers and database facilities, the telephone company turned to supplying lines for carrying data in digital form. As well as supplying the lines, the telephone company also established the limitations for their use. The basic designation for these lines are T carriers and they begin with the basic T1 carrier specified for the Bell System Voice Network.

T1 carriers are digital, leased twisted-pair lines, operating as 4-wire trunk lines, designed to handle 24 PCM voice grade channels operating at a maximum data rate of 64 Kbps each. Thus, the T1 line requires a bandwidth large enough to handle 1.544 Mbps ($24 \times 64$ Kbps). T1 lines are designed to interconnect stations that are up to 80 kilometers (km) apart, using regenerative repeaters every 1.6 km. The last repeater placed before the nearest switch station is to be no more than 0.8 km from that station. These repeaters reshape the digital data that has been distorted due to attenuation or pulse spreading caused by effects on the signals of the propagation delays inherent in the reactive nature of the cabling.

Each channel is sent, using time division multiplexing of eight bits per channel at a time. Since 24 channels are sent at once, a total of $8 \times 24$ bits, or 192 data bits, are sent per transmission period. An additional frame boundary bit (like a start/stop bit) increases the total to 193 bits per frame (transmission). Dividing the total data rate by the total number of bits per frame yields a frame rate of 8,000 frames per second.

The 193rd framing bits may have several interpretations based on the frame format used. The basic frame format is called D1 and is recognized by the alternating states of the 193rd frame bits of each succeeding frame. Two other common frame formats are the superframe (SF or D4) and the extended superframe (ESF), which are listed in Table 6-2 along with each one's use and framing bit interpretation. Every 24 channels make up a frame, which ends with one of two types of framing bits, the terminal frame (Ft) bit or the signal frame (Fs) bit. Furthermore, for a superframe, the states of the framing bits repeat every 12 frames. The terminal and signal frame bits alternate each frame with the odd-numbered frame bits being terminal bits and the even-numbered being signal bits. The pattern for the framing bits starting with frame number 1 is 100011011100 as shown in Table 6-2.

**TABLE 6-2**

## Superframe (SF) and Extended Superframe (ESF) Formats and Use or 193rd Frame Bit

| Frame Number | Superframe | Extended Superframe |
|:---:|:---|:---|
| 1 | Ft = 1 | Data-Link Control (DLC) |
| 2 | Fs = 0 | Cyclic Redundancy Check (CRC) |
| 3 | Ft = 0 | DLC |
| 4 | Fs = 0 | Framing Bits (F) = 0 |
| 5 | Ft = 1 | DLC |
| 6 | Fs = 1 | CRC |
| 7 | Ft = 0 | DLC |
| 8 | Fs = 1 | F = 0 * |
| 9 | Ft = 1 | DLC |
| 10 | Fs = 1 | CRC |
| 11 | Ft = 0 | DLC |
| 12 | Fs = 0 | F = 1 |

Sequence is repeated for every twelve frames with the single exception for the state of the framing bit for the extended superframe frame 8(*). The state of this bit alternates between 1 and 0 for each group of twelve frames.

For the extended superframe, the framing bits take one of three meanings. They are regular framing bits (F), serving the same function of terminal and signal framing bits; data-link control (DLC) bits, used for control and diagnostic activities; and cyclic redundancy check (CRC) error detection bits. Each of these bits appear as the 193rd bit of their particular frame. Framing (F) bits are relegated to every fourth frame, starting with frame number 4 and ending at frame 24 with the following bit pattern: 001011.

Data-link control (DLC) framing bits have the same function as the high-level data-link control (HDLC) protocol or other bit protocol bits. They occupy every odd number frame's 193rd bit position starting with frame number 1. Lastly, the CRC bits occupy the remaining frame bit positions using a total of 6 bits for each 24 frames. CRC-6 is used, which is a shorter form of the CRC-16 error detection method described in Chapter 3.

## BPRZ-AMI Encoding

The digital data form used on T1 lines is bipolar, return-to-zero, alternate mark inverted or BPRZ-AMI. The format for this digital form is illustrated in Figure 6-29. It results from a combination of digital formats. Zero logic levels are at 0V. Bipolar alternate mark inversion indicates that logic 1s are at +V and –V, altering between each level. Each logic 1 returns to a zero level midway through the bit period.

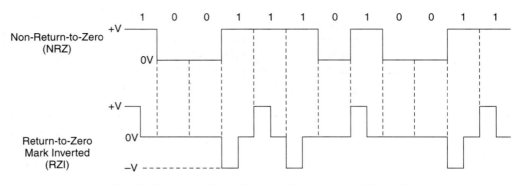

**FIGURE 6-29** Bipolar Return-to-Zero, Alternate Mark Inverted Digital Form

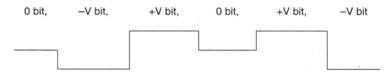

**FIGURE 6-30** Bipolar Violation

**bipolar violation (BPV)**
occurs when the voltage level
of two successive 1s is the
same when using alternate
mark inversion data format.

When using alternate mark inversion formats, two consecutive 1 bits with the same level (positive or negative) results in a condition called a **bipolar violation (BPV)** shown in Figure 6-30. Normally this condition indicates that an error has occurred.

A string of 32 consecutive 0s indicates that a sender is no longer in service. To avoid misinterpreting a long string of 0s if they are required in the data string, a logic 1 bit is inserted after 15 consecutive 0's. The receiving station is required to strip this added bit before interpreting the data.

**binary 8 zero suppression (B8ZS)**
scheme that prevents the
transmission of a long stream
of zeros.

Another method frequently used to avoid the occurrence of a long string of 0s is an encoding scheme called **binary 8 zero suppression (B8ZS),** which is illustrated in Figure 6-31. When the transmitter detects eight consecutive 0s, it checks the level of the logic 1 bit that preceded the eight 0s. If the level of that bit is positive, the 9 bits are encoded in accordance with Figure 6-31*a*. If the leading 1 bit is a negative level, then the encoding takes the form shown in Figure 6-31*b*. Notice that the difference between the two forms is that the logic 1 levels are inverted from one to the other.

Recall that a bipolar violation results when there are two consecutive positive or negative levels. However, if the patterns shown in Figure 6-31 are detected by a receiver set to decode B8ZS format, then the double-bipolar violation is recognized and decoded into a single 1 followed by eight 0s. Any other detected bipolar violations are assumed to be data errors.

## T1 Carrier

The T1 carrier is similar to the telephone system DS-1 or digital signal-1 carrier. DS-0 designates one of the 24 8-bit channels used to form the DS-1 or T1 channel. The In-

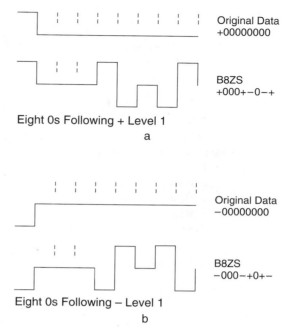

**FIGURE 6-31** B8ZS Encoding

ternational Telecommunications Union (ITU), the former Consultative Committee for International Telegraphy and Telephony (CCITT) standards organization, based in France, produced a similar digital carrier specification to apply to European digital communications systems. The significant differences in the specifications is an increase from 24 channels to 32 channels (30 voice grade and 2 for frame synchronization and signaling), which increased the bandwidth requirements to handle 2.048 Mbps, known as an E1 channel.

## Fractional T1 Service

**fractional T1 (FT1) service**
uses T1 type service to send fewer channels.

An addition to the T carrier system was made to facilitate users who wanted to have access to these lines but did not have sufficient data needs to fill 24 channels. This addition, called **fractional T1 (FT1) service,** uses conventional T1 technology to send fewer channels of information. User's data is inserted into alternating channels of a T1 frame. Unused channels are filled with filler data. As an example, suppose a user had a need to send a 384-Kbps video data sequence on a FT1 carrier. The user's data would fill 6 channels (64 Kbps per channel × 6 yields 384 Kbps). This data is placed in channels 1, 3, 5, 7, 9, and 11. Channels 2, 4, 6, 8, 10, and 12–24 are filled with filler data to complete the 24-channel frame. Since part of the frame is not being used for customer data, it is up to the customer to supply framing data so that a receiver can correctly interpret the data sent. The 193rd bit normally used for framing is still applied to the T1 frame, but it cannot supply specific framing information for the actual data contained in the fractional T1 frame.

## T1 Line Impairments

T1 carrier lines are susceptible to all the impairments and problems inherent to telephone lines plus a few of their own. Among these additional problems is that of **bipolar violation.** Since T1 carriers use alternate mark inversion, such an occurrence is generally in error. Shifts in clock frequency, jitter, and wander also cause data errors or framing misinterpretation to occur. A shift in the clock frequency by either the transmitter or receiver in a T1 connection causes data and/or framing bits to be sampled at an incorrect time. This causes a condition called a **slip** to take place. A slip is the loss or addition of a bit due to an error in sampling time.

Excessive jitter can also cause slips to be experienced. Most jitter is introduced in clock recovery circuits as they regenerate the clock signal from the incoming data transitions. Most of the time, the jitter is very slight and causes no problem. However, as components in repeaters and customer equipment begin to degenerate with time, jitter becomes worse until data or framing errors result. **Wander** is a form of low-frequency jitter (usually less than 10 Hz) and is caused by instabilities in the master timing source or from temperature variations experienced by the equipment. The effect of both jitter and wander is similar to frequency shifts. They cause data or framing bits to be incorrectly sampled, resulting in a data or framing error.

**slip**
loss or addition of a bit due to a skip in sampling time.

**wander**
low-frequency jitter in a delta modulator.

## T Carrier Hierarchy

Additional T carrier lines are available for systems requiring expanded use. Table 6-3 charts these lines and the specifications associated with each carrier. T2 lines carry 96 channels of digital information at 6.312-MHz rate on lines that can reach 500 miles distance with the aid of repeaters. Instead of using the 1 bit insertion following 15 consecutive 0's specified for T1, a false disconnect is avoided by using a method called **binary six zero substitution (B6ZS),** which is an abbreviated version of B8ZS. Any group of six consecutive zeros are replaced with one of two sequences shown in Figure 6-32.

**binary six zero substituion (B6ZS)**
another method used to prevent long string of transmitted zeros.

### TABLE 6-3

### 3 T Carrier Specifications.

| Carrier Type | Alternate Designation | Number of Channels | Multiplex Format | Rate (Mpbs) | Media Type |
|---|---|---|---|---|---|
| T1 | DS-1 | 24 | 24 Analog | 1.544 | Twisted Pair |
| T1C | DS-1C | 48 | 2 DS-1 | 3.152 | Twisted Pair |
| T2 | DS-2 | 96 | 4 DS-1 | 6.312 | Twisted Pair |
| T3 Radio | DS-3 | 672 | 28 DS-1 | 44.736 | 3A-RDS 11 Ghz |
| T4 | DS-4 | 4,032 | 6 DS-3 | 274.176 | T4M Coax WT4 Waveguide DR18 18 Ghz Radio |

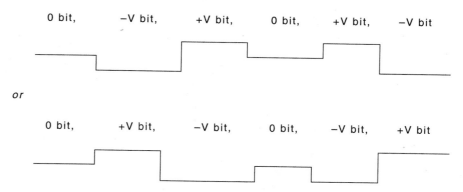

**FIGURE 6-32** B6ZS Format

Both sequences contain a bipolar mark inversion violation. When the receiver detects this violation, it replaces the sequence with 0s.

The remaining T carriers listed on Table 6-3 showing their specifications. Each carrier specification contains details specific to that T level. Basically, each is used to handle increased number of channels and each requires larger bandwidths to handle the increased number of channels and data rates at which they are operated. Different techniques exist for multiplexing the channels onto a single T carrier. One of the most common methods is time division multiplexing.

## Testing T1 Lines

A number of instruments are available for monitoring and testing T1 lines. One of these, the TM1 T-Carrier Monitor manufactured by Electrodata, Inc., is pictured in Figure 6-33. This handheld unit is used to monitor active lines without disrupting service. It can also be used to test an off-line transmitter to check its operation. The monitor is capable of automatically detecting the type of framing used (unframed, superframe, or extended superframe). Three LEDs in the upper right corner are used to indicate the framing type.

To be useful, a T1 test set must be able to decipher digital data in either standard **alternate mark inversion (AMI)** or B8ZS form. Additional indicators on the left side of the TM1 show the states of eight current data bits and four signal bits. The user is capable of selecting which one of the 24 channels to monitor, using two switches located at the bottom of the unit. Three types of errors can be detected by the monitor, bipolar violations (BPV), framing errors, and CRC data errors. Current indicators will flash as they occur, while a second set, called the history indicators, latch on the first occurrence of an error. These latched indicators are manually reset by pressing the switch located directly below the history LED column.

The two remaining indicators tell the user if the carrier becomes lost (loss of signal) and if the battery charge on this portable unit is becoming too low. As an emulator, the

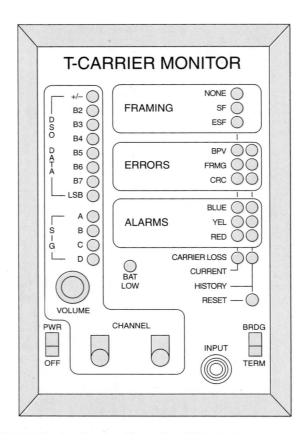

**FIGURE 6-33** TM1 T1 Carrier Monitor *(Courtesy of Electrodata, Inc.)*

TM1 Monitor is capable of generating a voice signal adjustable from +3 to –55 dBmDC coupled into 600 ohms termination. The frequency response of the output is held between + and – 0.2 dB of the original setting from 300–3 kHz, the telephone line bandwidth and between + and – 0.3 dB overall from 200 to 3.3 kHz. The device is small, easy to use, and helps the technician determine if there are any problems on a T1 line.

**Section 6.7  Review Questions**

1. What is the difference between T1 and E1 channels?
2. How many 8-bit channels make up a single T1 channel?
3. What is used to avoid misinterpretation of a long string of 0s?
4. What is the bandwidth of a T1 and an E1 channel?

## 6.8 COMPANDING

One additional method used to increase the volume of traffic on a given line is to reduce the number of digital bits that represent a signal to be transmitted without corrupting the signal that is replicated at the receiving end. It is desirable to compress the digital codes at the transmitter and then expand them back to their original form at the receiver. The acronym for this procedure is **COMPression/ExPANDING (COMPANDING).**

Companding can be done in analog or digital form. In analog companding, the signal to be digitized is companded first and then sampled and digitized. Digital companding reduces the digital code created after the signal has been digitized. Companding is also used to increase the signal-to-noise ratio of the coded signal.

**compression/expanding (COMPANDING)** of signals to reduce signal-to-noise effects of smaller signals.

### μ-Law

Analog compression follows one of two algorithms, depending on whether the system that does the companding was developed for North America and Japan or for everyone else. The North American μ- or 255-law is a system that divides the analog signal range into fifteen segments each eventually encoded into an eight-bit digital value. The actual voltage value ($Y$) for each segment is determined by the following formula:

$$Y = \frac{V \ln(1 + \mu v / V)}{\ln(1 + \mu)} \tag{6-4}$$

where $Y$ is the compressed value out of input voltage $v$ for a signal whose maximum peak value (and compressed value) is $V$. A graph comparing compressed voltage ($Y$) to input voltage ($v$) for peak values of +/– 1V is shown in Figure 6-34.

A coded PCM segment is shown in Figure 6-35. The first bit is the sign bit, where a logic one representing positive values and a logic zero, negative values. The next three bits designate the segment number 000 to 111 or 0 to 7. The fifteenth segment is the zero segment, which is half positive and half negative. The last four bits (at times, referred to as bits ABCD) determine the actual level within the range of a particular segment. Table 6-4 lists an example of the voltage levels represented by a compressed signal using μ-law. Note that the step size doubles with each segment and that the levels within a segment are divided into sixteen steps (the last four bits of the segment code).

The μ-law can be applied to digital compression of the signal. The analog signal is digitized and coded using 12 bits. The format for the 12-bit codes is shown in Figure 6-36. Using 12 bits to represent an analog signal improves the digital resolution when compared to 8 bits, due to increased number of quantized levels and reduced step sizes. This improvement in resolution increases the accuracy of the replicated signal. The first bit is the sign bit with the remaining eleven bits reflecting the magnitude of the sample. The form in the illustration shows eight codes for the twelve bits. Bits ABCD contain any one of 16 values from 0000 to 1111. $X$ values have actual digital values, which get discarded during the compression process.

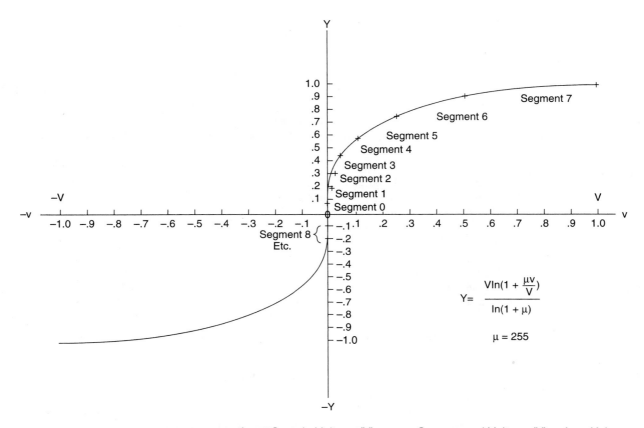

**FIGURE 6-34** Input Sample Voltage (V) versus Compressed Voltage (Y): μ-Law Using Unit Peak Value (V = 1.0)

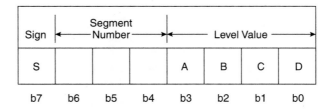

**FIGURE 6-35** PCM Code Example

Digitally, the 12-bit values are encoded into an 8-bit compressed code as follows:

1. Retain the sign bit as the first bit of the 8-bit code.
2. Count the number of zeros until the occurrence of the first 1 bit. Subtract the zero count from 6. This is the segment number.

**TABLE 6-4**

**Step Size and Maximum Voltage
per Segment for an Example PCM**

| Segment | Step (mV) | Maximum Voltage (Mv) |
|---------|-----------|----------------------|
| 0 | 0.5 | $16 \times .5 = 8$ |
| 1 | 1.0 | $8 + 16 \times 1 = 24$ |
| 2 | 2.0 | $24 + 16 \times 2 = 56$ |
| 3 | 4.0 | $56 + 16 \times 4 = 120$ |
| 4 | 8.0 | $120 + 16 \times 8 = 248$ |
| 5 | 16.0 | $248 + 16 \times 16 = 504$ |
| 6 | 32.0 | $504 + 16 \times 32 = 1,016$ |
| 7 | 64.0 | $1,016 + 16 \times 64 = 2,040$ |

| Segment | S | Data Code Bits (b11 → b0) |
|---------|---|---------------------------|
| 0 | S | 0 0 0 0 0 0 0 A B C D |
| 1 | S | 0 0 0 0 0 0 1 A B C D |
| 2 | S | 0 0 0 0 0 1 A B C D x |
| 3 | S | 0 0 0 0 1 A B C D x x |
| 4 | S | 0 0 0 1 A B C D x x x |
| 5 | S | 0 0 1 A B C D x x x x |
| 6 | S | 0 1 A B C D x x x x x |
| 7 | S | 1 A B C D x x x x x x |

x — Don't Care Bits

ABCD — Any Digital Value from 000–1111
Segment Value Code

S — Sign Bit   0 = +   1 = −

**FIGURE 6-36** 12-Bit Quantize Sample Codes

3. The first occurrence of a 1 is assumed during the expanding process, so it is set aside during compression.
4. Copy the next four bits (ABCD) into the 8-bit compressed code.

The compressed code consists of a sign bit, segment number, and the value of the ABCD bits.

### EXAMPLE 6-6

Code the 12-bit code 100001011010 into an 8-bit compressed μ-law code.

### SOLUTION

Retain the sign bit (1). Count the zeros until the first 1 occurs (4). Subtract from 7 (7 − 4 = 3 = 011). Discard that first 1 and get ABCD = 0110. Place them into the 8-bit format:

10110110.

Expanding back digitally, reverses the process:

1. Retain the sign bit.
2. Take the segment number, subtract from 7 and add that many 0s.
3. Make the next bit a 1.
4. The next bits are ABCD values.
5. Add a 1 and sufficient 0s to complete the 12-bit value.

The addition of the 1 and 0 bits following the ABCD bits sets a value in the middle between the actual values the remaining bits could have had. While this introduces some error, it is minimal since it occurs in the least significant bits of the 12-bit code.

### EXAMPLE 6-7

Expand the compressed code of Example 6-6.

### SOLUTION

Retain the sign bit (1). Subtract the segment number from 7 and insert that number of 0s. 7 − 3 = 4 = 0000. Replace the discarded 1. Append the ABCD bits followed by 10 (a 1 and the necessary 0 to complete 12 bits):

100001011010

Note that the last two bits of the original 12-bit code from Example 6-6 could have been any one of four combinations, 00, 01, 10, 11 (0, 1, 2, 3). This means that a maximum potential error of 2 could exist. The maximum percentage of error for 2 out of 1011000 = 2/88 = 2.3%. Since the original value of the two lowest bits was 10, there is no error in Example 6-6.

## A-Law

In Europe and elsewhere, the ITU standard for companding is the A-law, which uses two equations to compute compressed value, $Y$, depending on the peak value of $V$ in relation to constant $A$:

$$Y = \frac{Av}{V(1 + \ln A)}; \quad |v| < V/A \tag{6-5}$$

and:

$$Y = \frac{1 + \ln(Av/V)}{1 + \ln A}; \quad V/A < |v| < V \tag{6-6}$$

These formulas are used with systems that employ signals divided into 13-segment sections; each segment is encoded using a sixteen- to eight-bit compression conversion. The main significance in the difference between the use of the two codes is the facilitation of the different systems used in North America and Japan to those used in Europe and the rest of the world. It starts with the difference in T1 and E1 carriers, including the channel capacity of each, and ends with the size of the digital codes used to digitize the analog signals. The repercussions of these differences is to make interconnections between North American and European systems difficult.

---

**Section 6.8 Review Questions**

1. What is the benefit of using companding?
2. Give one reason that different companding schemes are used in the United States and Germany.

---

## 6.9 CODECS

**CODEC**
COder/DECoder used to convert voice signals to digital codes.

A **CODEC** is a device designed to convert analog signals, such as voice communications into PCM-compressed samples to be sent onto digital carriers and to reverse the process, replicating the original analog signal, at the receiver. The term CODEC is an acronym for CODer/DECoder signifying the pulse coding/decoding function of the device. Originally, CODEC functions were managed by several separate devices, each performing the tasks necessary for PCM communication. These functions are sampling, quantizing, analog-to-digital and digital-to-analog conversions, filtering, and companding. Today, all of these functions are available on a single CODEC IC chip like Intel's 2913.

The 2913 CODEC operates in one of two modes—fixed data rate or variable data rate. In the fixed data rate modes, the chip is capable of operating with one of three fixed-clock frequencies: T1's 1.544 MHz, ITU's 2.048 MHz, and also 1.536 MHz. In the variable data rate, the clock frequency can be set anywhere from 64 kHz to 4.096 MHz. The 2913 allows selection of either μ-law or A-law for the companding function.

On the transmit side of the 2913 CODEC, (Figure 6-37), the analog signal is fed into an operational amplifier, whose gain is set by external resistor components shown as R1 and R2 in Figure 6-35. If it is undesirable to use the amplifier, the signal can be entered,

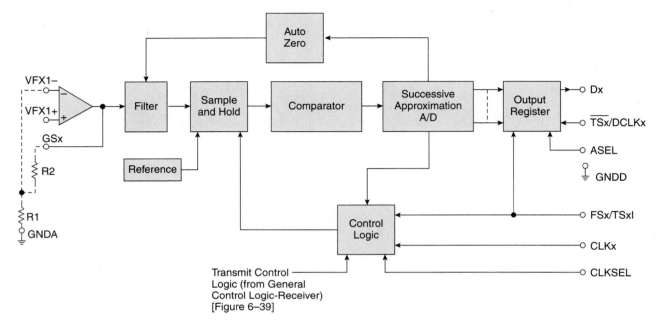

**FIGURE 6-37** CODEC Transmit Blocks

single-ended via the GSx pin and analog ground (GNDA). The gain (Av) for the op amp conforms to the standard op amp non-inverting gain formula:

$$Av = 1 + \frac{R2}{R1} \tag{6-7}$$

The output of the op amp is fed through an anti-aliasing filter to a sample-and-hold circuit. The low-pass anti-aliasing filter attenuates the sampling frequency −35dB. The bandpass filter following the anti-aliasing filter is designed to yield a flat response from 300–3kHz as illustrated in Figure 6-38. An additional high-pass filter rejects power line frequencies (50 and 60 Hz) and European railroads and ringing frequencies (17 Hz) and their harmonics.

Samples are held for encoding and μ-law or A-law compression in an internal sample-and-hold circuit. One additional unit, an auto-zero circuit is used to correct for any DC offsets that were created by the process to this point. The compressed PCM serialized data is sent out through the Dx transmit data line.

On the receive side of the CODEC (Figure 6-39), the serial data stream on the DR data receive pin is shifted into the input register and output in parallel to the digital-to-analog converter. The resulting analog signal is held in a second internal sample-and-hold circuit until the next coded sample is shifted into the input register. The decoded samples are then shaped and filtered before being sent out through a power amplifier. The filter is a bandpass filter with a sharp roll-off at 4,000 Hz. It has a flat response, like the transmit bandpass filter between 300 and 3 kHz.

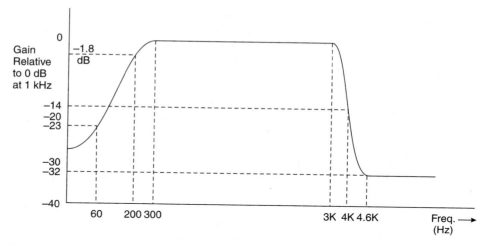

**FIGURE 6-38** CODEC Transmit Filter Characteristics

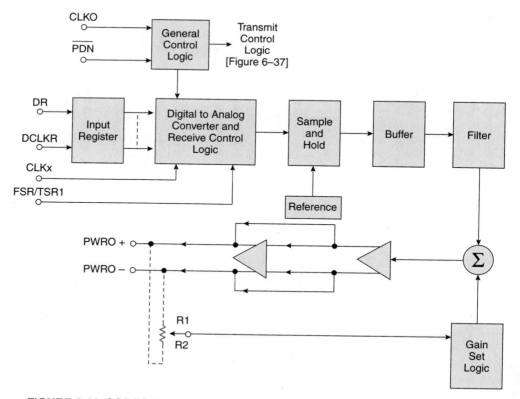

**FIGURE 6-39** CODEC Receive Blocks

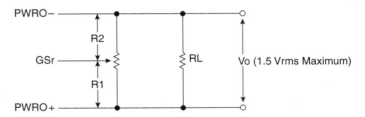

**FIGURE 6-40** Gain-Setting Resistors for Receive Power Amplifier

The maximum output power amplifier can be used as a single-ended circuit with the signal sensed from either PWRO– or PWRO+ (power amplifier out + or –) and analog ground (GNDA) or as a differential buffer by sensing the output between PWRO– and PWRO+. The maximum voltage level from the power amplifier is 1.50 Vrms. The actual output (Vo) between the PWRO outputs is a function of the gain (Ao) of the op amp circuit and can be set anywhere from the maximum of 1.5 Vrms to a minimum of 0.375 Vrms. The output level is set by the voltage level applied to gain set pin GSr. Maximum output is achieved by connecting GSr to PWRO–, while minimum voltage is achieved by connecting GSr to PWRO+. Voltage levels between these extremes can be accomplished with a potentiometer connected from PWRO+ to PWRO–, whose wiper is connected to GSr as shown in Figure 6-40.

The gain of the power amplifier is calculated by:

$$Ao = \frac{R1 + R2}{R1 + 4R2} \tag{6-8}$$

By reshuffling the formula, a ratio of resistor values based on the gain can be shown:

$$\frac{R1}{R2} = \frac{4Ao - 1}{1 - Ao} \tag{6-9}$$

The significance of the ratio of R1 to R2 is that the gain of the amplifier is limited to values that produce a positive ratio of R1 to R2.

---

**EXAMPLE 6-8**

Using the relationship of equation 6-9, determine the ratio for R1/R2 for maximum, minimum, and midrange gains of the power amplifier.

**SOLUTION**

A maximum gain of 1 brings the denominator of equation 6-9 to 0 making the ratio of R1 to R2 infinite. This gain is achieved by connecting GSr to PWRO– and disconnecting the resistor network. A midrange gain of 0.5 develops a ratio of 2 for R1/R2, that is, R1 is twice the value of R2. Lastly, the minimum gain occurs when the numerator of Equation 6-9 is brought to zero. This occurs at a gain of 0.25. Here the ratio of R1 to R2 is 0. Once again, this cannot be achieved using the resistors. They are removed and GSr is connected to PWRO+. For any gain between 1/4 and 1, the ratio of R1 to R2 can be calculated and used as a guideline for selecting values for the gain network.

This CODEC can operate using either the μ-law or A-law for companding. The selection is made utilizing the ASEL (A-law select) input. Connecting this input to Vbb sets the A-law mode. Connecting it to Vcc or GNDD selects the μ-law and causes the least significant bit of the transmit signaling frame to take the state of ASEL.

# SUMMARY

Multichannel transmissions allow many numerous communications links to be established simultaneously. Time division multiplexing assigns channels to time slots in a transmission stream while frequency division multiplexed systems fit each channel into a portion of the bandwidth of the transmitted signal.

Reducing noise problems, increasing data capacity, and/or increasing transmission speeds are among the considerations that led to the development of pulse code modulation techniques. These techniques involve sampling, quantization, and companding principals to achieve those goals.

Sampling and coding techniques are behind the modulation techniques described in this chapter. Several forms of pulse code modulation describe how higher data rates can be achieved by reducing the number of bits per sample through coding. Additional savings are achieved through adaptive modulations to reduce quantization step sizes. Delta modulation is useful when coding signals that do not change drastically from sample to sample.

Higher data rates using some of the techniques described in this chapter are achieved through V.34 and V.90 modems that have brought modem technology up to the 56-Kbps range.

# QUESTIONS

**Section 6.2**

1. Why is full-duplex frequency shift keying (FSK) an example of frequency division multiplexing (FDM)?
2. What is a guardband?
3. How are FDM groups created?
4. How are supergroups formed?
5. What is the frequency range for FDM group 4?
6. What is the purpose of carrier pilots in FDM?
7. What are the two largest terminal FDM groupings?
8. How do the groups in question 7 differ?

9. Which band of frequencies is occupied by a data under voice signal when sent with a mastergroup?
10. How are multiple channels sent on a single line using TDM?
11. What is the essential difference between STDM and STAT MUX TDM?
12. What is meant by pulse amplitude modulation (PAM)?

### Section 6.3

13. What is the difference between natural and flat-top sampling?
14. What is the purpose for using sample-and-hold techniques?
15. What is the Nyquist sampling rate when sampling a signal whose highest frequency is 1630 Hz and lowest frequency is 250 Hz?
16. A 3.6Cos(t) signal is naturally sampled at a rate of 56 kHz using 1.25 microsecond sampling pulses. What is the value of the reconstructed output signal?

### Section 6.4

17. What is the advantage of quantizing a signal's samples?
18. What is quantizing error?
19. What is the step size for a quantized signal whose peak voltages are +16.8V and −22.4V? There are 16 bits in the PCM code for each sample.

### Section 6.5

20. What happens to quantized samples in a PCM system?
21. Which type of analog-to-digital converter is commonly used for PCM encoding?
22. How does differential PCM differ from normal PCM?
23. What is the advantage of DPCM?

### Section 6.6

24. How does the sampling rate for delta modulation differ, in concept, from standard PCM sampling?
25. What is the basic idea behind delta modulation?
26. What is the difference between an adaptive delta modulator and a regular delta modulator?
27. When are step sizes changed when using adaptive delta modulation?

### Section 6.7

28. Give four characteristics of a T1 carrier.
29. How do E1 carriers differ from T1 carriers?
30. What type of multiplexing is used on T1 carriers?
31. Generally, what type of function does the 193rd bit perform for a basic T1 frame?
32. What use is made of the 193rd bits for superframes and extended superframes?
33. How many channels make up a superframe and an extended superframe?
34. Draw the BPRZ-AMI form for the data stream 11000111011. The least significant bit is on the left and the first logic 1 is at a +V level.
35. What is the purpose for using B8ZS encoding?
36. How does B8ZS differ from B6ZS encoding?
37. What is meant by a bipolar violation?
38. When are bipolar violations used on purpose?

39. Define *slip*. What causes slip to occur?
40. Define *wander*.
41. List at least five characteristics, data, or error indications that the TM1 T1 Carrier monitor is used to detect.
42. An integrated service requires 768 Kbps data to be sent on a fractional T1 carrier. Assuming that the first channel used to hold real data is channel 1, which channels will contain real data and which filler data?

### Section 6.8

43. Define slope overload.
44. Why is companding used?
45. Given a maximum voltage of 3.6V, what is the compressed value for an input voltage of 1.26V using μ-law? Using A-law?
46. Digitally compress this 12-bit code: 100101101101.
47. Expand the result from Problem 45. How much error is introduced? What is the percentage of error?

### Section 6.9

48. What is a CODEC and how does it differ from a modem?
49. What is the data rate ranges for the 2913 CODEC using variable data rate mode?
50. What is the bandwidth of the 2913 CODEC's transmit filter? Receive filter?
51. What is the ratio of R1 to R2 to achieve a gain of 2.5 for the input op amp of the transmit side of the 2913 CODEC?
52. What is the ratio of R1 to R2 to set the output power amplifier gain to .75?

## DESIGN PROBLEMS

1. Design, construct, and verify a three-channel FDM system. Each channel is limited to a 0–4 kHz bandwidth. Verify the circuit works by setting channel 1's signal frequency to 1 kHz, channel 2 to 1.8 kHz, and channel 3 to 2.5 kHz. This project can be separated into two parts with one group building the transmit side and the other the receive side. When the systems are completed, the three channel signals should appear at the received output.

2. Devise a TDM system to time multiplex three channels. Verify the system works by sending three short messages from transmitter to receiver at a slow enough data rate to check that they arrive correctly. Do not bother to packetize your data.

3. Design and construct a sampling circuit that creates sufficient samples of an input sinewave.

4. Design and construct a circuit that replicates samples into the original sinewave signal.

5. Design and construct a simple PCM coder that converts samples into a digital code.

6. Design and construct a PCM encoder that replicates the samples from a PCM type code.

7. Combine the projects from Problems 3 to 6 into a system.

8. Design a system built around the 2913 or 2914 CODEC chip. Verify that a signal driving the transmit side can be replicated at the output at the receive side.

# ANSWERS TO REVIEW QUESTIONS

**Section 6.1**

1. A channel is a single communication link.
2. Baseband systems are occupied by one channel.
3. Broadband systems have several channels sharing its bandwidth.

**Section 6.2**

1. Insufficient bandwidth.
2. Difference frequency.
3. Modulation of a carrier by a channel, group, etc.
4. Microwave and satellite communications.
5. Mixing received signal with same carrier frequencies used at sending site.
6. Less overall delay in receiving messages.
7. STDM—assigns permanent equal time slots that are there whether or not they are used.
   STAT MUX—assigns time slots as they are needed.

**Section 6.3**

1. Nyquist is used to determine the minimum sampling rate to replicate the sampled signal accurately.
2. Pulse amplitude modulation.
3. Natural sampling uses a sample's actual amplitude from beginning of sample to the end.
   Flat-top sampling takes an average of a sample's peak amplitude and uses that for the duration of the sample.
4. Fills in gaps between samples.

**Section 6.4**

1. Quantization is the breakup of a signal into a set of samples.
2. Quantization noise is the difference between original and replicated signal.

**Section 6.5**

1. Filter; sample; quantize; encode.
2. Quantization and encoding.
3. Maintain signal level between samples.
4. Code based on the difference between samples rather than the samples themselves.
5. Shorter codes.
6. Overall shorter codes.
7. Quantized step size.

**Section 6.6**

1. Samples are close together so changes can be one step size between samples. Requires only one bit to indicate if change is up one or down one step.
2. Slope overload is a large difference between original and replicated delta modulated signals.
3. Vary step size using adaptive delta modulation.

**Section 6.7**

1. T1 handles 24 channels per line.
   E1 handles 30 channels per line.
2. 24 channels.
3. Bipolar violation.
4. T1: 1.544 Mbps E1: 2.048 Mbps

**Section 6.8**

1. Companding allows binary codes to be compressed, reducing transmission time and bandwidth needs.
2. Reflects difference between T1 and E1 services.

# CHAPTER 7

# Data-Link Layer Protocols

## OBJECTIVES

After reading this chapter, a student will be able to:
* identify polling and selection messages.
* establish a communication link using the asynchronous or bisync protocols.
* establish and maintain a data link using SDLC.
* define the different SDLC frame types and their use.
* apply the HDLC protocol to peer-to-peer networks.

## OUTLINE

7.1   Introduction
7.2   Data-Link Sections
7.3   Character-Oriented Protocols
7.4   Bit-Oriented Protocols
7.5   Protocol Analyzers

**protocol** or **data-link protocol**
set of rules for successful communications between nodes.

**asynchronous protocol**
character-oriented asynchronous data-link protocol.

**binary synchronous control (BISYNC or BSC)**
character-oriented synchronous data-link protocol.

**synchronous data-link control (SDLC)** and **bit-oriented data-link protocol high-level data-link control (HDLC)**

## 7.1  INTRODUCTION

It was shown in Chapter 1 how characters are defined or represented by binary character codes for the purpose of digital data transfers. Alphanumeric (printable characters) and formatting characters (tabs, carriage returns, line feeds, etc.) determine how the actual messages appear on a terminal screen or printed page. Data-link characters are used to establish beginning and end of messages, set attributes applied to the message, acknowledgments of transmissions, error responses, and station identification. A set of rules that establishes which of these characters to use and how communications software responds to them is known as a **protocol.** The type of protocol used to establish a link between two stations in accordance with the second layer of the OSI model is known as a **data-link protocol.**

Four widely used data-link protocols that serve as a good representation of this type of protocol, are the character oriented **asynchronous** and **binary synchronous control (BISYNC or BSC)** protocols and bit-oriented **synchronous data-link control (SDLC)** and **high-level data-link control (HDLC)** protocols.

## 7.2 DATA-LINK SECTIONS

Besides establishing how a station reacts to data-link and other characters and which responses are given for certain conditions, protocols also determine the method of identifying the secondary station that a primary wishes to communicate with.

### Polls and Selections

**poll**
message sent by a primary station asking if a secondary station has traffic to send.

**selection**
message sent by a primary asking a secondary station if it is ready to receive traffic.

In a multipoint bus topology communications system, as shown in Figure 7-1, a single primary controls the communications interface between it and many secondaries. After the electronics system link is operational, the communications link is established by the primary using one of two types of messages. The primary can **poll** a specific secondary to inquire if that secondary has any messages to send to the primary or the primary can send a **selection** message to a secondary inquiring if it is ready to receive data messages (traffic). Software programs at the primary set polling and selection sequences so that each secondary, in turn, has an opportunity to communicate with the primary.

Secondary stations are required to respond, in some manner, to a primary's poll or selection. These responses to polls and selections depend on the state of the communications and peripheral links at the secondary station. The secondary could be in any one of three operational modes: *send, receive,* or *local.* The send mode is selected by a secondary that has traffic to send to the primary. The receive mode reflects a secondary's readiness to receive messages from the primary. Finally, the local mode indicates that a secondary is off-line and not able to handle or send traffic. In all three cases, the secondary is required to respond to the poll or selection.

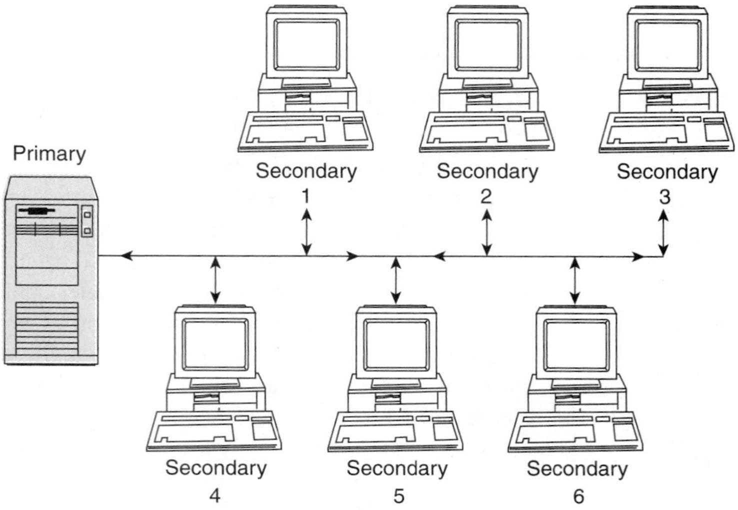

**FIGURE 7-1** Primary Bus Topology

The most common responses to a poll are:

1. The secondary is in the send mode and transmits an actual message.
2. The secondary acknowledges the poll, but is in the receive mode and informs the primary that it has no messages to send but is capable of receiving traffic.
3. the secondary is "off-line" (local mode) for maintenance or local use, in which case, it sends back an acknowledgment informing the primary of this condition.

The responses to a selection are similar, but have a slight variation:

1. The secondary is in the receive mode and sends a positive acknowledgment that it is ready to receive the primary's traffic.
2. The secondary is in the send mode and is not ready to receive messages, but has something to send.
3. The secondary is in the local mode and is not ready to send or receive traffic.

---

### EXAMPLE 7-1

Show the sequence that allows a message from secondary station 3 to be delivered to secondary station 2 in Figure 7-1.

### SOLUTION

In order for this message to be sent, the primary must first poll secondary station 3 to inquire if it has any messages to send. Secondary station 3 responds by sending the message it intends for secondary station 2. After the polling sequence is completed, the primary begins a selection sequence for the secondary stations that are to receive a message.

In this example, secondary station 2 is selected and asked if it is ready to receive traffic. After secondary station 2 responds positively, the primary sends the message it had previously received from secondary station 3. Secondary station 2 must respond with a status message after it fully receives the message from the primary.

---

Each protocol has a program included in it to handle polls and selections and the secondary's specific responses. In addition to a specific format of data for these functions, secondary addresses are included to determine which station is being polled or selected. Polling addresses are unique for each secondary station in the link, which means that the primary polls only one station at a time.

## Types of Secondary Addresses

**group address**
in a multipoint system, this is an address used to designate a number of secondaries, but not all secondaries.

Unlike polling addresses, selection addresses come in three different forms. Each secondary station has a *unique selection address,* similar to a polling address, allowing the primary to select it individually. Secondary stations may also be placed into groups, which include a number, but not all, of the secondary stations in the system. In this case, the group of secondaries is assigned a **group address.** These allow a number of stations to be

**broadcast address**
in a mulltipoint system, it is the address used by the primary to send a message to all of the secondaries in the system.

selected to receive the same message simultaneously. The last selection address type is called a **broadcast address.** Every secondary station is selected when this address is used, essentially making this address a group type address that is assigned to all secondary stations. The primary can send the same message to every secondary by first selecting them using the broadcast address.

## Data-Linking

Besides establishing the formats for polls and selections and secondary addresses, protocols also designate message formats and responses to data-link characters. Recall, from Chapter 1 that character codes such as ASCII contained three types of characters: alphanumeric, graphics, and data-link. The first two types dealt with printed characters and where and how they appeared on the screen or printed page. Data-link characters were used to establish the software communications link for character oriented protocols. These include functions such as *start of text (STX)* and *end of transmission (EOT).* Responses to these characters differ for each protocol, but overall, they perform the same necessary functions.

Bit-oriented protocols, on the other hand, recognize specific groups of bits or single bits to perform the data-link functions. A chief significance of data-link protocols is that they do not address the actual content of the data that they transport. They may assist in how the messages actually appear by contributing various forms of attributes, but there is no real concern over the format of the data area.

## Payloads

Data-link protocols may actually transport the frame of an upper level protocol within its data field. When the message arrives at its destination, it is taken by that receiver, which may use it to transport a message encapsulated within the framework that was delivered to it. Because the data transported by data-link protocols may not always be pure data, a more appropriate term for it is payload. The purpose, then of a data-link protocol is to establish a data-link between two stations and deliver a payload from one station to the other.

### Section 7.2  Review Questions

1. Define the purpose of poll and selection messages.
2. What are the three operational modes of a secondary station?
3. What are the three types of selection addresses? What is each used for?

# 7.3  CHARACTER-ORIENTED PROTOCOLS

Even though character-oriented data-link protocols are in use today, it is not the intention of this text to dwell on them, but to illustrate enough about them to understand their use in applications later on. The main thrust on details concerning data-link protocols is placed on the bit-oriented SDLC/HDLC protocols that are more widely used. The asynchronous and BISYNC protocols are used here to illustrate the formats and requirements for character-oriented protocols.

### Asynchronous Protocol

The asynchronous protocol has no requirement to establish bit and character synchronization between the sender and receiver. This task is performed through the nature of asynchronous data as shown in Figure 7-2. Recall that start and stop bits are used to frame asynchronous data characters. The line is held at a logic one (mark) condition, often referred to as idle line 1s, as long as the link is established and data is not being sent. The first character transmitted is detected by the receiver as a change from the idle line 1 to a 0 or space (start bit) condition. This establishes both bit and character synchronization. Bit synchronization is established beginning with the detection of the start bit and lasting for a time period dependent on the data rate (one bit time period being the reciprocal of the data rate). Character synchronization is established because the least significant bit of the character follows the start bit. Since there is no form of "preamble" message required to establish synchronization, actual messages are sent starting with the first character transmitted.

A polling message in the asynchronous protocol is recognized by the reception of a sequence of three characters:

$$
\begin{array}{ccc}
E & D & \\
O & C & R \\
T & 3 &
\end{array}
$$

**ASCII**
American standard code for information interchange.

**EBCDIC**
extended binary coded decimal interchange code.

**device control character 3 (DC3)**

Each character is recognized by reading its name vertically. They are written this way so that each character occupies only one word. That is, EOT is the acronym for the end of transmission character in the **ASCII** or **EBCDIC** code. Spaces between characters are used for clarity and are not included in the sequence. The EOT character is used as an alerting character. All secondaries will react to this character in anticipation of receiving a poll or selection message.

**Device control character 3** (**DC3**) further tells these stations that this message is a poll. Any other character immediately following the EOT character becomes part of a two-digit selection address. Finally, the third character (R in this case) is a secondary's polling address or, in the case of a selection, the second digit of the selection address.

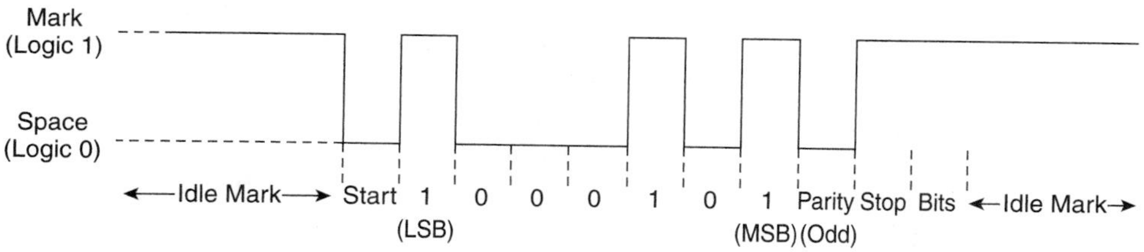

**FIGURE 7-2** Asynchronous ASCII Q

The secondary polled or selected is required to respond in one of three ways described earlier. If the secondary being polled is in the send mode, it sends its messages in response to the poll. In response to a selection, a secondary station in the send mode sends the character sequence * *.

A secondary that is in the receive mode responds to both polls and selections with the character sequence

A
\ C
K

Lastly, if the secondary is in the local mode, the sequence for the response is \ \.

After the communications link has been established through the poll or selection and the corresponding response, traffic can be sent and received between the primary and the polled or selected secondary station. Secondary stations are always required to respond, in some form, to a message sent by a primary. Failure to respond to the primary is taken to mean that the secondary communications system is not functioning properly and that station will be removed from the poll and selection sequence. In order to be restored to the sequence, a separate call must be made by the owners of the secondary station to the primary station to inform them that the secondary is ready to resume communications.

## BISYNC Protocol

**BISNYC**
character-oriented synchronous data protocol.

**BISNYC** or **BSC (Binary SYNchronous Control)** protocol is also a character-oriented protocol, but it is used for synchronous data. Synchronous data has no framing bits and relies on the receiver to regenerate a synchronizing clock from the incoming data stream. Once the receiver has recovered the clock, it must then establish bit and character synchronization. It does this by recognizing specific data character patterns sent by the transmitting station in the preamble to a message. This preamble and a BISYNC poll requires many characters and appears as:

```
P P S S E P P S S S S E P P
A A Y Y O A A Y Y P P D D N A A
D D N N T D D N N A A A A Q D D
```

The two leading PAD characters are alternating 1s and 0s, which are ideal for clock recovery since they are alternating logic levels at a fixed rate, unlike other data streams that have groups of 1s and 0s. Alternating 1s and 0s also allows for bit synchronization by a receiver since the received data is of a known sequence.

**SYN**
SYNchronization character.

The first pair of **SYN** characters aids the receiver to establish character synchronization by supplying a recognizable sequence of 1s and 0s, which forms the code for the SYN character to lock in on. The end of transmission (EOT) character is used here, as it is in the asynchronous protocol, to alert secondaries to a polling or selection message. The second PADs are groups of four 1s each that supply a short time delay to allow the secondaries to respond to the EOT character.

**station polling address (SPA)**

A pair of SYN characters reestablishes character synchronization in case it gets lost during the time delay. These are followed by a **station polling address (SPA)** and

**station device address (SDA)**

**ENQ**
enquire character.

**block check character (BCC)**
error detection character.

**station device address (SDA).** These establish a communication link with a specific device at a specific station. A SDA equivalent to the character " is recognized as a general poll of all devices at the station. The address characters are followed by an **ENQ (enquire)** character and two more delay PAD characters.

Usually, synchronous data does not use parity for error checking, thus reducing the number of additional bits to be sent with each character. Instead a CRC-16 error detection character is used as a **block check character (BCC)** for the purpose of detecting errors. For BISYNC polls and selections, this character is not generated. Instead, error detection is performed by duplicating significant characters. There are two SYN characters and two polling address and device characters. Each pair contains identical characters. Any errors in these characters prevents the secondaries from recognizing the poll and there is no response. The primary will retransmit the poll if it fails to get a response from a secondary within a specified time period.

A secondary, in the send mode, responds to a BISYNC poll by sending a message. The sequence used for responding when the secondary is in the receive or local mode and there are no problems at the secondary station is called a *handshake* and has the following sequence of characters:

```
S  S  E  P
Y  Y  O  A
N  N  T  D
```

If the secondary is experiencing some difficulty with any part of its system and it needs to inform the primary of this, it will respond to the poll with a sense and status message. This message informs the primary that the secondary cannot communicate with the primary. There are codes within the message that define most of the common reasons for the breakdown in communication.

Selection in asynchronous protocol is performed by sending an EOT followed by two selection address characters. These characters identify the secondary or secondaries selected. Selection in BISYNC is done by using a selection address in place of the polling address in the polling example shown above. This requires the definition of a unique set of polling addresses and a unique set of selection addresses by the BISYNC protocol. Device addresses are coded using the same characters as polling addresses, but since their position in the poll or selection message differs from polling address and selection address positions, mistakes are not made.

There are specific secondary responses to selections, which are similar to the ones used for polling. Additionally, there are responses designated to indicate the result of checking the cyclic redundancy check (CRC-16) block check character used for error detection.

## Asynchronous Text Message

Actual message formats for these protocols include data-link, formatting, and alphanumeric characters. Both protocols use similar formats, but the use of the characters differs significantly. How they affect the message as it appears on a terminal screen or printer page is also significantly different. It would take considerable time and space to reproduce and analyze text messages in both of these protocols.

STX or SOH (start of text or
start of heading)

For asynchronous, the message begins with a **STX** or **SOH** (**start of text** or **start of heading**) character. These characters, when sensed by stations that are not being polled or selected, cause these stations to become "blinded" to the message. They cannot receive them. The polled or selected station does read and interpret the message. Either starting character is followed by text, which may be any combination of alphanumeric and formatting characters.

Various applications respond to SOH characters in different ways. Some protect heading data from being altered at the receiving site. Some highlight the heading area while others do nothing different in response to SOH or STX. A heading is terminated by a STX character, which also denotes the remaining text as data rather than heading. STX text is terminated by an EOT (end of transmission) character, which also "unblinds" the other secondaries on the line.

Formatting characters, besides line feed and carriage return, are designated by a preceding *ESC (escape)* character. For instance, an ESC 3 highlights text that follows it while an ESC 1 sets a tab at the present cursor location. Numerous others are used and are defined by the asynchronous protocol specification.

An example of a asynchronous message:

```
 S  E  E                                          E     E
 T  S  R  S  3  D A T A S C O M M U N I C A T I O N S  S  4  N  O
 X  C  C           P                                 C     L  T
```

As mentioned above, STX indicates the beginning of the message and causes any receiver not previously polled or selected to be "blinded" or effectively taken off-line. ESC R sequence causes the cursor on a CRT to be placed in the home position and also clears the screen of anything on it. ESC 3 causes the following data to be highlighted. Data Communications is the actual text followed by an ESC 4, which removes the highlight attribute. *New line (NL)* is a combination of carriage return and line feed. Finally, the message is ended by an EOT character.

---

### EXAMPLE 7-2

In the message just shown, pick out the alphanumeric, formatting, and data-link characters.

### SOLUTION

Alphanumeric characters are those that are printable, which are : DATA COMMUNICATIONS. Formatting characters dictate how and where the message appears on the screen. They include all of the ESC n sequences. Finally, data-link characters are used to establish the software linkage and include STX and EOT.

---

What about message efficiency? In Chapter 1, it was shown that the inclusion of start and stop bits along with parity reduces the efficiency of transmission, since these are overhead or non-data bits. In its broadest meaning, transmission efficiency takes into consideration anything sent that is not purely information. For the asynchronous message above, start, stop, and parity bits are still noninformation bits, but now all data-link and graphics characters must also be included as non-information. They also reduce the efficiency of the message transmission.

There are a total of 30 characters transmitted in the example message. Using one start, two stop, one parity bit, and ASCII characters, there are 30 characters × 11 bits per character or 330 bits in the total message. The text bits in DATA COMMUNICATIONS, including the space character between the two words, account for 19 characters × 7 bits per character or 133 message bits. The efficiency of a message can be derived by dividing the number of information bits by the number of noninformation bits and multiplying the result by 100%:

$$133/330 \times 100\% = 40.3\%$$

## BISYNC Text Message

A BISYNC message format, shown in Figure 7-3, starts with a STX or SOH character similar to the asynchronous protocol. This is followed by a erase-write command character preceded by an ESC character. This sequence determines which portion of a message is being sent. Specifically, whether that portion is unmodified or modified (altered) text or has some other designated attribute. This is useful when data entered into a fixed form is being sent from one station to another. Since the body of the form is fixed (and therefore not modifiable), both stations already have that portion of the message. The only section that is new is the data entered into the form's blanks. By allowing the sending of modified data only, the message can be much shorter then sending both the body of the form and the entries into it.

The opening sequence, which is illustrated in Example 7-3 following this discussion, is followed by a write control character, which initializes the receiver terminal to the correct size and form of the message. The beginning location of the message, which is a memory buffer address associated with a specific terminal screen or printer location, follows the command sequence.

Actual text is next, which includes an assorted set of characters as before. Graphics indicators called **attributes,** which determine how the text looks, are defined by characters designated by a preceding **GS (group separator)** character.

BISYNC messages are sent as blocks of data. This allows larger messages to be broken into smaller units, or *blocks*, for more efficient transmission. An *ETX (end of text)*, *ETB (end of transmission block)*, or *ITB (intermediate end of transmission block)* character used to end a message block, is followed by a block check character (BCC) and trailing time delay PADs. ETX also designates this block as the last block of the message. ETBs and ITBs indicate that additional blocks of the message are due to follow.

**attributes**
those accents of a message that bring attention to a part of the message. They include bold face, underline, blinking, etc.

**GS (group separator)**

| PAD 55 | Preamble | Opening Data Link | —Text— | Closing Data Link | BCC | PAD FF |
|--------|----------|-------------------|--------|-------------------|-----|--------|

**FIGURE 7-3** BISYNC Message Format

ETX and ETB characters require the receiver to respond to the condition of the **block check character (BCC),** which is CRC-16 error detection characters, after they are received. ITBs are used to designate the end of the block that does not require an immediate response from the secondary. Instead, the secondary will respond to the condition of all ITB BCCs when it receives the ETX terminating the message.

ETBs and ITBs have the advantage that if an error occurs, only that block of the message needs to be retransmitted, instead of the entire message. ITBs have the advantage that a message can be broken into blocks as with ETB, but there is no time lost by the receiver responding to each block individually. Instead, all BCCs are acknowledged after receipt of ETX when ITBs are used. The receiving station can inform the sender which block is bad as part of that response.

---

### EXAMPLE 7-3

How does the message DATA COMMUNICATIONS appear using the BISYNC protocol? It is to be displayed on an 80-character screen and is modified data (data entered into a modifiable field by the user).

### SOLUTION

The actual characters are taken from numerous tables included in the BISYNC protocol and are shown here for illustration purposes. Assume the opening dialogue (poll and responses, etc., have already occurred). The message appears as:

```
P S S E P S S S E    D   G         G
A Y Y O A Y Y T S 5 O C S S S I D A T A S S S C O M M U N I C A T
D N N T D N N X C    1             P P


         D E B B P
I O N S C T C C A
         3 X C C D
```

The meanings of each character as defined by the BISYNC protocol are as follows:

First PAD—55 or AA is used to establish clock recovery and bit synchronization.
SYN SYN—establish character synchronization.
EOT—alerting character.
Second PAD—FF to allow time to respond to EOT.
SYN SYN—re-establish character synchronization.
STX—start of text indication.
ESC 5—Erase/write character—clears the screen and homes the cursor.
0—write control character—sets 80 character per line mode.
DC1 SS—signifies that the message will be stored in video memory starting at the location associated with the home position on the screen.
GS I—the group separator indicates that an attribute character (I) follows. Attribute I causes the field of the message to follow to be intensified and protected from being altered.

DATA—Data followed by a space—actual text.

GS SP—group separator and attribute character space—removes intensify and protect from the next field of the message.

COMMUNICATIONS—actual text.

DC3 inserts the cursor at the current video memory location (placing it following the S in communications in the message).

ETX—indicates the end of the message. Block check characters are expected next.

BCC—two block check characters to hold the results of

CRC-16 error detection are next.

PAD—time delay PAD (FF hex) allows the receiver to check the BCCs for any indication of an error.

---

What about the efficiency of this message? With synchronous data there are no framing or parity bits within each character, so all seven bits of each ASCII character are data bits. There are still only 19 information characters in DATA COMMUNICATIONS, but there are a total of 42 characters in the message. Since all bits are character bits, the efficiency of a synchronous message is the total information characters divided by total messages characters multiplied by 100%:

$$19/42 \times 100\% = 45.24\%$$

It's not much more efficient than the asynchronous version, but keep in mind, that this example is a short message, which is not the usual case for data transmissions. The difference becomes much greater as message length grows. Actual data transmissions will include hundreds to thousands of characters. As an example, compare the efficiency of an asynchronous and synchronous 256 character message. Assume the same number (23) of graphics characters (ESC and GS/DC characters) are used in both messages. The efficiency of sending synchronous data is:

$$\text{efficiency} = \frac{256}{279} \times 100\% = 91.8\% \text{ for BISYNC}$$

You will soon note that the efficiency of synchronous data transmissions will far outdistance that of asynchronous. The reason is simple enough. Each character added to an asynchronous transmission is 63.6% efficient by itself (7 ASCII bits divided by 11 total using parity and two stop along with a start bit). Each additional synchronous character is 100% efficient (all data bits) by itself. The asynchronous message retains its 63.6%, not counting data-link characters as noninformation bits. Once you take into account the additional non-data characters as overhead as well, you quickly realize that the efficiency drops well below 63.6%.

## Transparency

One additional overhead character is used in the BISYNC protocol to facilitate a function called *transparency*. There are occasions when data messages may contain numerical values or other characters that have the same codes as the data-linking characters. An example application is the sending of an inventory to the accounting department of a huge store. It is very possible that the count of any number of items might appear as link control characters. To avoid this, a delimiter character (DLE) precedes the start of text (STX) character of the message. All characters within the message that are intended to be data-link characters are preceded by the **DLE** character. The receiver then interprets only those characters following DLE as data-link characters. All others are treated as information or formatting characters.

**DLE**
data-link escape character or delimiter character.

This description of the character data-link protocols is a stripped down version with the purpose of illustrating the concept of character interpretation in establishing communications and defining the actual message. There are a number of detailed functions and code tables, responses, and requirements that these character-oriented protocols have. Furthermore, information and further details about these character- oriented protocols can be obtained from the individual protocol specifications themselves.

---

**Section 7.3  Review Questions**

1. How are polling and selection messages recognized using both character-oriented protocols?
2. Compare how messages are transmitted by both protocols.
3. How is error detection accomplished by both protocols?
4. Why are BISYNC transmissions more efficient than asynchronous protocol- based transmissions?

---

## 7.4  BIT-ORIENTED PROTOCOLS

Synchronous data-link control (SDLC) and the enhanced version high-level data-link control (HDLC) are bit-oriented data-link protocols that provide a means to break messages up into smaller sections, called *frames*. This ability lets a primary station interleave messages to several secondaries by sending each a frame of its message in turn, rather than sending entire messages at one time. As the communications continues through a network, each secondary is linked to the primary in a continual sequence until all the frames are sent. The secondaries then reassemble their messages from the frames they have received from the primary. To do this, there is a requirement to maintain the sequence of the message frames so that a secondary can correctly reassemble them.

### SDLC/HDLC Protocols

Being bit-oriented means that the data-linking processes are established by the particular grouping of bits rather than by characters such as STX or SYN. SDLC was conceived by IBM to be used with the EBCDIC character code solely. Additionally, IBM selected b0 in a group of eight bits (byte) to be designated as the most significant bit and b7 as the least significant bit of the byte. This is the reverse of standard notation that most of us are familiar with. For example, converting the hexadecimal numbers 36 to binary by IBM appears as:

$$
\begin{array}{ccccccccc}
 & b0 & b1 & b2 & b3 & b4 & b5 & b6 & b7 \\
36 = & 0 & 0 & 1 & 1 & 0 & 1 & 1 & 0
\end{array}
$$

while in the more common standard notation, it would appear as:

$$
\begin{array}{ccccccccc}
 & b7 & b6 & b5 & b4 & b3 & b2 & b1 & b0 \\
36 = & 0 & 0 & 1 & 1 & 0 & 1 & 1 & 0
\end{array}
$$

To reduce the confusion that this may cause in the remaining discussion, this text will retain the more common notation of b0 as the least significant bit. Do keep in mind, though, if you look into the SDLC specification in more detail, that the IBM notation will be used. Fortunately, most upper-level protocols are designed to be similar to the enhanced HDLC version, which among other things, returns to the standard notation of b0 as the least significant bit.

## SDLC/HDLC Frame Format

The basic format for SDLC and HDLC messages is shown in Figure 7-4. Each message transmission is framed by a starting and ending *delimiting flag,* which is recognized as six ones between two zeros, or 01111110, or 7E in hexadecimal form. Do not forget that SDLC and HDLC are bit-oriented protocols, so while it is easier to refer to six 1s surrounded by 0s as 7E, it is still the sequence of the eight bits that are being referred to. If this sequence of six 1s and two 0s appears *anytime* or *anywhere* in an SDLC/HDLC message, the SDLC/HDLC protocol recognizes it as a flag, intended or not. The reason for this is that this protocol is bit-oriented.

How is this a problem? Suppose in the body of text, the EBCDIC character G is followed by a U. The EBCDIC code for G is 11000111 and U is 11100100. Together, the bit sequence becomes 1100**011111110**0100. Notice that a 7E sequence (boldfaced) is created in the middle of the data sequence. SDLC/HDLC reacts to this combination as a flag and would end the current message at this point. Equally as well, since a flag sequence does not follow this sequence, it will also assume that this ending flag is also the beginning flag of the next frame!

To avoid the problem of misinterpreting a message sequence as a flag, a method of transparency called zero stuffing or zero insertion is employed to make unwanted 7E sequences transparent to the applications software. A zero is inserted following five consecutive ones anywhere in a message except for intended flag sequences. The receiver removes any zeros it detects following five consecutive ones before interpreting the message contents.

| 7E Flag | Secondary Address | Control Field | Text Field | Frame Check Sequence | 7E Flag |
|---------|-------------------|---------------|------------|----------------------|---------|

**FIGURE 7-4** SDLC/HDLCFrame

**EXAMPLE 7-4**

For the EBCDIC message sequence for GUTTER, where are zeros inserted to avoid mistaken flags?

**SOLUTION**

the binary sequence for the message GUTTER is:

11000111111001001110001111100011110001011101001

Zeros are inserted following five consecutive 1's. The message sent becomes (with inserted zeros boldfaced to aid in showing their location):

1100011111**0**1001001110001111**0**00011110001011101001

Notice that a zero is inserted following the second group of five 1s even though they would not be interpreted as a flag, which requires six consecutive 1s. The part of the program that detects five consecutive ones makes no distinction based on the next bit. This makes the insertion and removal of the extra zero universal and comparatively easy.

---

You can use as many consecutive 7E sequences as desired before beginning a message. The SDLC protocol will not interpret a beginning flag until the bits following that flag no longer look like a flag. With consecutive flags, an ending 0 of one flag can be used as the beginning zero of the next flag. So these are both two consecutive flags:

0111111001111110   and   011111101111110

As stated earlier, the ending flag of one frame can be used as the beginning flag of the next frame, which makes for a more efficient transmission of consecutive frames.

Two additional control types of information are supplied by a specific count of consecutive ones. Between 7 and 15 consecutive ones followed by a zero defines an abort sequence. A station may experience problems in the middle of a transmission sequence, and rather than continue to the end, will send an abort to disrupt the transmission. More than 15 consecutive 1s is recognized as a return to an idle condition which maintains a link without any additional messages or frames being sent.

## SDLC/HDLC Address Field

The next field in the message is the secondary address field. All messages between a primary and a specified secondary contain that secondary's address in this field. This identifies the station that the primary is communicating with. In SDLC this field is eight-bits wide, allowing for 256 different addresses.

In HDLC the address field is unlimited. Each address must be in multiples of eight bits (bytes) and can be any total length. A 1 in the most significant bit of an address byte designates that byte as the last byte of an address. All other bytes in the address have a 0 in the most significant bit position. Effectively, there is no limit to the number of secondaries used with the HDLC protocol.

The significance of the unlimited address field is not just to allow for a larger number of stations on a single network, but rather to facilitate addressing for a peer-to-peer network. Peer-to-Peer networks do not have a primary. All stations have equal access to the network and any station can communicate directly with any other station. The ability to do this type of communication requires the address field to contain information about both the source and destination stations.

In a primary/secondary system, one end of the communication is always the primary, so SDLC only provided a method to identify the secondary station. HDLC address fields can be divided, by application software, into subfields containing a source address and a destination address that is used to identify both ends of the link.

## SDLC Control Field

Following the address field, is the control field which defines what type of message is being sent. There are three different message frames used in SDLC and HDLC. They are information, supervisory, and unnumbered frames. Figure 7-5 shows the control field for an information frame, which is identified by the least significant bit (b0) of the control filed being in a logic low state.

**poll/final flag (PF)**

A **poll/final (P/F)** bit identifies the last frame sent by either the primary or secondary. For SDLC, the last information frame sent by a primary is a polling frame (P/F = 1) that requires an immediate response from the secondary station. Other frames sent by the primary are nonpolling (P/F = 0), which do not require a secondary to respond.

The last frame sent from a SDLC secondary station is a final frame (P/F = 1), while the other preceding frames are not final frames designated by P/F bit = 0. The SDLC primary is not required to respond to a final frame from a SDLC secondary station. It is used to inform the SDLC primary that the SDLC secondary station is sending its last frame.

**frame number being sent (NS)**

**frame number expected to be received next (NR)**

The remaining bits in the control field of an information frame are interpreted in groups of three for SDLC. One group contains the **frame number being sent (NS)** and is composed of bits b3, b2, and b1. Bits b7, b6, and b5 yield the **frame number expected to be received next (NR)** from the responding station. These frame numbers are used to keep track of the number of successful frames sent and received. An example best illustrates how they work.

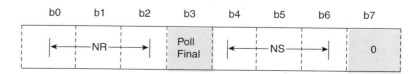

**FIGURE 7-5** Information Frame Control Field

---

### EXAMPLE 7-5

Starting with a reset condition, how does the frame numbering sequence operate to keep track of frames?

### SOLUTION

For now, a reset condition sets all the NS and NR sequence number bits to zero. Initially, we will assume that the primary sends a single polling frame. The control frame, in binary (LSB on the right) is 0001000. NS and NR are both zero indicating that this is the first frame sent (NS = 000) and that the primary expects frame number 000 (NR = 000) as the first frame the secondary will use when it replies to the primary's poll. A high in b4 (P/F) makes this a polling frame and a low in b0 designates this as an information frame.

The secondary replies by sending three information frames:

00100000
00100010
00110100

Realize that each of these control fields are part of a complete message and are extracted for illustration purposes so that we can concentrate on the sequence of frame numbers in the control field. All of the control fields have a NR of 001, telling the primary that the secondary expects frame number 001 as the next frame to be sent by the primary. This is expected since the primary only sent one frame with an NS of 000 making its next frame, in a correct sequence, frame number 001.

The P/F bit of the first two fields is low, informing the primary that additional frames follow. The last frame has a high P/F in its control field indicating that this is the end of the current transmission. The last field is the final frame sent by the secondary in this sequence. The control field NS numbers indicate that the secondary has sent frames 000, 001, and 010.

Now, suppose the primary receives the first two frames without any problem, but something goes wrong with the third frame. The primary wants this frame retransmitted. The primary also has five frames to send. To accomplish this, the primary sets its NR to 010. This tells the secondary that the primary received frames 000 and 001 all right, but did not receive number 010 or detected an error in the data field of that frame. In either case, it wants the secondary to retransmit this frame. The primary's control fields for the next five frames are:

01000010
01000100
01000110
01001000
01011010

The secondary, when it finally replies, starts by sending frame 010 and any additional frames it has ready to send.

The one drawback of this method of automatic request for retransmission occurs when an error is detected in an early frame. For instance, if the secondary had sent five frames and the primary detected an error in the very first frame, it will request the secondary to start its response with the bad frame and continue in sequence, retransmitting all five frames, even if the following four had originally been received without a problem.

To illustrate what happens in the case of an early detected error, let's suppose this system really gets fouled and the receiver does not get frame number 011 correctly from the primary. The secondary has two frames to send besides the one to be retransmitted. It will send a NR of 011 with the three frames it sends. The primary will respond by transmitting frames 011, 100, and 101 again. When using SDLC, there is no way for a sending station to detect if frames following a bad frame were good or bad, so it must retransmit all frames starting from the bad frame until the last frame is sent. The sequence, first from the secondary and then the primary:

sent by the secondary:

> 01100100
> 01100110
> 01111000

sent by the primary:

> 10100110
> 10101000
> 10111010

The secondary is required to reply to the poll to complete the dialogue.

---

For SDLC, a maximum of seven consecutive frames can be sent by a station without a response from the receiving station. When frame number seven (111) is followed by the next frame, the NS and NR numbers wrap around to 000 and resume counting. The reason for limiting the maximum number of consecutive frames to seven is to avoid confusion concerning two or more frames with the same NS number. If this should occur, there would be a problem in maintaining the correct sequence. A request to retransmit frame 010 in response to a sequence of more than seven frames in a row would be confusing. Which frame 010 is to be retransmitted, the first one or the second one?

## HDLC Control Field

In the HDLC control field, both NS and NR fields are expanded to seven bits, increasing the maximum consecutive frames to 127. A logic zero state of the LSB of the control frame is still used to indicate an information field and the bit between the NS and NR numbers is still the P/F bit, but it is interpreted to refer to the last frame sent by the current source station whether that station initiates the communication or is the responding station.

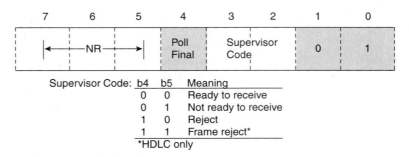

| 7 | 6 | 5 | 4 | 3 | 2 | 1 | 0 |
|---|---|---|---|---|---|---|---|
| | ←——NR——→ | | Poll Final | Supervisor Code | | 0 | 1 |

Supervisor Code:

| b4 | b5 | Meaning |
|----|----|---------|
| 0 | 0 | Ready to receive |
| 0 | 1 | Not ready to receive |
| 1 | 0 | Reject |
| 1 | 1 | Frame reject* |

*HDLC only

**FIGURE 7-6** Supervisor Frame Control Field

## Supervisory Frames

SDLC and HDLC supervisory frames are recognized by $b0 = 1$ and $b1 = 0$ in the control field as seen in Figure 7-6. There is no NS number. In place of the NS number is a code is used to define the type of supervisory frame being sent. Supervisory frames do not include text within the message. All supervisory information is conveyed by decoding bits b2 and b3. NR number field (b7, b6, and b5) and poll/final function of b4 remain as before. As with the information control field, for HDLC, the NR field is expanded to seven bits to maintain correct sequence numbers.

Supervisory fields are most commonly sent by a station when it has no information to send, but needs to reply to a sending station. For example, a primary may have twenty frames of information to send and the secondary only two. After the secondary sends its two frames, it uses supervisory "ready to receive" formats in response to primary polling frames to indicate that it is still online and in a state capable of continuing to receive traffic, therefore maintaining the dialogue while the primary completes sending its information.

The "not ready to receive" function informs the primary that the secondary is off-line and incapable of receiving future frames.

The "reject" code informs the primary that the frames sent starting with the one indicated by the NR number will have to be retransmitted because an error of some sort was detected in the indicated frame.

Because HDLC has so many more frames that it can send without a response from the destination station, that protocol includes an additional supervisory code for "frame reject." This is used by HDLC to indicate a specific bad frame. In the earlier scenario, if a bad frame is detected, the receiving station indicated this by sending the appropriate frame number in the NR number bits. The sending station was then required to retransmit all frames beginning with the bad frame. With HDLC, if only one frame is bad, that frame number can be indicated using the "frame reject" supervisory message. The sending station is, then, only required to retransmit the faulty frame and then can resume sending frames following the last one it originally sent.

## Unnumbered Frame Formats

The third frame type is the unnumbered frame of Figure 7-7 and is recognized by bits b0 and b1 both being high. Bit b4 still signifies polling and final frames as before. The re-

(*Unnumbered Frame Function Code)

**FIGURE 7-7** Unnumbered Frame Format

**TABLE 7-1**

## Unnumbered Frames

| Binary Code | Hex Equivalent P/F: 0 | 1 | Frame Name | Sender | Description |
|---|---|---|---|---|---|
| 000*0011 | 03 | 13 | UI | Both | Information |
| 000*0111 | 07 | 17 | SIM | Primary | Set Initialization |
| 000*0111 | 07 | 17 | RIM | Secondary | Request Initialization |
| 000*1111 | 0F | 1F | DM | Secondary | Disconnect Mode |
| 010*0011 | 43 | 53 | RD | Secondary | Request Disconnect |
| 010*0011 | 43 | 53 | DISC | Primary | Disconnect Command |
| 011*0011 | 63 | 73 | UA | Both | Acknowledge |
| 100*0011 | 83 | 93 | SNRM | Primary | Set Normal Mode |
| 100*0111 | 87 | 97 | FRMR | Secondary | Frame Reject |

*P/F bit.

maining bits in the control field define the type of unnumbered frame being sent. NR and NS number fields are not used with unnumbered frames, which means that they are not sent in a sequence to form a larger message, but are complete in themselves.

These frames are used by the primary to send commands to the secondary and by the secondary to send status messages or to request an action from the primary. Table 7-1 details the control field codes and their related unnumbered frames. Hexadecimal equivalents of the control field are used with the P/F bit (b4) high to indicate a polling or final frame condition in one column and low, for all other frames, in the other. The next columns give the unnumbered frame mnemonic, who originates the frame, and a brief description of the frame function. Essentially, request frames are initiated by the secondaries and command frames by the primary station. The reason for this lies in the very core of a multipoint system—the primary always has control of the system and is the only station that can issue commands.

If a secondary wishes to go to a local mode, it must first request the primary to disconnect it from the line by sending a RD (request disconnect) frame to the primary. The primary then issues a DISC (disconnect) command, which allows the secondary to go off-line. The secondary can be placed back online when the primary sends a

SNRM (set normal response mode) frame. NR and NS numbers are reset to 000 with the SNRM command. When a primary polls a secondary that is off-line, the station is still required to respond to the poll. In the local or disconnected mode, it does this by sending a DM (disconnect mode) status frame. The secondary can also request to be placed back online by sending a RIM (request initialization mode) in response to a primary's poll. This is sent in place of the DM status message. A primary would then respond with an SNRM, which re-initializes the secondary's NR and NS numbers and places that secondary back online.

When communication is first established with a secondary, the primary issues a SIM (set initialization mode) command. This command's function is to initialize the secondaries NR and NS numbers to 000 so that frames can be correctly tracked. In this manner, the primary controls the start of the dialog with a secondary station. SIM differs from SNRM in that SIM will also initiate a secondary's initialization program while SNRM will only set the NR and NS numbers to 000 and place the secondary back online.

The remaining frames include an unnumbered information (UI) frame, unnumbered acknowledge (UA) frame, and a frame reject (FRMR) frame. UI is used whenever text outside of normal traffic is to be sent between the primary and secondary. Details about problems one station is having or a test message are examples of the data used with an UI frame. UA frames are used by the secondary or primary as a means of acknowledging an unnumbered transmission.

While HDLC still utilizes many of the unnumbered frame formats, the application of them is slightly different. Because there is no primary station on the network, the commands are generally issued by the station that initiates the communication link. This is particularly true for SIM in order to assure that both stations in the link start with NS and NR numbers at zero.

## Unnumbered Frame Reject

Frame reject is used by the secondary to inform the primary why the secondary is rejecting the frame sent by the primary beyond a normal detection of an error in the information field. The format for a frame reject frame is shown in Figure 7-8. The flag, secondary address, control field (containing the FRMR code), frame check sequence, and ending flag comprise the standard SDLC/HDLC frame format. In the text field, instead of char-

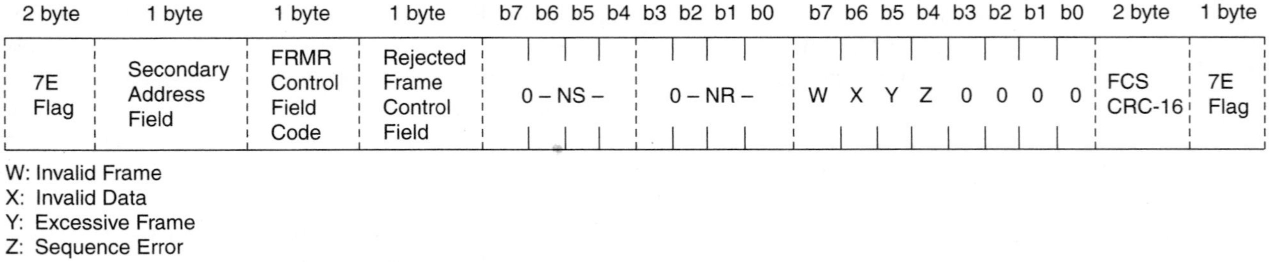

| 2 byte | 1 byte | 1 byte | 1 byte | b7 b6 b5 b4 b3 b2 b1 b0 | b7 b6 b5 b4 b3 b2 b1 b0 | 2 byte | 1 byte |
|--------|--------|--------|--------|-------------------------|-------------------------|--------|--------|
| 7E Flag | Secondary Address Field | FRMR Control Field Code | Rejected Frame Control Field | 0 – NS – | 0 – NR – | W X Y Z 0 0 0 0 | FCS CRC-16 | 7E Flag |

W: Invalid Frame
X: Invalid Data
Y: Excessive Frame
Z: Sequence Error

**FIGURE 7-8** SDLC Frame Reject (FRMR) Format

acter code information, a specific bit format is employed to identify the problem. Sixteen bits are used to fit the format shown in Figure 7-8. The NR and NS numbers in the frame reject text area, are those frame numbers at the point when the frame was rejected. A high state in one of the status bits specifies the reject cause. These causes are problems other than error detection requiring simple retransmission. Those are resolved using the NR/NS system detailed earlier.

The invalid frame status bit indicates that the frame received by the secondary could not be identified. This is used if a station sends an unnumbered code in the control field, which is not a code included in the protocol. The destination station cannot interpret the frame so it sends an unnumbered reject with the invalid frame bit set high.

The invalid text status bit is used when text is sent with a field that does not allow text. For instance, if text is included with a supervisor frame, this frame would be rejected and the invalid text bit would be set.

The excessive frame status bit indicates when too many consecutive frames are sent without a poll frame (more than 7 for SDLC and 127 for HDLC).

Finally, sequence error status bit indicates when a NR number does not agree with the transmitted NS number sequence. For example, three frames are sent by primary with NS numbers for frames 011, 100, and 101. The secondary responds with a ready to receive supervisor frame setting the NR number to 110. The primary starts its next sequence by sending a frame with NS number of 111 instead of an NS number equal to 110 as it was expected to do. The secondary is confused because it expected the primary to send frame number 110 first. The secondary has no idea why the primary did not follow the correct sequence. In response to the polling frame, the secondary sends a frame reject. In it, the secondary includes the NS frame number (111) that the primary sent and sets the sequence error bit of the frame reject text field high informing the primary that it transmitted out of sequence.

## Remaining SDLC/HDLC Fields

The next field in SDLC and HDLC frames that allows text is the text field, which follows the control field. Text is generally sent in groups of bytes using EBCDIC character code for SDLC and any character code for HDLC. Applications of the HDLC protocol as utilized in numerous other protocol suites, often find complete upper-level protocols placed as payloads in the text field.

Error detection is accomplished using CRC-16 or CRC-32 algorithms. The error characters are computed on the data following the starting 7E flag up to, and including, the last character of text. The CRC bytes, called a *frame check sequence (FCS),* are placed following the text and preceding the ending 7E flag. The ending delimiter flag completes an SDLC/HDLC frame sequence.

---

EXAMPLE 7-6

How does the SDLC message Help! appear transmitted as final SDLC frame number 101 sent by a secondary station whose address, in hexadecimal, is 3E? Assume the primary already sent three frames numbered 000, 001, and 010.

**SOLUTION**

Start by determining the contents of the control field. This message is an information frame, so bit 0 is low. The NS number is 101 to indicate the frame being sent. The NR number is set to 011 since the primary already sent frames 000, 001, and 010, making frame number 011 the next expected frame to be transmitted by the primary station. The P/F bit is set high, indicating that this is a final frame. Placing this information into the correct control field bits produces the control sequence 01111010 or 7A in hexadecimal.

To clarify where each piece of information is placed, hexadecimal bytes and the fields of the frame are separated by a space. This is not the way it is transmitted. Actual frames do not have the spaces. They are shown here in this form for clarity only. The message is:

7E   3E   7A   C8   85   93   97   5A   FD   C2   7E

The first 7E is the starting frame flag. This is followed by the secondary's address, 3E. 7A is the control field derived above. The next five hexadecimal bytes are the EBCDIC code for Help! in hexadecimal: CB 85 93 97 and 5A. The hexadecimal characters FD C2 is the actual frame check sequence (FCS) CRC-16 bytes computed for the data, starting with 3E and ending with 5A. The FCS is followed by the 7E ending flag.

## HDLC Operating Mode Enhancements

Since HDLC is used primarily with peer to peer networks, it uses two modes of operation to facilitate activity between the stations on the network. The *asynchronous response mode (ARM),* specified in the HDLC protocol, allows a station to respond to any frames, removing the restriction of responding to poll frames.

When applied to a primary controlled system, the final (P/F) bit no longer indicates a final frame from the secondary. It now indicates when the secondary is responding to a poll frame. This bit is low for secondary responses to nonpoll frames. This added mode gives the SNRM (set normal response mode) command more meaning. When this command is issued, besides resetting NS and NR numbers and placing a secondary online again, it also places that secondary in the normal response mode instead of the asynchronous response mode. In the normal response mode, the link acts like it did in SDLC with the primary controlling all functions.

An additional unnumbered command, the set asynchronous response mode (SARM) does the same functions as the SNRM except it places the secondary into the asynchronous response mode. If it is desirable to reset NS and NR numbers without having to worry about which response mode is in use, then an unnumbered REST command can be used. This command resets the sequencing numbers but leaves the secondary in its current operating mode.

With the *asynchronous disconnect mode (ADM)* of HDLC, the secondary can initiate a disconnect (DISC) and go off-line at any time. It no longer requires the secondary to request a disconnect and then wait for a command to be sent from the primary. The secondary is still required to issue a *disconnect mode status (DM)* in response to a primary's poll frame if the secondary is off-line (disconnected). This is beneficial in a peer-to-peer system to allow any station to go off-line at any time. When another station attempts to establish communication with that station, it would respond with a disconnect status (DM) message.

---

### Section 7.4  Review Questions

1. Compare the differences between SDLC and HDLC.
2. Compare the difference between SDLC and BISYNC data-link protocols.
3. What is the general purpose of each of the three SDLC frame types?
4. How is flag transparency achieved in SDLC?
5. When is the unnumbered frame reject message used?

---

## 7.5  PROTOCOL ANALYZERS

**protocol analyzer**
test equipment used to monitor network activity.

One of the primary digital and data communications test instruments in use today is the **protocol analyzer.** This necessary test set is used to monitor the performance and help determine problems on the local network. Many analyzers have been developed, some to test specific types of networks and others to be more universal. One of the latter analyzers is the Hewlett Packard HP4954A series Protocol Analyzer (Figure 7-9). This set has a monitor screen, keyboard, internal 20-Mbyte hard drive, and accessible 3 1/2″ floppy drive. Additionally, there are a number of LED indicators on the front panel. Connectors are provided on the back of the set to allow access to various networks. These

**FIGURE 7-9** HP4954A Protocol Analyzer *(Photo Courtesy of Hewlett Packard)*

include connections for RS232C and RS449 serial interfaces, ISDN port, X.25, and SNA (system network architecture) interfaces. An additional RS232C connector is provided for connection to a printer used for making hard copies of analyzer results. ISDN and X.25 are discussed in future chapters.

The HP4954A series was developed to be used with X.25, SNA networks, and systems that use high-level data-link control (HDLC), synchronous data-link control (SDLC), BISYNC, and asynchronous data-link protocols. The type of test being performed on the network is controlled through the keyboard using "soft keys" defined by previously written or user-developed programs. The analyzer incorporates a version of the C programming language called DataCommC to allow user-specific applications to be written and executed.

To analyze a system's performance, protocol analyzers use statistical methods to record and display the amount of data being transferred and the number and types of errors experienced during any given transmission. These errors include bit errors (parity, cyclic redundancy check [CRC], and longitudinal redundancy check [LRC]), framing errors, and loss of carrier. A portion of the data being transmitted can be captured and displayed on the screen at various selected points. These points are triggered by detection of framing elements, errors, or after a specific number of bits have been sent. The screen then displays the data, along with interpretation of certain data-linking bits or characters. When capturing data based on an error, the position of the error is shown, assisting the user in analyzing and determining the cause of the error.

Protocol analyzers are capable of data and error simulation, which allows the user to inject a specific data pattern with or without errors. Capturing the data on the receiving end, the user can determine what errors occurred, as well as where they occurred. Since the data sent are known, any deviations from the data will be actual errors. Errors can be injected to test the error-detection or error-correction capabilities at the receiver.

Figure 7-10 shows a HP4954A being used to emulate a device entry into a network. The operator can select the operation to be run on the system by keying in the selection through the keyboard or by running emulation programs from a floppy disk set into the drive. The HP4954A series analyzers are capable of emulating data terminal equipment (DTE) or data communications equipment (DCE) using any number of data-link protocols. The same types of networks that can be tested by the analyzer can also be accessed under emulation. The emulation facility aids system developers, who can test the response of the system to a device planned to be added to it. Customized applications can be tested before they are adapted to the network, saving a lot of development and network application test time.

As networks have become more complex and interwoven, the need for more complex analyzers has also risen. Much analysis is now performed using software programs that simulate network traffic and check the results of that simulation. The advantage of using this software is that network lines are uninterrupted and the analyzing "equipment" is internal to the system. It is also readily accessible and usable. Software systems can also be easily upgraded to meet the demands of evolving networks. These software analyzers can be tailored toward a particular network layout by using configuration commands that come with them.

**FIGURE 7-10** With the X.25 and SNA Network Performance Analyzer, Easy-to-Read Graphs Let You Quickly View and Assess Such Key Performance Statistics as Line Utilization, Overhead, Response Time, and Network Errors *(Photo and caption courtesy of Hewlett Packard)*

### Section 7.5  Practice Questions

1. What are the functions performed by a protocol analyzer?
2. What is the benefit of being able to emulate a network access planned for a peripheral device?
3. What are the advantages of using a software protocol analyzer program?

## SUMMARY

Data-link protocols were developed to establish a standard set of rules to govern the establishment and maintenance of a communications link. For character-oriented protocols, numerous data-link characters, such as STX and ESC, are used to define when messages begin and how the text is to be handled. Methods for error detection are also specified within the protocol. Bit-oriented protocols rely on specific groups of bits to define message beginnings and ends, definition of each message unit (frame), and a method for flow control (frame number sequencing).

## TABLE 7-2

**Protocol Summary**

| Protocol | Data Type | Data-Link Form | Transmission Size | Overhead (Nonmessage/Nongraphics) |
|---|---|---|---|---|
| Asynchronous | Asynchronous | Character | Message | Parity/Framing Bits/Data-Link Characters |
| BISYNC | Synchronous | Character | Block | Preamble/Data-Link Characters/CRC (BCC) |
| SDLC/HDLC | Synchronous | Bit | Frame | Flags/Control Field/CRC (FCS) |

Data-link protocols set prescribed rules and procedures for originating calls and for station responses. Message formats and sequences and error detection methods are also included within the protocol specifications. The four most common types of protocols are the character-oriented asynchronous and BISYNC protocols and the bit-oriented SDLC and HDLC protocols. Table 7-2 summarizes these protocols, including a brief comparison of each.

# QUESTIONS

### Section 7.2
1. Define protocol.
2. Why are data-link protocols necessary?
3. What is the difference between a poll and a selection?

### Section 7.3
4. Of the four data-link protocols discussed in this chapter, asynchronous, BISYNC, SDLC, and HDLC, which one(s):
   a) is used with asynchronous data.
   b) uses CRC-16 exclusively for error detection.
   c) uses EBCDIC exclusively as a character code.
   d) can send messages in blocks.
   e) can send messages in frames.
   f) is bit-oriented
   g) is character-oriented.
   h) makes no distinction between a poll and a selection.
   i) uses frame numbers to indicate retransmission needs.
5. Under which circumstances must a secondary respond to a primary's message?
6. What is the response, in the asynchronous protocol, to a poll if the secondary is in the send mode? the receive mode? the local mode?
7. How is bit and character synchronization achieved using the asynchronous protocol?
8. How is bit and character synchronization achieved using the BISYNC protocol?
9. What is the purpose of the following asynchronous characters?

a) DC3  b) /ACK  c) //  d) **
10. What is the purpose of the following BISYNC characters?
    a) STX  b) DC1  c) EOT  d) GS  e) SYN
11. What is a BISYNC handshake message used for?
12. What is the sequence of characters in the BISYNC handshake message?

**Section 7.4**

13. When is a 7E flag sequence recognized as a flag in an SDLC message?
14. What is used to prevent a 7E sequence from being mistaken for a flag?
15. Stuff zeros in this message wherever appropriate. The message is framed correctly with one start and one ending flag. The least significant bit is at the left and is the first bit transmitted.

> 011111100011101010110011111110000111100111111100001111110

16. In SDLC, what is indicated by a sequence of ten consecutive ones followed by a zero?
17. In SDLC, what is the meaning of more than 15 consecutive ones on the line?
18. What types of frames can be sent with SDLC?
19. Which SDLC frame type never includes a data field?
20. What are the six fields of an SDLC information frame?
21. How are information frames recognized by the SDLC protocol?
22. Which address is used in the SDLC address field?
23. What is the maximum number of stations that can be used with a SDLC protocol?
24. A SDLC frame is sent from the primary to the secondary with a control field of 10110010 (least significant bit to the right).
    a) What type of frame is it?
    b) Is it a poll frame?
    c) What is the frame number being sent?
    d) What is the frame number of the expected response frame?
25. Three frames are sent by the primary with the last frame's control field being 11010110. The secondary responds with a first frame control field of 01001100.
    a) Is this response correct?
    b) What must the primary do in response to this frame?
26. Which station in a multipoint environment sends unnumbered command messages?
27. Which station in a multipoint environment sends unnumbered request messages?
28. What is the effect of sending an unnumbered set initialization mode command? Which station does it effect?
29. How does a SIM command differ from a SNRM command in SDLC?
30. How does a SNRM command differ in the SDLC and HDLC protocols?
31. How do the fields in an HDLC frame differ from an SDLC frame?
32. How many consecutive SDLC frames can be sent without an acknowledgement from a secondary station?
33. How many consecutive HDLC frames can be sent without an acknowledgement from a destination station?
34. Contrast the number of stations that can be used with HDLC compared with SDLC.

35. Which supervisor frame is included in the HDLC protocol and not the SDLC protocol? What is that frame used for?

36. How do the receivers in a multipoint environment know when an HDLC address is completed?

37. Develop the HDLC message for a single frame with the following requirements:
    a) Information frame to station address 3CD4F9 (in hex).
    b) Frame number 39 (decimal) is being sent and frame number 78 (decimal) is expected as the next frame from the secondary.
    c) This is the final frame in the sequence.
    d) EBCDIC characters are used for the data field.
    e) The message is: HDLC is a more versatile protocol.
    f) CRC-16 is used for error detection.

38. What are the types of rejects that can be specified in an unnumbered frame reject message?

39. What is the purpose of the HDLC asynchronous response mode?

40. What can the HDLC asynchronous disconnect mode provide that SDLC protocol does not have?

41. Compare the efficiency of sending a 512-byte message using asynchronous, BISYNC, and SDLC protocols. Assume EBCDIC character code is used for each message. For the character-oriented protocols, 30 bytes are used for graphic characters (including ESC, NL, DC, GS, and attribute characters). BISYNC sends the message in three even blocks and SDLC uses three frames to complete the transmission. CRC-16 is used for error checking for the synchronous protocols. Asynchronous data uses one start, one stop, and even parity.

**Section 7.5**

42. Explain some of the uses a protocol analyzer can be put to.

43. List several advantages of a software analyzer program.

---

# ANSWERS TO REVIEW QUESTIONS

**Section 7.2**

1. Polls—inquires if secondary station has any messages to send.
   Selections—inquires if a secondary is ready to receive messages.

2. Send, receive, and off-line (or local).

3. Individual—specific station address.
   Group—addresses several, but not all secondary stations.
   Broadcast—addresses all secondary stations.

**Section 7.3**

1. Asynchronous poll:   E  D          selection:   E
                        O  C  polling              O  selection
                        T  3  address              T  address

   BISYNC uses pairs of SPA (station polling address) or SSA (station selection address).

2. Asynchronous data-link protocol uses start bit for character and bit synchronization, BISYNC uses double SYNC characters and a preamble.
   Asynchronous transmits in full messages, BISYNC in blocks
3. Asynchronous uses parity for error detection, BISYNC uses CRC algorithm.
4. Less overhead.

### Section 7.4

1. 

| Item | SDLC | HDLC |
|------|------|------|
| NS/NR Numbers | 3-bit | 7-bit |
| Address Field | 1-byte | Unlimited |
| Character Code | EBCDIC | Any |
| Supervisor Frame | 3 Modes | 4 Modes |

2. 

| Item | SDLC | BISYNC |
|------|------|--------|
| Data-Link | Bit | Character |
| Message Format | Frames | Blocks |
| Link Enquiry | Poll Only | Poll/Selection |
| Error Detection | Frame Check Sequence | Block Check Character |
| Message Start | 7E Flag | SYN SYN |
| Message End | 7E Flag | ETB, ITB or ETX |
| Transparency | Zero Stuffing | DLE Character |

3. Information frames—sequenced data.
   Supervisor frames—maintain link without data.
   Unnumbered frames—commands and requests.
4. Zero insertion or stuffing.
5. Unnumbered frame reject used to resolve problems that occur that cannot be handled using supervisor frame reject mode.

### Section 7.5

1. Monitor system performance and analyze problems and data errors.
2. Analyze response of system to peripheral before it is included in system.
3. Network lines are not interrupted and analyzing "equipment" is internal to system.

# CHAPTER 8

# Network Architecture and Protocols

## OBJECTIVES

After reading this chapter, the student will be able to:
- Understand how local area networks are configured and specified.
- Apply the IEEE 802.3 protocol to a bus network.
- Apply token ring protocol to ring networks.
- Describe how network interface cards allow PCs to have access to networks.
- Interconnect networks using bridges, gateways, routers, and servers.

## OUTLINE

8.1   Introduction
8.2   Networks by Size
8.3   IEEE 802.3 and Ethernet
8.4   IEEE 802.4 Token Bus
8.5   IEEE 802.5 Token Ring
8.6   Network Interface Cards
8.7   Interconnecting LANs
8.8   IEEE 802.6 Metropolitan Area Network
8.9   X.25 Packet Switch Protocol

## 8.1 INTRODUCTION

**instilude for electronic and electrical engineers (IEEE)**

Several organizations have become responsible for generating the numerous standard specifications used in the data communications and networking field. One such organization, the **institute for electronic and electrical engineers (IEEE)** created the 802 committee in 1980 to draft standards for local area networks (LANs). By 1984 they

257

generated their first standard based on Xerox and Digital Equipment Corporation's (DEC) Ethernet bus network. This standard is known as either IEEE 802.3 or CSMA/CD, for carrier sense multiple accesses with collision detection (more on this standard later). Within the 802 group are many subgroups, each responsible for developing standards for specific LAN types or subgroupings. These groups and the titles of their responsibilities are:

802.1: Spanning Tree standard for bridges used to interconnect similar LANs based on the OSI model.

802.1a: Network management architecture.

802.1b: Network management protocols.

802.1D: Bridging standard.

802.1d: Source routing standard.

802.1g: Remote bridge standard for wide area networks (WANs).

802.1p: Prioritization of MAC layer bridges.

802.1Q: Virtual LANs (VLANs).

802.2: Logical Link Control (LLC).

802.2i and j: Acknowledged connectionless LLC service.

802.3 and above contain Media Access Control (MAC) specifications.

802.3: Carrier Sense Multiple Access with Collision Detection. (CSMA/CD) bus network (Ethernet).

802.3i: CSMA/CD over twisted-pair and fiber optics.

802.3u: Fast Ethernet.

802.3x: Gigabit Ethernet flow control.

802.3z: Gigabit Ethernet.

802.3ab: Gigabit on category 5 UTP cabling.

802.4: Token Bus Network.

802.5: Token Ring Network.

802.5j: Fiber optic token ring.

802.6: Metropolitan Area Network (MAN).

802.7: Broadband Networks.

802.8: Fiber optic networks (covered in Chapter 12).

802.9: Integrated voice and data service (ISDN) (covered in Chapter 10).

802.9a: Isochronous Ethernet (ISONET).

802.11: Wireless LAN.

802.12: Fast Ethernet.

802.14: Integrated services over private cable.

802.16: Broadband Wireless Access.

In this chapter, we will examine a number of these specifications as they apply to networks in use today. Other sections of the 802 group are covered in later chapters that are concerned with specific areas of networking. This chapter starts with the oldest, and still highly popular, local area network specified in IEEE 802.3, the CSMA/CD or Ethernet network. This is followed by discussions on other IEEE 802 specifications that are not included in other chapters.

# 8.2 NETWORKS BY SIZE

**local area networks (LANs)**

Originally, the 802 committee was formulated in 1980 by IEEE to develop standards for **local area networks (LANs),** which comprised a number of communicating stations interconnected into a single system. This system is small and usually contained within a single building. As networks grew in application, additional subcommittees under the 802 heading were formed to create additional standards. These standards included applications for **metropolitan area networks (MANs)** and **wide area networks (WANs).** A MAN is a system that combines several LANs interconnected over a limited area such as a city or regional area. Wide area networks (WANs) extend the coverage to large sections of a country. As networking became involved in worldwide coverage, additional network classifications appeared. The most common classification for a world wide network is the **global area network (GAN).** One last network classification has become popular by use to accommodate networks that were larger than LANs but not quite large enough to be truly a MAN. Since these networks generally interconnected various sites on college and university campuses, they became known as **campus area networks (CANs).**

**metropolitan area networks (MAN)**

**wide area networks (WANs)**

**global area network (GAN)**

**campus area networks (CANs)**

A network that interconnects work stations and personal computers (PCs) within a single manufacturing plant is a good example of a LAN. A network interconnecting library, student resource areas, classrooms, and other facilities within a single college campus illustrates the makeup of a CAN. MAN as aptly represented by a network of automated teller machines (ATMs) for a local banking concern. A case can also be made for extending the ATMs into neighboring states for larger banking concerns, forming a WAN. One very common application of a GAN is the well-known Internet or World Wide Web (WWW) network.

## Logical Link Control

A subcommittee of IEEE 802, designated as 802.2, took on the responsibility of formulating a basic set of standards for address allocations and frame formatting of messages used for networking. This committee developed two sublayers to address these concerns, the **logical link control (LLC)** and **media access control (MAC)** sublayers. The LLC layer is designed to enable the exchange of data across a transmission channel controlled by a **media access unit (MAU).** LLC and MAC work together to formulate an ISO data-link layer protocol, which enables network layer traffic the ability to deliver reliable, errorfree information across an established communication link.

**logical link control (LLC)**
exchange of data across a channel.

**media access control (MAC)**
data-link control across media.

**media access unit (MAU)**
controls data exchange on LLC.

The responsibilities attributed to the LLC sublayer are those outlined for the data-link layer, and include establishing and terminating a communication link, supplying a frame format for the payload, detecting and correcting errors, and maintaining control over the traffic flow or dialog between the end stations in the link.

The establishment of a communication link at the LLC level can be made in one of two ways—via either connection or connectionless services. When using a connection-oriented service, denoted as LLC1, the data-link connection is established before data is sent between the communicating stations. This allows transfer of data over an established connection. In contrast, a connectionless service, known as LLC2, establishes the connection path as the first frame is being transmitted. Address information within the frame is checked by the media access units, which then determine the best available path to complete the connection between sending and receiving stations. Once the path is determined, further communication uses it until both stations are finished sending and receiving data.

Connection service operation reduces overhead in the message frames, but requires a dedicated connection be established and maintained even when there is no traffic. Connectionless service operation requires additional overhead to assist in determining the communication path taken by the frames, but it does allow transmission paths to be released for other use when there is no more messages to transmit.

The basic format of a logical link control frame is shown in Figure 8-1a. It is no mistake that it closely resembles an HDLC format, because LLC is a data-link level protocol. Destination and source addresses, labeled **destination service access point (DSAP)** and **source service access point (SSAP)** define the location of the end points in communication link for either connection-oriented or connectionless service. Sending and receiving sequence numbers, in the information and supervisor control fields of the LLC frame, serve the same purpose that NS and NR numbers did in the HDLC protocol—they maintain frame sequencing. The remainder of the control field identifies the frame type in the same manner as HDLC.

If the LLC frame is used to encapsulate a higher-level protocol within the payload field, 802.2 provides for a means, known as the *sub-network access protocol (SNAP)*, to identify that protocol. SNAP formats are recognized by DSAP and SSAP values of 10101010, or AA in hexadecimal. In response to these addresses, the LLC protocol will interpret the payload as a SNAP format. The first three bytes of the SNAP header (Figure 8-1b) are a organization code established by the 802.2 specification. This is followed by a 2-byte type code that identifies the protocol used in the data area of the payload. Actual source and destination addresses are included in the protocol format of the data area in the payload.

**destination service access point (DSAP)** and **source service access point (SSAP)**
end points in communication link.

## Media Access Control

The media access control (MAC) sublayer defines how different stations can access the transmission medium. It takes into consideration, the **topology** in use (bus, ring, star, etc.), and the protocol being used. It includes data-link functions of addressing, error detection, and framing. The LLC and MAC fields are combined with data and additional fields required by an upper-level protocol to form the final message format. They are used to define the use of that message, its source and destination stations, and the networks that message will travel through to complete the data-link.

**topology**
physical or logical description of a network. Comes in three basic forms: bus, ring, or star (hub).

### Section 8.2  Review Questions

1. What was the original purpose for forming the IEEE 802 committee?
2. Describe the differences between LAN, WAN, MAN, CAN, and GAN.
3. Which OSI model layers does the 802.2 standard apply to?

| DSAP 8 | SSAP 8 | Control Field 8: Unnumbered 16: Numbered | Information Field 8m | CRC–32 |
| --- | --- | --- | --- | --- |

DSAP–Destination Service Access Point

| I/G | E | D | D | D | D | D |
| --- | --- | --- | --- | --- | --- | --- |

I/G: 0 = Individual   1 = Group   D: Destination Address
E: 0 = Normal Address   1 = IEEE Defined Address

SSAP–Source Service Access Point

| C/R | E | S | S | S | S | S |
| --- | --- | --- | --- | --- | --- | --- |

C/R: 0 = Command   1 = Response   S: Source Address
E: 0 = Normal Address   1 = IEEE Defined Address

Information Field: m is an Integer

CRC–32 = 32 Bit Cyclic Redundancy Check

a

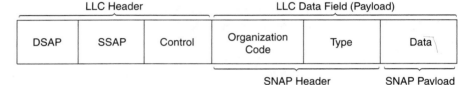

LLC Header | LLC Data Field (Payload)

| DSAP | SSAP | Control | Organization Code | Type | Data |
| --- | --- | --- | --- | --- | --- |

SNAP Header         SNAP Payload

DSAP = Destination Service Access Point
SSAP = Source Service Access Point
SNAP = Sub-Network Access Protocol

b

**FIGURE 8-1** IEEE 802.2 Frame Format (a), LLC SNAP Format (b)

## 8.3  IEEE 802.3 AND ETHERNET

Ethernet is a specification for a local area network using a bus topology. It was developed as a joint effort by Xerox Corporation, Intel Corporation, and Digital Equipment Corporation in 1973. The IEEE 802.3 committee formalized a standard based on, and closely resembling, Ethernet called **carrier sense multiple access with collision detection (CSMA/CD).** The standard details specifications for a bus network in reference to the first two layers of the OSI model.

Originally, the Ethernet system was a local area network (LAN) designed to exist within a **bus** topology using 50-Ω coaxial cabling to interconnect the network stations. The cable would run the needed distances within and between a building's rooms and floors. Today, many Ethernet systems incorporate fiber optic and unshielded twisted-pair copper cables to link stations together. This section will concentrate on the original

**carrier sense multiple access with collision detection (CSMA/CD)**
Ethernet bus access.

**bus**
system topology that uses a common set of lines that each station taps into.

coaxial cable application of Ethernet. Improvements to the original standard are examined in later sections of this chapter.

## IEEE 802.3 Physical Layer

**nodes**
local area network access point.

Figure 8-2 illustrates an IEEE 802.3/Ethernet topology. Stations at specific locations within the Ethernet network, called **nodes,** are attached to the coax bus by using taps. Specifically, a coax segment of 500-meters maximum length can support up to 100 nodes. These nodes are to be no closer than 2.5 meters apart within a segment to reduce the possibility of standing wave signals on the cable. Local repeaters are used to interconnect two 500-meter segments. The maximum distance, from end to end, between any two nodes on the network is specified at 1,500 meters, using these local repeaters. An additional untapped 1,000 meters can be used between segments with remote repeaters to extend the maximum end-to-end distance between nodes to 2,500 meters. Repeaters are devices that regenerate signals that may have been attenuated as they travel on the bus. Figure 8-2 shows how these limits are applied to an Ethernet LAN. L is used to designate a local repeater, R for a remote repeater, and T for a tap or node location.

Both the Ethernet and 802.3 specifications state that a maximum of 1,024 nodes, are permitted on a single Ethernet network. At this point you must realize that these nodes could be used to connect to devices that would allow other networks to be connected to an Ethernet network. We will explore these devices later in the chapter.

---

### EXAMPLE 8-1

What is the maximum number of nodes that can be serviced by a basic Ethernet system using a single remote and a single local repeater?

### SOLUTION

Three tappable segments can be interconnected along with an additional untapped 1,000-meter segment as shown in Figure 8-2. Each tappable segment allows 100 nodes to be attached to it for a grand total of 300 nodes. Three nodes are used to connect to the repeaters, leaving 297 nodes available to connect to network stations.

---

Two basic types of coax cable are specified in the IEEE 802.3 standard for the Ethernet network. They are designated 10BASE5 or thicknet and 10BASE2 or thinnet. The 10 in each designation indicates that the cable is designed to carry data at a 10-Mbps rate as defined in the 802.3 specification. Thicknet is used to connect the nodes together and for untappable extender segments. Thinnet is used between the node taps on the thicknet cables and the stations connected to the network at a maximum distance of 185 meters between node and station. Thirty stations can be connected to a thinnet cable serviced by a single node. Thicknet can manage 100 nodes per 500-meter segment as stated in the previous paragraph, with a minimum of 2.5 meters between nodes.

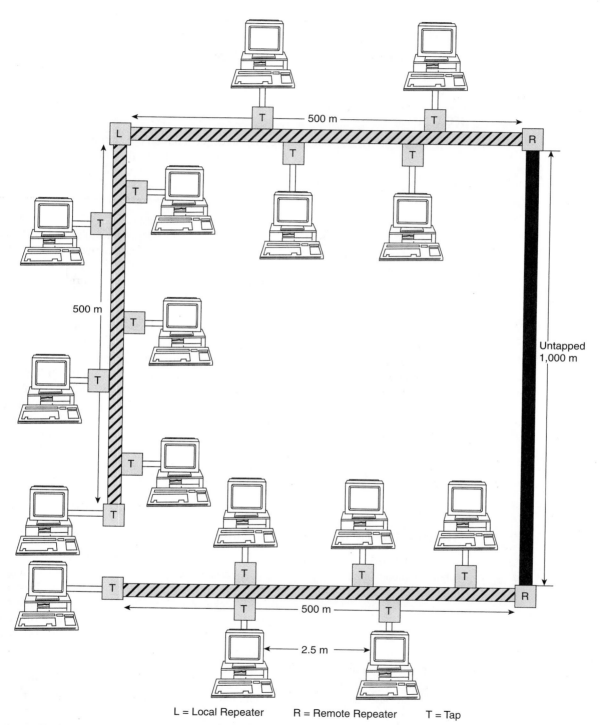

L = Local Repeater    R = Remote Repeater    T = Tap

**FIGURE 8-2** Ethernet Topology Using Coaxial Cable

**EXAMPLE 8-2**

Each station connected to the nodes in Example 8-1 is replaced with a thinnet cable supporting the maximum number of stations. What is the total number of stations in the network?

**SOLUTION**

In Example 8-1, 297 nodes were used to connect directly to stations. If these were all replaced with thinnet cables, each supporting 30 stations, then the total number of stations would jump to $30 \times 297 = 8,910$. This would violate the maximum limit of 1,024 set by the 802.3 specification. We will see why the limit is imposed shortly.

A subcommittee, IEEE 802.3i was convened to add fiber optic cables and unshielded twisted-pair (UTP) copper cable specifications to the CSMA/CD network standard. UTP cables are known as 10BASET and have a maximum length of 100 meters per segment. Fiber optic cables with larger bandwidths and less attenuation concerns are used to a maximum length of 4.5 kilometers (KM). The different types of cabling are illustrated in Figure 8-3.

Because of the higher bandwidth associated with fiber optic cables, Ethernet networks employ them as backbones to interface several networks into a single system. A backbone is an interconnecting cable or system whose sole purpose is to deliver traffic from one network to another over a specified distance. This transfer is done at higher data rates than found on the networks themselves. This is to assure that internetwork traffic does not present heavy delays to the delivery of messages. There are no stations connected to a backbone. The end nodes are connected to the devices that collect the traffic and deliver it to the network. When employed in this manner, the fiber optic portion of the network is configured as a star or hub topology, while each connecting network maintains a bus topology as shown in Figure 8-4. The hub is a centralized Ethernet switch that directs traffic amongst the bus networks connected to it.

## Manchester-Encoded Ethernet Data

Ethernet networks send digital data on the bus at a rate of 10 Mbps using an encoding scheme called Manchester encoding. Manchester encoding ensures transitions (changes of logic level) for every bit sent to aid in the job of recovering the synchronizing clock by the destination station. The data stream and a synchronizing clock signal are combined, as shown in Figure 8-5, to produce a mixed data stream that includes the inverse of the data bit state in the first half of each bit period followed by a clock transition midway and a clock level for the remaining time of the bit time. In the data stream shown in Figure 8-5, the first data bit in the original stream is a logic 1.

As shown below in the Manchester-encoded stream, the logic level for the first half of the original bit time is low. At the midpoint of the first original data bit, the Manchester-encoded stream transitions to a high state and remains that way until the next data bit

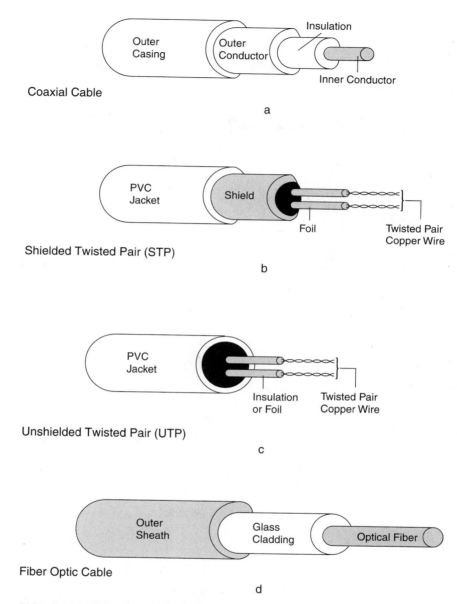

**FIGURE 8-3** Cable Types

begins. The second bit is a 0, so the Manchester stream shows a high level for the first half of the bit time, then a transition in the middle to a low state, where it remains until the third bit begins. Notice for each of the original data bits, that there is a transition in the Manchester-encoded stream midway through the bit time. This insures that a clock can be recovered at the receiver by using these midpoint transitions.

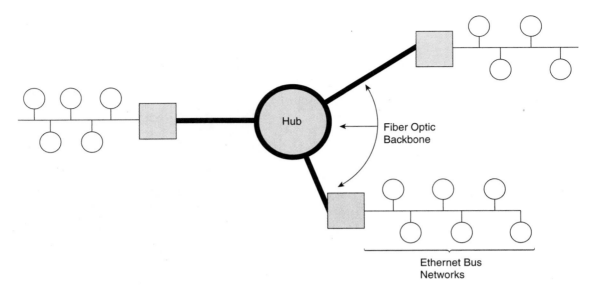

**FIGURE 8-4** Ethernet Fiber Optic Backbone

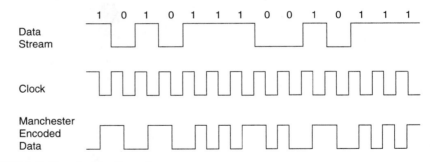

**FIGURE 8-5** Manchester Encoding

Figure 8-6*a* is a circuit for generating a Manchester-encoded data stream. The original data and synchronizing clock are fed into the two inputs of an exclusive OR gate. The high level of the clock signal causes the output of the exclusive OR to be the inverse of the input. Since the clock changes state to complete its alternation, the exclusive OR output will change state also, right at the midpoint of the data input.

Figure 8-6*b* is the decoder for a Manchester stream of data. As you can see, it isn't any different from the encoder, simply another exclusive OR! The encoded data and the recovered synchronizing clock are fed to the exclusive OR inputs. Here, the logic function of the exclusive OR works to recreate the original data stream that was decoded by the source station.

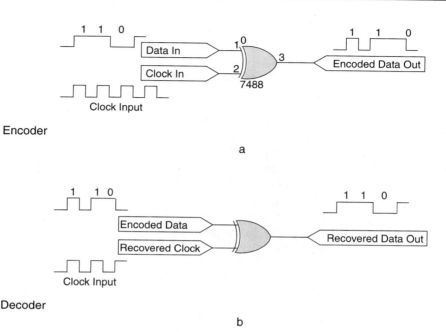

Encoder

a

Decoder

b

**FIGURE 8-6** Manchester Circuits

## Access to an Ethernet LAN

**baseband**
systems that use a single
communication channel.

**broadband**
system that uses multiple
channels on a single
communication line.

Ethernet uses a **baseband** half-duplex exchange at a 10-Mbps rate. Baseband designates a single channel system in contrast to a multichannel **broadband** system. All Ethernet nodes have equal access to the common single channel configured as a bus topology. This access is achieved using a system called carrier sense multiple access with collision detection or CSMA/CD. Stations desiring access to the network monitor the bus for the presence of a carrier signal from another station for a minimum time period equivalent to a round-trip propagation time between nodes at the maximum distance used on the network. This is the time it would take a signal to travel from a node to the furthest end of the network and back again. For the maximum allowed distance on an Ethernet bus, this propagation time is 8.6 microseconds ($\mu$S). This time period assures that no signal is present for a round-trip propagation time between stations 2,500 meters apart. The time would be less for networks whose furthest nodes are closer than the maximum distance allowed.

If a carrier is sensed, the station desiring access to the bus must continue to monitor and wait for the bus to be free. Once the bus is detected as being free of any carrier signal for the 8.6 $\mu$S period, the monitoring station can then begin to access the line. The station continues to monitor the line for an additional period equal to the round-trip delay, after it begins to transmit its message to detect the possibility of a collision with another station's transmission.

There is the possibility that more than one station has been waiting for use of the bus. In that case, two transmitted signals may be placed on the bus at the same time once both stations sense the bus is clear of any carrier signal. The two stations have equal right to access the bus once it is free. Eventually, these two signals will meet and collide causing both messages to become garbled. The sending stations have been monitoring the bus for just this possibility.

Once the collision is detected, both stations back off and send a jamming signal, composed of 8 bytes of alternating 1s and 0s, onto the bus. The jamming signal indicates to all stations on the network that a collision has occurred and that these two stations are going to attempt to regain access to the bus. The jamming signal is detected by the other stations as an active carrier, preventing them from gaining access to the bus until the collision is resolved. The two stations that experienced the collision wait a random amount of time before attempting to regain access to the network. The randomness of this delay very nearly assures that another collision will not occur. One station will access the bus before the other.

The amount of time delay is determined by an algorithm called *truncated binary exponential K back off* and is equal to a value from 0 to $2^K$ time slots. $K$ is the current number of attempts by a station to gain access to the bus. This value has a maximum of 10, even though a station is allowed 16 tries before it is disconnected from the network. A time slot is the time it takes for a signal to propagate to the end of the network and back. For Ethernet, the period is 51.2 $\mu$S. After timing out, a station must again monitor the bus for 8.6 $\mu$S to determine whether the bus is free of the other station it collided with. The timed-out station can attempt to seize the bus and transmit its data.

Since the time delay is random, there is still a possibility that both time delays could be close enough to cause a second collision. In that situation, the process is repeated in the hope that the randomness of the delay time would not allow a third collision. In some applications, the delay can be less random to allow for a priority to be established. In that event, after the collision is detected and the colliding stations have backed off, the station with the higher priority will time out long before the other station. The higher priority station then is guaranteed access to the network. This does reduce the sense of "peer" stations on the network, since the nature of being a peer system is that all stations have equal initial access rights to the bus.

To reduce the problem of bus contention and aid in keeping the back-off algorithm within a reasonable time frame, 802.3 and Ethernet specifications limit the total number of stations on a single network to 1,024 units.

## The Data-Link Layer

Specifications for CSMA/CD are applicable to OSI model physical and data-link layer. For the data-link layer, Ethernet uses a form of the HDLC protocol as the basis for its frame format as shown in Figure 8-7. The IEEE 802.3 committee settled on a similar format for the CSMA/CD frame shown in Figure 8-8. Both frame formats begin with a preamble sequence comprised of seven bytes of sequential binary bit patterns of alternating ones and zeros. Each four-bit pattern is represented by the hexadecimal character, A, which is a binary sequence of 1010. The eighth byte of the preamble, which the 802.3 committee designated as a start frame delimiter (SFD) is AB or 10101011. The preamble

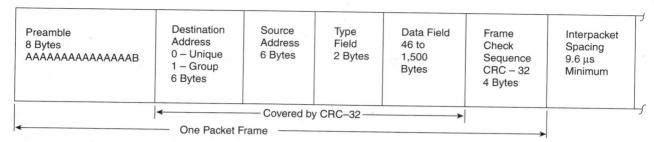

**FIGURE 8-7** Ethernet Frame

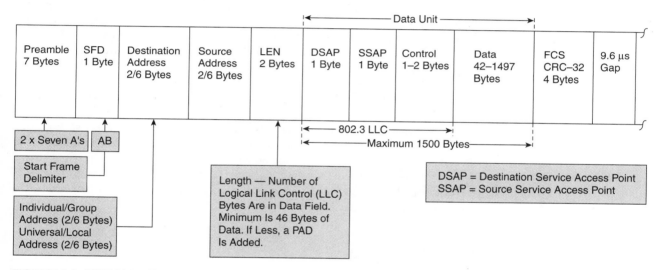

**FIGURE 8-8** IEEE 802.3 Frame

is used to establish clock recovery and bit/character synchronization for synchronous data used on the Ethernet bus network.

The preamble is followed by two address fields, one for the destination location of the message and one for the source location. The use of two addresses is facilitated by the unlimited size of an HDLC address field. Two addresses are required because all stations on the Ethernet network are peer stations and have equal access to the bus. Since any station can initiate a data link, it is necessary to identify both the source and destination locations for that link. Ethernet frames allocated 6 bytes each for both addresses while 802.3 specifies 2 or 6 bytes, depending on the type of address in use (individual or group).

The 2-byte length (802.3) or type (Ethernet) field is an extension of the concepts of the HDLC control field. In addition to the functions of identifying frame types (information, supervisor, or unnumbered) and frame numbers, the type/length field also contains information identifying which one of 247 possible data formats, selected for a specific application, is being used in the data field.

The data field itself can contain from 46 to 1,500 bytes of information depending on the application in use. The 802.3 specification identifies a number of subfields within the information field for logical link control (LLC) functions. They are a destination and source access point and control field. The access points identify subnetworks within a multiple Ethernet system that uses some form of interconnecting backbone such as the fiber optic network briefly mentioned in the previous section.

The last field in the Ethernet and 802.3 frames is the frame check sequence field that uses a cyclic redundancy check or CRC-32 error detection method to monitor the frame between the end of the preamble and the start of the CRC-32 field itself. The constant used for this CRC-32, in hexadecimal, is 104C11DB6. A minimum of 8.6 microseconds spacing is set aside as a gap between frames to allow the receiving station time to compute and verify the CRC-32 frame check sequence.

Messages are divided into smaller units called *packets* and encapsulated into the Ethernet or 802.3 frame. The sending station is responsible for selecting source and destination addresses as well as access points in the case of a multinetwork environment. It then follows the procedure outlined for accessing the network bus and eventually sends the packets out toward the receiving station.

Frame numbers within the type or length field are used by the receiving station to reassemble the packets into a complete message. The use of packets allows several stations to use the bus in a shared time manner. One station might send two packet sections of a message and then yield use of the bus to another station that would send a number of packets before yielding the bus for another station's use. Eventually, the original station will regain access and send additional packets. Determining the maximum number of packets or available access time is incorporated in the management software used by the stations connected to the network.

## Ethernet Interface Device

**serial interface adapter (SIA)**
Manchester encoder/decoder chip.

Advanced Micro Devices (AMD), located in northern California, has developed and marketed a chip set that manages and interfaces to an Ethernet network. One unit of the chip set is the **serial interface adapter (SIA),** which functions as a Manchester encoder/ decoder and collision detection mechanism. The encoder operates with a crystal-driven, 20-MHz oscillator (Figure 8-9). The logic within the oscillator block also divides the 20 MHz by 2 to produce a 10-MHz clock (TCLOCK). The two clocks create transitions within each data bit to satisfy the Manchester coding format. The Manchester encoder combines the transmit data (TX data) with the clocks to produce the Manchester-encoded signal. The encoding process continues as long as transmit enable remains high (usually the duration of the data stream). Figure 8-10b shows the transmit timing diagram for the serial interface adapter. Data are sent out differentially on transmit (XMT) + and – lines. The transmit select signal determines whether transmit mode 1 or its inverse, transmit mode 2, is being used. In mode 1, the idle time voltage between XMT +/– is V, while in mode 2, it is a high (logic 1) level.

The receive side of the serial interface adapter includes a carrier detect and Manchester decoder circuits. Incoming data on the differential receiver (RCV) + and – inputs are fed to both circuits. The front of the decoder circuit uses a phase lock loop (PLL) to cap-

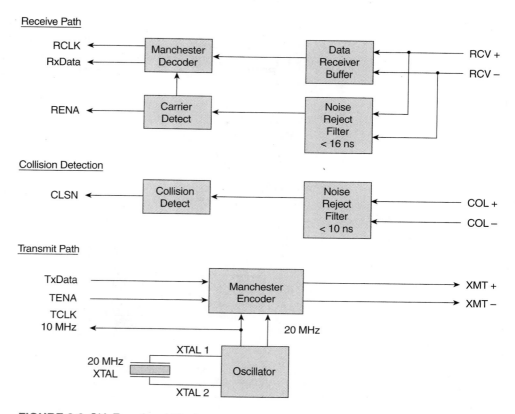

**FIGURE 8-9** SIA Functional Blocks

ture clock (RCLK) from the preamble within 6 bit times. Sensitivity to the front end is such that a differential signal greater than −300 mV is adequate for detection. As long as the signal does not drop below −175 mV once detection and clock recovery have begun, the receiver will detect the incoming data stream and output the decoded data on the RX line. A carrier detect circuit generates a receive enable (RENA) to indicate that the receiver is detecting the data. Transitions less than 16 ns are rejected by the carrier detect circuit to prevent false starts due to line noise. The relationship between the signals is illustrated on the receive timing diagram of Figure 8-10a.

**collision detection signal (CLSN)**

A third functional block within the serial interface adapter receives a collision indication on COL + and − and generates a **collision detection signal (CLSN)** to the processor in the system. Again, signals on the differential input must be greater than −300 mV to start and not fall below −175 mV. Transient signals less than 10 ns are rejected by a noise reject filter preceding the collision-detection circuit (Figure 8-9). CLSN remains active as long as the collision-jamming signal remains on COL +/−.

Another unit in the AMD chip set is the actual processor called a *local area network controller for Ethernet (LANCE),* which creates the Ethernet packet form, computes the CRC-32 frame check sequence, and responds to the enable and collision-detection signals

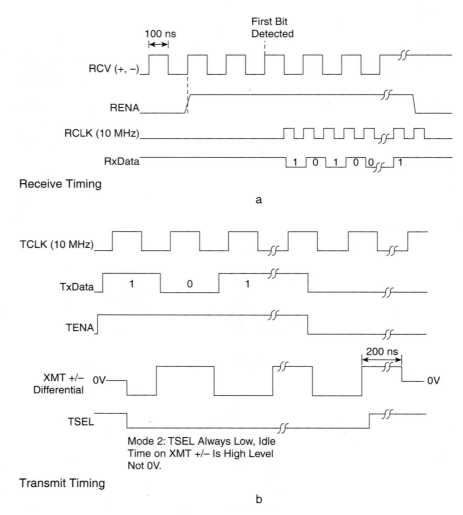

**Receive Timing**

a

**Transmit Timing**

b

**FIGURE 8-10** SIA Receive and Transmit Timing

from the serial interface adapter. The processor also strips the data from the received string and checks the CRC-32 of the received data for errors.

## Fast and Gigabit Ethernet

The basic Ethernet or IEEE 802.3 specification, based on coax bus topology and CSMA/CD access, was written during the early days of network technology. Portions of that specification have become obsolete or too restricting for the changes in technology that have occurred since the inception of the IEEE 802 committee. To facilitate improvements in networking technology, more subcommittees have been formed to adapt older networks to newer technologies, and thereby, creating new and current networks. One such committee, 802.12 drew up specifications for a 100-Mbps Ethernet network capable of operating at long distances using fiber optic cables.

Basic CSMA/CD or Ethernet networks transmit and receive data at 10 Mbps and operate over coax cable. Other medium such as unshielded and shielded twisted pairs of copper wire and fiber optic cables have been added by the 802.3i specification. A separate subcommittee, the 802.12 group considered increasing the data rate of Ethernet transmissions to improve the overall data throughput (the amount of data sent and received) on the network based on *fast Ethernet* technology. The result of this committee's work was to draft a standard for 100-Mbps Ethernet networks.

The faster rate resulted in changes in cable specifications, network topology, and access methodology. Attempts were made to retain CSMA/CD over twisted-pair cabling under a specification designated 100BaseT. 100 is used to specify the 100 Mbps data rate and BaseT to indicate the use of twisted-pair wiring. Limitations of 250 meters for the network span with no more than 100 meters between stations forced 100BaseT networks to be small. To increase the network size, additional specifications were added by the 802.12 committee, to allow 500-meter multimode or 2-kilometer single mode fiber optic cable to be used. Discussion on the details of fiber optic cable appear later in Chapter 11.

**star**
topology that uses a central controlling station called a hub to direct traffic on the network.

To improve access to the network, the 802.12 committee adopted an alternate access method and topology. A **star** topology with a 4,000-foot maximum diameter is specified to assist in controlling the access to the network. Collisions at 100 Mbps occurred more often than at 10 Mbps, influencing the 802.12 committee to alter the access method from a collision-detection to a collision-avoidance operation. The hub of the star determines if there is activity on the bus portions of the star spokes and controls which station has access to a given bus segment at any one time. The hub uses a round robin arbitration scheme, which effectively maintains each station as a peer station still having equal access to the system. This is the basis for the **carrier sense multiple access with collision avoidance (CSMA/CA)** access scheme used for 802.12 fiber optic fast Ethernet networks.

**carrier sense multiple access with collision avoidance (CSMA/CA)**

A subcommittee under the 802.3 blanket, called 802.3z has drafted a specification for a faster gigabit Ethernet network. The original idea behind this version of the popular network protocol was to develop a compatible system for an Ethernet backbone operating at half duplex. This backbone would interconnect numerous 10- or 100-Mbps networks together, transferring packets between them at a gigabit rate. This collapsible backbone connects several hubs into a central high-speed repeater so that traffic can be moved between the hubs quickly.

One restriction the higher data rate imposed on the backbone was a 18-meter distance limit. Additionally, because of the difference in data rates between the subnetworks and the backbone, the backbone carrier would have to remain on for longer periods of time than required by the message length to assure that the receiving hub or node absorbed the entire message. This is referred to as *carrier extension* and it wastes bandwidth required to accommodate the additional carrier time. One proposal to overcome this waste is to transmit packets on the backbone in bursts. The first packet would use carrier extension, but immediately following packets would not since they would occupy each other's extension time. This does require more management capabilities to sort out packets at the destination nodes.

A second proposal made by the 802.3z committee is for a full-duplex, point-to-point switched gigabit Ethernet network. It would use optical fibers to connect backbone switches and high traffic network controllers or servers at distances up to 2 kilometers. Incoming and outgoing messages would use separate cabling to facilitate full duplex operation and avoid collisions.

In general, the 802.3z committee's work centered on a specification for the gigabit Ethernet to run over various types of media. They also set requirements for encoding, code conversion where needed, and various media interface schemes. The work has thus far resulted in a three-tiered hierarchy for Ethernet switching involving interfacing 10-Mbps, 100-Mbps and gigabit Ethernet networks. The lowest tier specifies 10-Mbps Ethernet switches uplinking to a 100-Mbps hub used to connect work group clients and servers to the network. The middle tier uses 100-Mbps Ethernet switches to uplink to a gigabit hub to connect interdepartmental servers and lower tier switches. The final tier connects gigabit Ethernet networks to other networks to connect enterprise servers and second tier switches to the gigabit backbone and to other systems.

Ethernet has come a long way from its inception as the first local area network to assume a dominate role in data communications networking. It remains the most popular network protocol for LANs and for internetworking using switches and backbone interfacing.

### Section 8.3  Review Questions

1. What is used as the medium for the original 802.3 specification?
2. What is the maximum length of tappable and untappable segments?
3. What types of medium were added to the CSMA/CD specification by the 802.3i amendment?
4. What is the benefit of using Manchester encoding?
5. What must a station do first before it can gain access to an Ethernet bus?
6. When a collision is detected, what do both stations do immediately?
7. What is the purpose of the back-off algorithm?
8. Which data-link protocol is used as the model for an Ethernet frame?
9. What is the purpose of the preamble?
10. How does the 802.3 frame compare to the 802.2 LLC frame?
11. What data rates do the 802.12 and 802.3z specifications address?
12. How does CSMA/CA differ from CSMA/CD?
13. What is the chief use for gigabit Ethernet?

## 8.4  IEEE 802.4 TOKEN BUS

**token**
short message that precedes transmission of data packets on a token ring or bus network.

An alternate method used to access a bus network is outlined in the IEEE 802.4 specification. It employs **token** passing to control which station has the right to bus access. A token is a short message that specifies the station currently using the network and the next station that has access to the network after the current station is finished using it.

While a station possesses the token, it can send messages out on the network and read messages already present. Tokens may be held for a maximum specified time before they must be passed on to the next succeeding station. The procedure as specified in 802.4 follows this process:

1. Each station knows the predecessor and successor station address in the token sequence.
2. The token is passed from station to station in descending order.
3. The station currently possessing the token adds its traffic to the data stream. It then modifies the token to include the successor's station address and the token is passed to it.
4. After sending the token frame, the sending station monitors the line to detect whether the successor station has received the token. This is verified by the successor station sending a valid frame following the token. This frame can be an information frame or a modified token frame. The latter case occurs if the successor station has no messages to add to the existing information frames. In that case, the successor station merely modifies the token and passes it along to its successor station.
5. Each station that receives the token checks the messages attached to it and reads any messages assigned to that station. After the message is read, a flag is set signifying that the station got its message, the token is remodified, and passed along to the next station.
6. The token is passed to each station in the sequence until it returns to the original station. Once the original sender detects its message returned, it checks to see if the destination station for that message has read it, and if so, removes that message frame, adds any new ones, reads any messages destined for it, again modifies the token, and passes it along as it did before.

This process is repeated as long as the system remains active. If the source station detects that the destination station did not read the message, it leaves the message in the queue, adds any new messages, modifies the token, and passes all of it to the next station. After three consecutive appearances of an unread message, the source station assumes that there is a problem with the destination station and removes the message from the queue.

## Fault Recovery

There is the possibility that the sending station that passed the token does not detect a valid frame following the token frame sent. In that situation the sending station attempts to retransmit the token to the successor station in the hope that the problem is transient. If the successor station again fails to return a valid frame, the sending station sends a "who follows" frame containing the successor's address. The next station in the sequence following the intended successor station responds with a "set successor" frame, which contains its address in the data field to allow it to become the next station in the sequence, bypassing the nonresponding station. The sending station now modifies the token with this address and passes the token to the new station, bypassing the original successor station. This is one method of *fault recovery,* which maintains network communications despite the failure of a particular station.

The "who follows" frame is tried a second time if there is no response to it. Failure to receive a response the second time causes the sending station to try a new tactic by sending a "solicit successor" frame. This frame is also sent at regular intervals by each station on the network. It is used to allow a new station to enter the token sequence. Any operational station responding to the solicit successor frame will receive an amended token. It is possible for numerous stations to respond to the solicit successor frame, particularly if it was sent because of a lack of response to who follows frames. In this event, a resolve contention sequence is initiated, causing all responding stations to wait for a random period of time before responding a second time. This is repeated until there is only one station left responding to the solicit successor frame. That station becomes the new successor frame and the token along with its appended traffic is sent to it.

### Token Bus MAC Frame

Figure 8-11 details the token bus formats for media ccess control (MAC) frames, which are specified in IEEE 802.4 to indicate how tokens are formed to allow access to the medium and how data are included following the token. There are two basic formats, one for a **free token** without data and one for an information frame that follows a modified token. As with most local area network protocols, these frames are bit-oriented and very loosely follow HDLC format guidelines.

**free token**
short packet with no data packets attached to it.

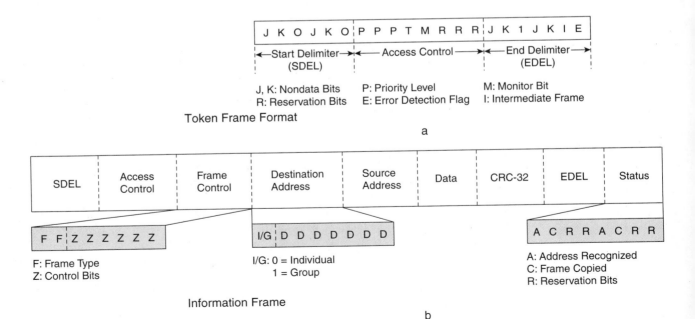

FIGURE 8-11 Token Bus Frame Formats

The token consists of an access field surrounded by beginning and ending delimiters (flags). Within the access field, the first 3 bits establish the priority level of the token. There are eight priority levels (000–111). Stations with equal or higher priority can accept and pass the token, while lower-priority stations must wait until a lower-priority valued token is passed to them. Lower priorities are assigned by a higher-priority station when it modifies the token before passing it along. The monitor bit (M) assures that a frame doesn't continuously circulate in the system. A station receiving its own message with the monitor bit set, recognizes that the frame has been received by the intended destination station, removes the message from the data stream, and resets the monitor bit in the token. Reservation (R) bits are used for requesting token access at the present priority level. The token frame is followed by the information frame, which contains detailed data about the information being sent. It begins with a preamble containing data about the frame type(F bits) and control information (Z bits). The frame types are listed in Table 8-1.

Destination address, which can be individual (first bit low) or group (first bit high) and source address fields are next in the frame sequence. Actual information is next, followed by the frame check sequence, which uses CRC-32 process for error detection. The last field in the media access control frame is the frame status field. Bit A in this field, called the address recognized bit, is set when the receiving station acknowledges that this frame was received and copied. The originating station can then remove the frame from the data stream upon detecting the set state of this and the monitor (M) bit in the token frame. The R bits are reservation bits used in similar manner to the reservation bits in the token frame.

## Differential Manchester Encoding

Discussion of the delimiter fields has been delayed because of the need to mention the encoding format used by an IEEE 802.4 token bus. It is a modified Manchester encoding scheme called diffrential Manchester encoding. Recall from the Section 8.3.2 that Manchester encoding is utilized to assure a transition in the center of each data bit to facilitate clock recovery. This is achieved by inverting the first half of each data bit followed by a second inversion midway through the bit time. Differential Manchester encoding still performs this task.

Figure 8-12 shows both forms of Manchester encoding. The problem with the original Manchester encoding method is that the actual data level is detected as a level from

---

**TABLE 8-1**

**Frame Identifier Bits**

| F | F | Frame Type |
|---|---|-----------|
| 0 | 0 | MAC Frame |
| 0 | 1 | LLC Frame |
| 1 | 0 | Not Used |
| 1 | 1 | Not Used |

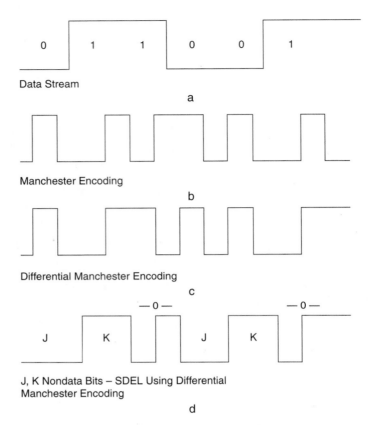

| 0 | 1 | 1 | 0 | 0 | 1 |

Data Stream

a

Manchester Encoding

b

Differential Manchester Encoding

c

J, K Nondata Bits – SDEL Using Differential
Manchester Encoding

d

FIGURE 8-12 Manchester encoding Forms Used for IEEE 802.×

the beginning to the middle of each data bit. Differential Manchester encoding shifts the detection to the beginning edge of each bit period. Zero logic levels are detected by the occurrence of a transition (or change) of logic state at the beginning of a bit period. A logic 1 is detected by the lack of a change at the beginning of a bit time. The level of the bit is then inverted at the midpoint, as a clock transition similar to regular Manchester-encoded bit streams, to assure that a transition occurs there for every data bit.

---

### EXAMPLE 8-3

Compare the encoded data stream for Manchester and differential Manchester encoding for the bit stream: 011001.

### SOLUTION

This is the data stream illustrated in Figure 8-12. Note that for both Manchester encoding forms, each second half of each bit time is inverted from the logic level of the first half of the bit period. The differences lie in the data logic state detection. For regular Manchester, the actual data level appears inverted in the first half of the bit time. For differential you have to observe the beginning of each period to see if there is a level change from the previous bit. A level change indicates that the current bit is a logic 0. No change indicates a logic 1.

The last section of Figure 8-12 shows the format of the start delimiter sequence. J and K bits are referred to as nondata bits and have no transitions at their midpoints. Using differential Manchester encoding, nondata J bits are held low, while nondata K bits are constantly high. Since these do not conform to standard Manchester encoding format, they identify the sequence as a starting delimiter, or flag, indicating the beginning of a frame.

### Section 8.4  Review Questions

1. How does a token bus avoid bus access contention between two or more stations?
2. What is the main difference between Manchester and differential Manchester encoding schemes?

## 8.5  IEEE 802.5 TOKEN RING

**ring**
topology that connects all stations into a ring format. Messages are passed from one station to the next, in an orderly fashion, around the ring.

Applying the concept of token passing to a **ring** topology is similar in the basic concept of token passing for a token bus topology. The main difference in a ring is that tokens and messages are passed around the ring to each station in the ring in a fixed sequence. Recall that in a token bus, the sequence can be modified using "who follows" type messages. For single ring topologies, there is nothing provided for fault tolerance should a station fail. Such a failure will cause the network to cease communicating until that station is brought online again. Later in this section we will focus on a dual-ring network designed to overcome this problem.

As with a token bus, messages that have been detected as received and read by the originating station are removed from the queue. At first when the network comes active, the process of passing traffic around the ring is initiated by the circulating of a free token around the ring. A station desiring to send a message to another station on the ring must grab the free token from the ring and append the message to the token. The token is modified to indicate that it is no longer free and the entire frame is sent to the next station in the ring.

### Token Ring Formats

The free token format for the token ring is shown in Figure 8-13. This frame includes starting and ending delimiters, optional priority, monitor, and reservation data. The priority level establishes which stations on the ring may have access to the free token. The

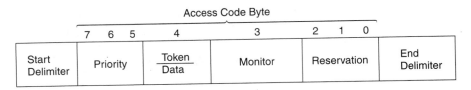

**FIGURE 8-13** Ring-Free

| Start Delimiter | Access Code | Control Field | Destination Address | Source Address | Data | Frame Check Sequence | CRC-32 | Address Recognized | Frame Copied |
|---|---|---|---|---|---|---|---|---|---|

8 Bit Frame Status

**FIGURE 8-14** Busy Token

monitor bit is used by a node point whose singular concern is to monitor the line for faults and errors. The reservation bits are used to allow higher priority stations to reserve token use, while a lower priority station has access to the token. The use of reservation bits will become more apparent after the free token has been captured.

A station requiring use of the ring looks for a free token. Upon detecting one on the ring, the station captures the token and immediately amends it to a busy token as shown in Figure 8-14. The busy token contains the same information as the free token with some additional fields. As well, a bit in the access code is changed to indicate that this is a busy token. The additional fields include destination and source address fields, control field information, which identifies the type of message being sent, data or payload, and a frame check sequence (CRC-32) field for error detection. Following the frame check sequence is a status field, which contain bits that identify the frame as being received (ADDRESS RECOGNIZED) and read (FRAME COPIED) by the destination station.

## Token Ring Sequence

Once the sending station creates the busy token, it sends it to the next station in the ring. The message is passed from station to station until it returns to the originating station. If the destination address was recognized by the intended receive station and the message was read, the status bits at the end of the token will be correctly set. In this case, the originating station, satisfied that its traffic was received, strips the message from the token, amends the busy bit back to free, and places the free token back onto the ring unless there are additional messages in the queue from other stations on the ring. In that case, the token will remain a busy token and the remaining frame will be passed along with the token to the next station on the ring.

Once all the messages have been removed from the queue, the token becomes a free token again, free to allow any station on the ring—when it detects the free token—to capture it and append traffic to it. If a higher-priority station decides it wishes to have access to

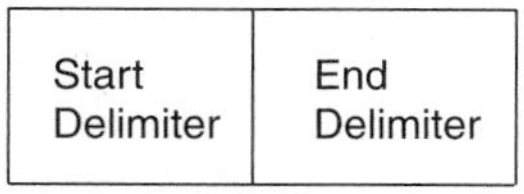

| Start Delimiter | End Delimiter |
| --- | --- |

**FIGURE 8-15**  Abort Sequence

the line, it must amend the busy token to indicate its desire. The higher-priority station achieves this by setting the reservation bits to its own priority. When the originating station returns the free token to the ring, the priority level is changed to the higher-priority to prevent lower-priority stations from gaining access before the higher-priority level station can capture the free token. Some token ring protocols maintain a hold timer, which limits the time that a station can hold a token before returning it to the ring as a busy or free token.

If the originating station receives its message without the address recognized or the frame copied bits set, it will return the busy token back to the ring in an attempt to resend it to the intended destination. As with the token bus, the attempt to resend the message is done a few times before the source station determines that there maybe something wrong with the destination station. At that point, the source station would cease trying to send the message and remove it from the queue.

In an extreme case, if there is trouble detected on the ring or any reason to abnormally terminate a transmission and free up the token, an abort sequence shown in Figure 8-15, consisting only of a start and end delimiter is sent.

## IBM Token Ring

International Business Machine' (IBM) Token Ring Network is implemented through the use of a Texas Instrument chip set designated TMS380 LAN. The network and the chip set actual use a form of star topology coupled with a logical ring as shown in Figure 8-16a. The logical transfer of data uses token passing and a contiguous ring with traffic being passed from one station to the next as if the actual topology was a physical ring. The traffic is actually sent from one station to the hub and then from the hub to the next station. Routing and management of the network functions are under control of the central hub of the physical star network. Stations connected in the ring are connected through a wire concentrator that acts as a normally closed switch. The hub controls when the switch is on, attaching a node to the ring; and when it is off, disconnecting a node from the network. The main advantage of using the logical ring/physical star configuration is that when a station fails, it can be bypassed (Figure 8-16b) so that the remaining stations can still pass traffic through the ring.

The IBM token ring/TMS380 LAN operates at 4 Mbps on twisted-pair wire or fiber optic cable. With the twisted pair, two pairs of lines are connected to each station. One is a downlink and the other is an uplink to allow data to be passed in a full-duplex four-wire environment. This is also known as a dual-ring configuration (Figure 8-17) and can be used to aid in fault tolerance. If a station fails, the system can revert to half duplex, using the second ring to "back around" the bad station as shown in Figure 8-18. The data is routed up to the failed station on one ring. The failure is detected and the data is routed

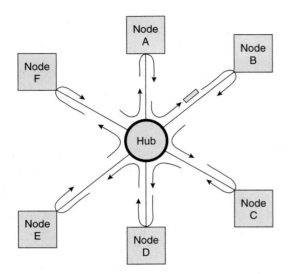

FIGURE 8-16a Logical Ring

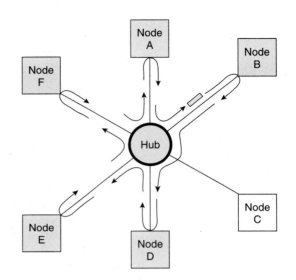

FIGURE 8-16b By-Passing Bad Node C

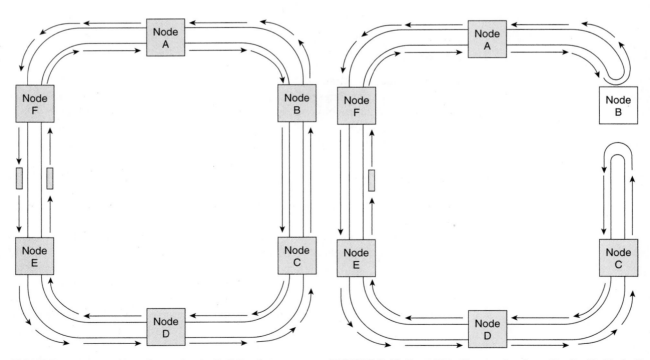

FIGURE 8-17 Dual Ring Operating in Full Duplex

FIGURE 8-18 Dual Ring Recovering from Fault with Node B

back along the second ring until it again reaches the failed station. Here the traffic is returned to the first ring to resume the ring sequence.

Token passing is accomplished using the same free token and message frame formats discussed for the token bus. The main difference is that tokens are passed from station to station around the ring rather than in a random sequence as used by token bus topology. Also, centralized control by the hub station in the star assures orderly, collision-free accesses to the network as well as fault tolerance.

The IBM token ring incorporates a network manager, which is responsible for network security, network configuration changes (station or node insertion and deletion), data requests and responses, and reconfiguration control. The star hub is an ideal source for the network manager control to emanate from. Access to the network manager functions from a node are available in two forms, an authorized mode and a nonauthorized mode. Nonauthorized mode makes all management functions accessible to a node while authorized mode accesses limit the functions available to a specified node on the network.

---

**Section 8.5  Review Questions**

1. How does a token ring differ from a token bus?
2. What happens if a station fails on the ring?
3. What is used in the token format to detect that the message has been successfully received and read?
4. What are the advantages of a logical ring using a physical star topology?
5. What are the two methods used for fault tolerance in a logical ring?
6. How is full duplex achieved in the IBM token ring?

---

## 8.6  NETWORK INTERFACE CARDS

**network interface card (NIC)**
interface PC to network.

To interface a personal computer to a local area network requires the use of a **network interface card (NIC)**, also known as a network adapter card, which plugs into a PC expansion slot. The NIC serves as a physical interface between a PC and a specific network. There are NICs for Ethernet, token ring, and a myriad of other network types that are discussed in later chapters. NICs operate at the media access control (MAC) sublayer of the data-link protocol to assure an operating interface operation. As such, they must adhere to the data-link protocol being used and the type of medium employed. For example, an Ethernet Adapter Card has to be able to respond to CSMA/CD access procedure and connect to a coax cable.

**attachment unit interface (AUI)**
connect thicknet to Ethernet NIC.

Several types of connectors are provided for NIC cards, depending on the medium requirements for the network. Figure 8-19 shows three of the most common types of connectors. A BNC connector is used to connect thinnet-type coax cable. Its bigger brother, thicknet coax, is connected using an interface cable and an **attachment unit interface (AUI)**

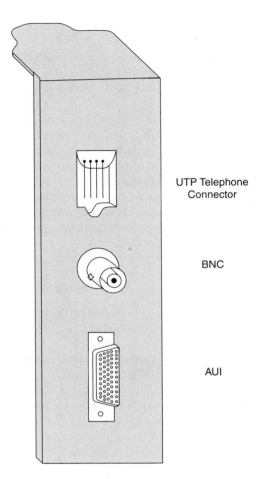

UTP Telephone
Connector

BNC

AUI

**FIGURE 8-19** NIC Line Connectors

15-pin connector as shown in the figure. Lastly, unshielded twisted-pair (UTP) cables are connected through standard telephone-type connectors.

On the PC interface side, the NIC card has to be compatible with the PC's internal bus structure. In order to operate as a component of the computer system, NIC drivers are required to provide application support for the card's operation. These small programs assure that the data entering from the network through the NIC card can be stored in the computer's RAM memory and data from that memory can be sent through the NIC back to the network. There are two primary ways a computing system can manage the transfer of data between its memory and the NIC card. The first is for the central processing unit, or CPU (usually a microprocessor), to execute a series of instructions to move data between memory and the card. For large amounts of data, this is not efficient. It also takes time away from the CPU to do other tasks.

**direct memory access (DMA)**

An alternate and more preferred method to transfer data between a computer's memory and the NIC card is to use a **direct memory access (DMA)** controller. When data is ready to be moved between the card and memory, the CPU grants bus mastership (or control) to the DMA. The device provides memory address and NIC access information directly to each unit. Once they are both accessed, data flows directly from one to the other. There is no requirement to decode and execute instructions like a CPU. As such, the transfer occurs at rates that are up to 100 times faster than those done by a CPU.

---

### Section 8.6 Review Questions

1. What is the main purpose for a NIC?
2. What type of access would a token ring NIC have to handle?
3. Why are DMAs preferred when transferring large blocks of data?

---

## 8.7 INTERCONNECTING LANS

As the need to increase the size of networks grew, there arose a need to be able to interconnect local area networks to form larger networks like MANs and WANs. These networks are interconnected using hardware/software systems called bridges and gateways. Bridges are used to connect two similar networks together, while gateways allow systems with differing protocols or message formats to be interconnected. Controlling traffic flow, maintaining fault monitoring, and adding and deleting stations from these new larger systems requires more and more sophisticated system management.

### Bridges

A bridge is a physical circuit that allows two networks to be interconnected as shown in the IBM token ring application in Figure 8-20. The type of data transferred between any two LANs connected by a bridge must have the same format. Notice that all of the networks in Figure 8-19, including the backbone, are token ring networks.

The actual protocols that define the message content can be any usable protocol as long as the source and destination stations use the same protocol. As such, bridges are protocol-dependent. For instance, a station on one of the token ring local area networks, can send a message through a bridge to a station on a second network as long as that station also uses token ring topology. In similar manner, Ethernet networks can be bridged together allowing a station on one network to send an Ethernet-type packet to a station on another Ethernet local area network through the bridge.

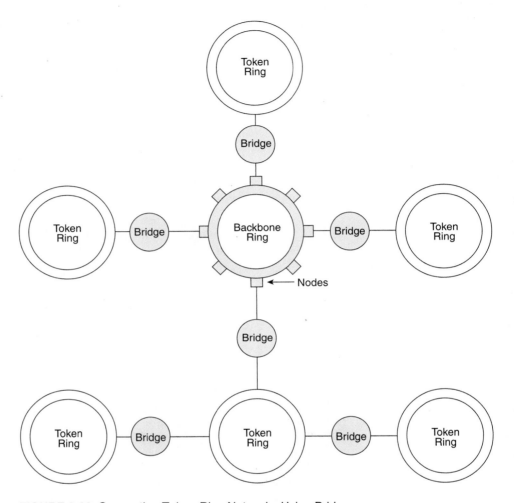

**FIGURE 8-20** Connecting Token-Ring Networks Using Bridges

### Address Tables

A bridge works by comparing source and destination addresses within the message packets against an address table. If a destination address is on the same network as the source address, the bridge ignores it, allowing the message to continue onto the destination station within the same network. However, if the destination address is on the other local area network connected to the bridge, the bridge copies the message onto the second network so that it can reach its destination. This process is called *filtering* and *forwarding,* since messages destined for the same LAN as the source are "filtered" from the bridge, while traffic destined for the other network is "forwarded" to it.

Address tables are formed by one of two methods. One method is to have the table sent to a bridge by a system manager. Station addresses and associated network numbers are stored directly into the bridge's memory. This method requires the manager to have beforehand knowledge about the system configuration in order for the table to be formulated. A second method for creating the address table is called *learning*. The bridge sends out a broadcast message asking the various nodes on the two LANs to identify themselves. As each station on the network responds to the broadcast request, the bridge builds the address table. This method requires the software at the bridge to perform the task and also requires time for the table to be built. In either case, once the address table is constructed, the bridge can begin the filtering and forwarding process.

Some bridges have additional filtering capabilities that allow it to filter broadcast or multicast packets so that a situation called a **broadcast storm** does not occur. Essentially, broadcast messages are meant to be received by all stations on the network. Unfiltered systems could experience broadcast packets crisscrossing through the bridge until the network becomes completely saturated and quits. A bridge with the capability of sensing a broadcast message would pass it along between the two LANs so that the broadcast message is made available to all the stations on both networks. Attempts by the broadcast packet to return to the first network through the bridge are filtered out.

**broadcast storm**
overload of traffic on a network due to the issuing of too many broadcast messages.

## IEEE 802.1 Spanning Tree Protocol

One concern when using a bridge to interconnect two local area networks involves the loss of network activity in case of a failure of the bridge. The IEEE 802 standard addresses this problem in the 802.1 specification by supplying a solution to this concern. That solution is called the spanning tree protocol and requires a second bridge to be used for back up purposes. The primary bridge used to handle traffic has a secondary parallel bridge connected across it. Under normal operation, the primary bridge passes all the messages between the two LANs. If the primary bridge fails, the backup bridge takes over the handling of inter-LAN messages until the primary bridge is repaired.

The spanning tree protocol details the conditions under which the backup bridge takes over and how the takeover is to be accomplished. The main disadvantage of the spanning tree method is the inefficient use of the backup bridge, which is effectively idle as long as the primary bridge remains healthy. A process called load balancing can be applied to a system to make use of the backup bridge under normal operation. The traffic load is divided between both bridges, which reduces the volume on the primary bridge. In the event that either bridge fails, the remaining good bridge takes on the full traffic load until repairs can be made to the failed bridge.

Another use for the spanning tree protocol is to allow multiple networks to operate with more than one interconnecting loop. This is an extension of the concept of a backup network, except that it implies that the interconnections, rather than a network are being used for backing up the system. In both cases, bridges negotiate for the best route through the networks, establishing a tree format from the routing. The highest-priority bridge becomes the root bridge of the tree network as can be seen in Figure 8-21. Each succeeding bridge determines which of its ports points to the root bridge, designating that port as the root port of the bridge. These root ports are set in a forwarding mode always sending traffic to the root bridge. Bridges then select additional ports that point away from the root bridge to move data from the root bridge through the local bridge and on to the connecting networks of that local bridge. Local bridges can also have root functions to subnodes that are subordinate to them.

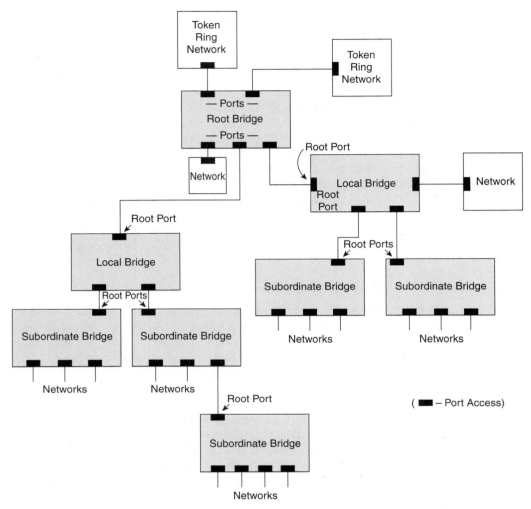

**FIGURE 8-21** Spanning Tree Topology

## Gateways

Gateways are similar in function to bridges in that they connect two networks together. The main difference between bridges and gateways is that gateways can interconnect networks with differing message formats or protocols. A bridge does not perform any protocol translation in transferring packets between networks. Gateways, on the other hand, are designed to make protocol translations between two dissimilar LANs. For instance, a token ring station can be made to communicate with an Ethernet station through a gateway. Recall, that with a bridge, any protocol can be passed through the bridge, but for communication to be successful, the source and destination stations are required to use the same protocol. When using a gateway, the gateway reads the message and determines the source protocol. Next, after comparing the destination address with its address table, the gateway determines the destination station's protocol and location. The source message's format is converted (or translated) into the destination protocol format and the gateway forwards the translated message to the network that has the destination node.

## Routers

Routers are packet switch type devices that operate at the OSI network layer and are sometimes referred to as network layer relays. They are used to interconnect networks to networks and backbones as shown in Figure 8-22. Routers direct traffic from any of the local area networks (LAN) in the diagram to any other LAN in the system. They do this by maintaining routing tables, which identify source, destination, and neighboring router addresses.

Messages are routed through the system using any number of routing protocols. These protocols initially provide a means for the routers to query the system and "learn" about a system's configuration. From the responses to the query message, routers will build the routing tables. An example of a routing protocol that is used with the TCP/IP protocol suite that underlies the Internet is the **routing information protocol (RIP).** Details about TCP/IP and the Internet are in Chapter 10. Other networks that use RIP are NOVELL NETWARE and Xerox National Systems.

Under RIP, after routers are initialized, they periodically send a request routing information and also await responses they use to update their routing tables. Other non-information upkeep-type messages that routers broadcast under RIP are presence and network configuration change messages. The presence message is just a reminder to adjacent routers that the router is still active in the system and the network configurations it supports has not changed. The change in configuration message also validates the router's presence, but it also indicates that some change has taken place in the networks it handles and adjacent routers need to update their tables accordingly.

When a message is sent from a host in a LAN, for instance LAN-1 in Figure 8-21, the nearest router accepts the message and examines its address information. If source routing is employed, the router will "read" the next address in the path selected by the sender of the message and pass it along to that node. This might be another router or node on a network. Source routing does not allow much flexibility in the way packets are routed through the system. A delay can be experienced if a selected point in the path is busy or congested.

Virtual routing allows routers to find the best path through the system to deliver the message from source to destination. The possible destinations that routers can be expected to handle are:

1. To the router itself, in which case the router accepts and processes the message.
2. To a node on the same LAN. Here the router forwards the message back to the LAN it came from.
3. To an adjacent router, which will determine the next point in the system to send the message.
4. To a node on another LAN, which the router forwards the message to.
5. To an unknown destination, in which case the message is discarded and an error message returned to the source node.

**routing information protocol (RIP)**
routing protocol used with TCP/IP.

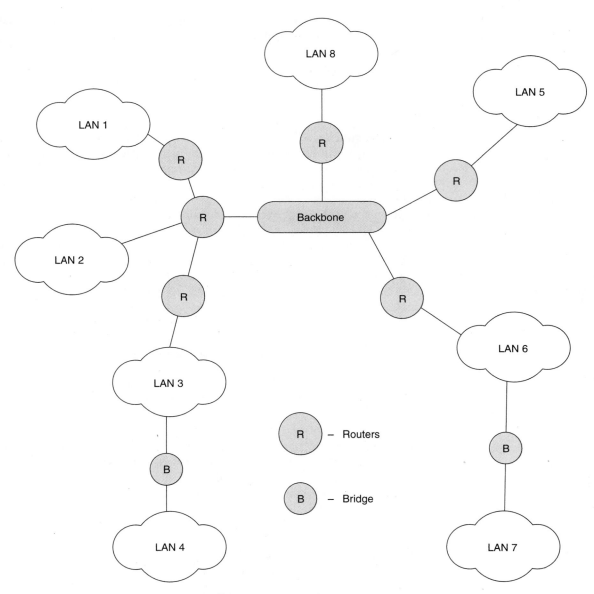

**FIGURE 8-22** Routers

Each point in a path that a message takes is called a *hop*. Routing protocols like RIP attempt to deliver the packets with as few hops as possible. When the system becomes congested with a large amount of traffic, packets that start to take too many hops and still not reach their destination are often discarded in an attempt to reduce the congestion. Notification of dropped packets is not frequently done. After all, the purpose of dropping a packet in the first place is to reduce the traffic load. RIP allows 15 hops before it discards a packet.

## Servers

**server**
protocol translation between
elements on a network.

At the network layer, a software/hardware system called a **server** is used to perform the local translation between a user's protocol and the common network. These servers (like SNA nodes) come in many variations, depending on the service they are required to perform. For example, there are file servers used to handle data file storage and retrieval. Printer servers provide printer resources to all the personal computers, terminals, or mainframe computers on the network. A similar service is provided by fax servers allowing faxes to be sent from a user's desktop computer through the network to the fax machine.

Communications servers provide means for users to communicate to available channels through the network. There are also database servers to handle database functions and archive or backup servers to assure against lost files or databases. A special type of file server known as an application server store and provide access of application programs for network users. Lastly, there are e-mail servers to handle the interfacing of e-mail messages from any e-mail client on the network.

### Section 8.7  Review Questions

1. What is the main difference between a bridge and a gateway?
2. How does a bridge protect a network from experiencing a broadcast storm?
3. What function can a gateway perform that a bridge cannot?
4. How does the spanning tree protocol provide for fault tolerance when a bridge fails?
5. How are backup bridges used to keep them from idling when the network has no faults?
6. What OSI level do routers function at?
7. What do routers use to keep track of LAN node addresses?
8. What term refers to a point in a system that a packet is sent to?
9. What is the prime purpose of a server?

## 8.8  IEEE 802.6 METROPOLITAN AREA NETWORK (MAN)

**distributed queued dual bus (DQDB)**

Local area networks are usually restricted to a single site, such as a building or a floor or room in a building. Metropolitan area networks (MAN) expand network coverage to include several buildings or sites within a limited area. These sites are usually located within a city's limits or limited to a specific region that may include a number of cities. The IEEE 802.6 addresses the management of MAN networks through the use of the **distributed queued dual bus (DQDB)** protocol.

## DQDB Physical Layer

For the physical layer of DQDB, this protocol specifies a dual-bus topology to carry data in forward and reverse directions. The forward direction bus carries data while the reverse direction handles queuing and control information.

Scheduling counters are employed to keep track of data frames called slots as they pass along the bus system. Bus contention for the purposes of moving slots of data is handled using a requesting mechanism. Up/down counters keep accurate count of data requests for traffic being made by stations on the bus. They count up for each request made and down for each request satisfied. A separate down counter notes the number of empty time slots that pass in the opposite direction. These counters monitor how busy the bus is at any given time.

When a station on the network is ready to transmit, it sends a request causing the up/down counter to increment. The down counter decrements as each empty time slot passes by to satisfy outstanding requests from other stations. When the down counter reaches zero, the requesting transmission occurs on the next empty (available) time slot. The up/down counter decrements at each instance that a time slot is used to satisfy a transmission request.

## DQDB Frames

A DQDB frame, shown in Figure 8-23, contains an access control field, which includes a busy bit indicating whether the slot is being used or not. The access control field also contains four request bits, one for each of four priority levels. When the station makes a request, it can also assign the priority level of the request by setting the appropriate request bit.

The access control field is followed by a segment header field, which contains addressing and sequencing information for the payload portion of the DQDB slot. The payload field itself contains data and upper-level header information within a 48-byte field size. The total frame size is 53-bytes, which makes DQDB compatible with the asynchronous transfer mode (ATM) cell size discussed in Chapter 10.

**Section 8.8  Review Questions**

1. What protocol is used to manage IEEE 802.6 MAN networks?
2. What types of information are carried on the dual-bus network using the DQDB protocol?

| Access Control 1 Byte | Segment Header 4 Byte | Segment Payload        48-Byte |
|---|---|---|

**FIGURE 8-23**  DQDB Slot Format for Data

3. What are the frames in DQDB called?
4. What are DQDB data frames set up for 48-byte information and 5-bit header format?

# 8.9 X.25  PACKET SWITCH PROTOCOL

**international telecommunications union (ITU)**
standards group for networks.

**International telecommunications union (ITU)** group's X.25 protocol is a commonly used network access protocol that complies with the lower three layers of the open systems interconnection (OSI) model. The X.25 protocol is a packet switch local area network (LAN) protocol that incorporates existing standards to specify network requirements.

## Physical Layer

ITU V.24, similar to IEEE RS232C, addresses the physical layer of the OSI model within the X.25 framework. The data-link protocol specified by X.25 is HDLC. The high-level data-link (HDLC) protocol frame used is essentially the same as described in Chapter 7. HDLC extended addressing is used in the address field to accommodate the larger number of stations in a network using X.25.

## Data-Link Layer

Part of the data field is designated as a packet field. This portion is used to identify the packet type so that it may be correctly interpreted. In all other aspects, the HDLC protocol remains unchanged in its application by X.25.

## Network Layer

**packet assembler/disassembler (PAD)**
interface non-standard X.25 networks to X.25.

X.25 also specifies the manner in which non-X.25 networks can interface to an X.25 network. This is accomplished using a **packet assembler/disassembler (PAD).** Essentially, the PAD protocol reads the nonstandard data and translates them into X.25 packets, which are then routed to their destinations on the X.25 network. The process is reversed when an X.25 station sends data to the nonstandard network.

PADs address the network level of the OSI model. In addition, at this level X.25 specifies how packets are formed and exchanged using a virtual circuit. Virtual circuits are those that are specified logically, but not directly connected physically. There are two classes of virtual circuits, **switched virtual circuit (SVC)** and **permanent virtual circuit (PVC).** The differences are as obvious as their names. A permanent circuit is one that is always connected between stations; while a switched circuit uses connections that are switchable between stations. It is necessary to remember that virtual circuits are logical

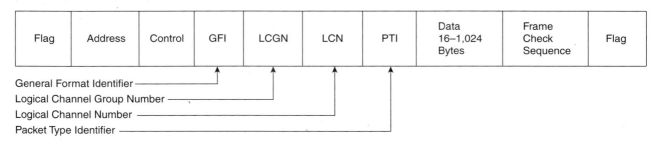

| Flag | Address | Control | GFI | LCGN | LCN | PTI | Data 16–1,024 Bytes | Frame Check Sequence | Flag |
|------|---------|---------|-----|------|-----|-----|--------------------|---------------------|------|

General Format Identifier ——————
Logical Channel Group Number ——————
Logical Channel Number ——————
Packet Type Identifier ——————

**FIGURE 8-24** X.25 Packet Frame

and not physical entities. The actual physical connections (even for permanent virtual circuits) are most likely to be different each time a link is established.

The network level packet (Figure 8-24) is formatted in the same manner as a basic HDLC packet. The beginning of the information field contains a packet header, which is used to designate the logical channel and packet identifications. It begins with a *general format identifier (GFI)*. Among other information, this field sets the size of the packet frame numbers (NS—the number of the current frame being sent and NR—the number of the next expected frame to be transmitted by the current receiving station).

**logical channel identifier (LCI)**

**logical channel group number (LCGN)**

**logical channel number (LCN)**

The general format identifier is followed by the **logical channel identifier (LCI),** which is comprised of a **logical channel group number (LCGN)** and **logical channel number (LCN).** The logical channel identifier is capable of identifying a specific channel from amongst 256 groups of 16 channels each.

The control fields of the network packet (Figure 8-25) identify the type of packet in use. The types are data packet (same as HDLC information control field), flow control packet (HDLC supervisor packet), and supervisor packet (conceptually the same as the HDLC unnumbered control field). For the data packet, a *M, or more* bit replaces the function of the poll/final (P/F) bit used with HDLC. A logic one state of the M bit indicates that there is more data packets to follow.

## Flow Control

The flow control frames differ from HDLC supervisor frames due to the lack of the P/F bit. Three of four supervisor frames used in HDLC are used in X.25 network flow control packets. They are ready to receive, not ready to receive, and reject with the one HDLC supervisor frame not used being frame reject.

## Supervisor Frames

Supervisor frames for the X.25 network layer use the unnumbered format but are totally different from the unnumbered frames used by HDLC. The X.25 frames include CALL REQUEST, CALL ACCEPTED, CALL CONNECTED, INCOMING CALL, CLEAR INDICATION, CLEAR CONFIRMATION, INTERRUPT, INTERRUPT CONFIRMA-

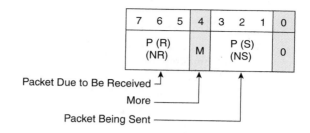

Data Packet

a

Flow Control

b

Supervisory

c

**FIGURE 8-25** X.25 Control Fields

TION, RESET REQUEST, RESET CONFIRMATION, RESTART REQUEST, and RESET CONFIRMATION. These satisfy the needs of the X.25 protocol in a manner in which the HDLC unnumbered frames could not. This layer is used to form control and data packets and establish the connections on a virtual circuit. Also, keep in mind that most of these local area networks use peer stations throughout. That is, there is no controlling primary station in the network.

### Section 8.9  Review Questions

1. Name a standard and a protocol that apply to X.25 physical and data-link layers.
2. Which OSI layer does an X.25 PAD comply with?

## SUMMARY

The IEEE 802 suite of standards specifies the requirements for various types of networks and their components. These specifications can be used in the planning of local area networks, metropolitan area networks, or wide area networks based on bus, star, or ring topologies. These standards, out of necessity, address one or more layers of the OSI model, making both the OSI model and the IEEE 802 standards compatible.

Beginning with LANs, IEEE 802 defined the basic building blocks in logical link layer and media access control specifications. Detailed network operations based on bus and ring topologies are defined in CSMA/CD, token bus, and token ring documents. Others include metropolitan area networking and isochronous or integrated services. As technology improved network capabilities, additional standards were added to the group. Fiber optic applications, gigabit Ethernet backbones, and wireless communications are examples of recent technology that has now been addressed by various 802 standards.

The whole idea and purpose behind setting any standards is to facilitate the process of communicating between users, be it through a point-to-point system or a complex arrangement of networks. In the next set of chapters, we will explore the protocols and networks that handle most of our data traffic and communications.

## QUESTIONS

**Section 8.1**

1. Which of the 802 subsections address
   a) media access control (MAC) layer specifications
   b) MAN networks
   c) Ethernet Protocol
   d) Spanning tree
   e) Token ring
   f) Wireless LAN
   g) Gigabit Ethernet
   h) Data transfers using private cable companies
   i) Isochronous Ethernet
   j) Logical Link Control (LLC)
   k) CSMA/CD

2. Which data communications areas are addressed by the following IEEE 802 subcommittees?
   a) 802.2
   b) 802.3
   c) 802.1z
   d) 802.4
   e) 802.5
   f) 802.11
   g) 802.12

**Section 8.2**

3. Define the difference between a LAN and MAN.
4. Specify the amount of territory covered by each of the following network types:
   a) LAN
   b) MAN
   c) CAN
   d) WAN
   e) GAN
5. What aspects of networking is addressed by the MAC specification?
6. What aspects of networking is addressed by the LLC specification?

**Section 8.3**

7. List five Ethernet/802.3 specifications that apply to the OSI model physical layer.
8. What are the Ethernet distance limitations for tappable and untappable segments?
9. How many nodes can be tapped into a single segment?
10. What are the minimum and maximum distances between nodes on an Ethernet network?
11. What are the three main types of network topologies?
12. Which network topology is used with Ethernet networks?
13. Which data-link protocol is the basis for the Ethernet data-link process?
14. What are the purposes for the preamble preceding Ethernet packets?
15. What is the data rate for the original 802.3 specification?
16. Encode the following unipolar data stream into a Manchester-encoded data stream.

011010111010010001

17. What data rates are specified by 802.12 and 802.3z?
18. How does CSMA/CD resolve the problem of line contention?
19. How does CSMA/CA differ from CSMA/CD?
20. What is the difference between Ethernet address fields and HDLC address fields?
21. What are two fundamental enhancements addressed by the 802.3i specification?

**Section 8.4**

22. What is used in token formats to assure that a delimiter form does not appear elsewhere in a message?
23. What type of token is used to allow a station to access a token bus network?

**Section 8.5**

24. How does an originating station on a token ring know if its message has been received and copied by a destination station?
25. What does the originating station do when it detects its message has been sent and copied?
26. Give an advantage of a token ring network compared to an Ethernet bus network.

27. Give a disadvantage of a token ring network compared to an Ethernet bus network.
28. Encode the following unipolar data stream into a differential Manchester-encoded data stream.

011010111010010001

### Section 8.6
29. What is the purpose of a network interface card (NIC)?
30. What are the NIC's specific tasks?
31. How do DMAs improve performance?

### Section 8.7
32. What do network bridges do?
33. How do network bridges differ from gateways?
34. What is the purpose of a router?
35. What is the primary purpose of servers?

### Section 8.8
36. Which IEEE standard addresses metropolitan area networks?
37. What are slots used for?
38. How is contention resolved using DQDB?
39. How are dual buses employed by DQFB?

### Section 8.9
40. How many total channels are available to be selected from using the X.25 LCI field?
41. What are the differences between X.25 and HDLC information frames?
42. What are the differences between X.25 flow control and HDLC supervisor frames?

# RESEARCH ASSIGNMENTS

1. The original Ethernet protocol has undergone quite a number of changes, some of which were outlined in this chapter. Write a paper on the current crop of Ethernet protocols and systems. Highlight improvements over older Ethernet protocols and systems.

2. The IEEE 802 committee keeps expanding as the need for defining standards for new aspects of networking arise. Research what the committee (and its subcommittees) are working on today. Specify what area of networking the new specifications will address and why they are needed.

3. Select a NIC network adapter card and report on how it interfaces a personal computer to a local area network.

4. Explore the latest internetworking devices and explain how and where they are used.

# ANSWERS TO REVIEW QUESTIONS

### Section 8.1

1. IEEE 802 originally formed to create standards for Local Area networks.
2. LAN—limited to single building or room.
   WAN—national network.
   MAN—regional network.
   CAN—links several sites like university campus networks.
   GAN—international network.
3. Physical and data-link layers.

### Section 8.2

1. Coax cable.
2. 500 meters for tappable and 1,000 meters for untappable segments.
3. Fiber optic and unshielded twisted-pair copper cabling.
4. Aids in clock recovery.
5. Station desiring access must first monitor the bus for a time equivalent to one round-trip propagation time.
6. Ceases transmitting message and starts transmitting a jamming signal.
7. Back-off algorithm is used to resolve contention between two or more stations trying to access the bus at the same time.
8. HDLC.
9. Clock recovery, bit and character synchronization.
10. The main difference is in the contents of the control fields. 802.2 control fields are the same as HDLC control fields. 802.3 control fields contain length information for the LLC (802.2) control field used in the payload (data unit) field of the 802.3 frame.
11. 802.12: 100 Mbps and 802.3z: 1 Gbps.
12. CSMA/CA protocol seeks to resolve bus contention before transmission begins to avoid collisions rather than trying to detect them after transmission has begun.
13. Backbone.

### Section 8.3

1. Uses a token to control when a station can access the bus (free token) or when the bus is in use (busy token).
2. Manchester encoding uses level sensing for data bit information. Differential Manchester encoding senses data states at the leading edge of the bit time period.

### Section 8.4

1. Token ring networks circulate messages from one station to the next. Token bus allows any station to access the bus.
2. When a station in the ring fails, the communication link is broken and further communication is stopped.
3. Address recognized and Frame copied status bits.

**Section 8.5**

1. If a station fails, the hub can bypass it and maintain the communication link.
2. Method 1: use a hub to bypass the failed station.
   Method 2: use dual rings—the secondary ring is used when a station on the primary ring fails.
3. Using dual rings, messages can be sent in two directions simultaneously using one ring for the forward direction and the other for the reverse direction in a full-duplex manner.

**Section 8.6**

1. A NIC interfaces a personal computer to a network.
2. Token passing.
3. Faster transfers.

**Section 8.7**

1. A bridge connects like networks, a gateway connects dissimilar networks.
2. By filtering broadcast messages that try to cross back through a bridge.
3. Protocol translation.
4. Backup bridge.
5. Load balancing.
6. Network.
7. Routing tables.
8. Hop.
9. Sharing resources.

**Section 8.8**

1. DQDB—distributed queued dual bus.
2. Data in the forward direction and queue/control information in the reverse direction.
3. Slots.
4. Handle ATM traffic.

**Section 8.9**

1. V.24 and HDLC.
2. Network layer.

# CHAPTER 9

# Integrated Services and Routing Protocols

## OBJECTIVES

After reading this chapter, the student will be able to:
- See how ISDN is used to transport integrated voice and data service.
- Define the different rates of service supplied by ISDN.
- Understand how a PBX is used to handle integrated voice and data services.
- Determine how messages are routed through multiple networks using the Asynchronous Transfer Mode (ATM) and Frame Relay routing protocols.
- Decipher the different levels of quality of service made available by routing protocols.
- Tell the difference between connection and connectionless service.

## OUTLINE

9.1   Introduction
9.2   Integrating Services
9.3   Broadband ISDN
9.4   IEEE 802.9: Integrated Voice and Data Services
9.5   Digital Subscriber Line (DSL)
9.6   Private Branch Exchange
9.7   Asynchronous Transfer Mode (ATM)
9.8   Frame Relay

## 9.1  INTRODUCTION

**integrated services digital network (ISDN)**
digital communications system that transports voice and digital data.

The data communication field is in a continual state of flux. New standards and communication network facilities are constantly being introduced and implemented. Many of the older standards are being updated to accommodate higher data rates and new technology advancements. Some of the newer technology of the past years has become the backbone of today's communication systems. Among these are the **integrated services digital network (ISDN),**

301

**asynchronous transfer mode (ATM)**
routing protocol that uses fixed size packets called cells.

**frame relay**
routing protocol that uses variable size packets called frames.

the **asynchronous transfer mode (ATM),** and **frame relay.** These are used to transport and route integrated services such as voice, data, and video information through the public phone systems (ISDN) and among numerous digital networks (ATM and frame relay).

Additionally, ISDN uses numerous established local area networks (LANs), wide area networks (WANs), metropolitan area networks (MANs), and other network configurations, interconnecting them via common public carriers. The benefits of using an integrated network such as ISDN include line sharing, high data throughput, easy identification of data packets, and error recovery.

The establishment and termination of calls are done the same regardless of the originating station's configuration. ISDN, being a digital network application, has become a popular method for accessing the Internet. Being digital, it allows for quicker connection and downloading of files from the Internet.

---

**Section 9.1  Review Questions**

1.  What is the main purpose of ATM and frame relay?
2.  Why is ISDN being used to facilitate connections to the Internet?

---

## 9.2  INTEGRATING SERVICES

ISDN data services integrate voice, data, and video information onto a single channel. There are a number of types of ISDN data channels used for this purpose, which are detailed in the former International Consultative Committee for Telegraphy and Telephony (CCITT), now International Telecommunications Union (ITU) specifications for ISDN networks. The I series of standards describe ISDN principles, service capabilities, network characteristics, network interfacing, internetwork interfacing, and maintenance principles. The Q series specify the actual protocols used. The actual breakdown of ITU ISDN standards is listed in the Table 9-1.

### ISDN Channels

**bearer (B) channel**
64-Kbps ISDN data channel.

ISDN services are transported on two different types of channels, the **bearer (B) channel** for data, voice, and video information at 64 kbps; and the Signaling (D or Delta) channel, used for control signaling or low-speed packet switching transmissions, operating at 8 or 16 Kbps. Specific uses of the D channel include call setup and termination and system maintenance control. Use of the packet-switched data on the D channel provides a means for telemetry information to be passed along while integrated data travel on the B channel. Telemetry data include low scan alarm, energy monitoring, and security data.

### ISDN Service Rates

**basic rate interface (BRI)**
defines an ISDN network service consisting of two B channels and one D channel.

**primary rate interface (PRI)**
defines an ISDN network service consisting of 23 or 30 B channels and one D channel.

Two established forms of ISDN service are commonly available—the **basic rate interface (BRI)** specified in ITU I.420/Q.931, consisting of two B channels and one D channel (2B + D) and a **primary rate interface (PRI)** specified in ITU I.421/Q.921. In North

**TABLE 9-1**

**ITU Specification Standards**

| Specification | Topic |
| --- | --- |
| I.113 | Vocabulary of Terms |
| I.121 | General Broadband Aspects |
| I.211 | Service Aspects |
| I.311 | General Network Aspects |
| I.321 | Protocol Reference Model |
| I.327 | Functional Architecture |
| I.420/Q.931 | Basic Rate Interface (BRI) |
| I.421/Q.921 | Primary Rate Interface (PRI) |
| I.430 | Physical Layer Interface |
| I.431 | PRI Physical Layer |
| I.440/441/Q.920 | Data-Link Specifications |
| I.450/451/Q.930 | Network Layer Specifications |
| I.610 | Operations and Maintenance |

America, Japan, and Korea, the primary rate interface consists of twenty-three B channels and one D channel (23B + D). In Europe and the rest of the world, the primary rate interface uses thirty B channels and one D channel (30B + D). Basic rate interface systems require bandwidths that can accommodate 144 Kbps to handle the two 64-Kbps B channels (B1 and B2) and a 16-Kbps D channel. Primary rate interface systems, because of the larger number of B channels used, need bandwidths of 1.544 Mbps for North America, Japan, and Korea, and 2.048 Mbps for everyone else.

**terminal adapter (TA)**
ISDN unit used for converting non-ISDN messages into ISDN format.

Data, in any form originating from a subscriber (customer or user), are converted into ISDN form by a **terminal adapter (TA),** transported onto the common carrier (telephone system), and delivered to the destination station. On the receiving side, the process is reversed, returning the data to the form usable by the receiving network customer. Because of the use of terminal adapters, the sending and receiving customers do not have to use the same type of networking or protocols at their sites. Essentially, the communication methodology is transparent to the user. What happens to the messages along the way is insignificant to the user as long as the data are successfully communicated.

## ISDN Topology

**terminal equipment**
ISDN physical user network entry equipment. TE may also be used for Terminal Endpoint.

ISDN units and interfaces throughout the network, shown in Figure 9-1, are defined by function and reference in the network. Users may access the network through one of two categories of entry terminal devices, **terminal equipment 1 (TE1)** and **terminal equipment 2 (TE2).** Terminal equipment 1 units contain standard ISDN interfaces and require no data protocol translation. Raw data are entered and immediately configured into ISDN protocol format. TE1s are illustrated in Figure 9-1 as a terminal and digital telephone

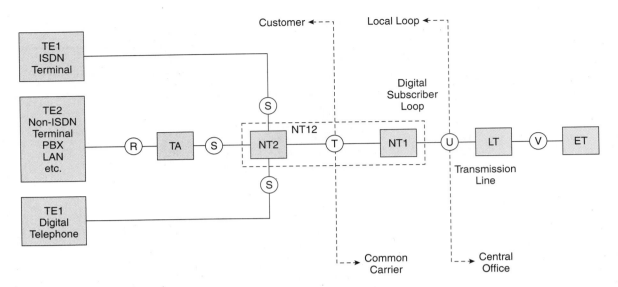

**FIGURE 9-1** ISDN Connections and Reference Point

supplying data to the system at reference point S. Non-ISDN terminals are classified as TE2 and are connected to the system at reference point R via physical interfaces such as RS232C or X.21.

Terminal adapters (TA) are required to perform the required translation between the non-ISDN data protocol and the ISDN network protocol formats before supplying data to the S reference point, as shown in Figure 9-1. Input data at the S reference point is formatted into the necessary B and D channels before being applied to the common carrier. Terminal adapters also support analog phones and facsimile by use of a 3.1-KHz audio bearer service channel. These audio signals are digitized and formatted into ISDN channels and sent onto the network.

The S reference point indicates interface lines that are presently in ISDN format. They supply the 2B + D data format for the ISDN basic rate to network termination 1 (**NT1**) units. These units supply the physical interface between the customer and the common carrier at reference points designated by the letter T. As such, these units are often called *Customer Exit/Entry Nodes.* An additional Network Termination 2 (NT2) unit may be used to terminate several S point connections, providing local switching. NT2s may also include protocols for data-link and network layers of the open systems interconnection (OSI) model to complete the interface between the common carrier local loop and the central office (U reference point). NT2s also perform 2-wire to 4-wire conversion for ISDN basic rate access to the common carrier.

The U reference point is the 2-wire media interface point between NT1 and the central telephone switching office. U interfaces are a 160-Kbps digital connection that uses echo cancellation to reduce noise. The most common signaling method used on U interfaces is known as 2B1Q (2 Binary 1 Quarternary), which provides 2 bits per symbol at the 160-Kbps transfer rate. Quarternary refers to four levels that can be represented as symbols using 2 bits. In effect, 2B1Q refers to 2 bits per quarternary symbol.

U interfaces are terminated at the office by a line termination (LT) unit, which incorporates physical layer interfacing between the central office transceiver and the local loop lines. The LT unit is connected at reference point V to an exchange termination (ET) unit, such as a PBX that routes the data through a local network to an outgoing channel or to a central office user.

## Additional ISDN Channel Types

The preceding sections discussed two types of ISDN channels, the B and D channels, used by subscribers. Other types of channels are included in the ISDN specification to provide additional services to subscribers. Specific ISDN services are selectively provided by telephone and networking companies. These additional ISDN channels and services include the following:

A channel—conventional analog voice.

E channel—packet switched using 64-Kbps, similar in function to a D channel.

H0 channel—six B channel circuit switch at 64-Kbps per channel used for high-speed facsimile, video, data, or signal imaging.

H4 channel—hyperchannel uses broadband ISDN at 139 Mbps over fiber optic cable.

H11 channel—24 B channel switch at 64 Kbps per channel requiring two primary rate interface services for a total data rate of 1.536 Mbps on the channel.

H12 channel—European version of H11, using 30 B channels instead of 24 for a total data rate of 1.92 Mbps.

The H4 channel is actually developed by time division multiplexing ISDN channels as shown in Figure 9-2. Thirty B channels (64-Kbps plus guard band) are multiplexed to form a 2.048-Mbps H1 channel. Four of these are multiplexed, increasing the bandwidth to H2 channel of 8.448 Mbps. In turn, four of these are multiplexed, creating an H3 channel of 34.348 Mbps. Finally, four H3 channels are multiplexed to create the H4 channel at 139.266 Mbps, which includes 135 Mbps of channel information plus guard bands between channels to prevent crosstalk. The total number of channels sent on the multiplexed H4 channel is 1,920 B channels at 64 Kbps each.

The common T1 channel described in Chapter 8 can be implemented on ISDN using various combinations of channels described earlier. The most direct definition is supplied by the H11 channel, which is designed principally for the twenty-four channel T1 carrier. However, T1 can also be implemented using 23 B + D or 24 B channels, 3 H0 + D, or 4 H0. The main difference in these is whether or not the communications link requires a signaling (D) channel.

Many ISDN services are not available because the cost of maintaining them is too high to justify the need for their use. Decisions to include a particular service in an ISDN offering by a provider is very individualized and selective. Additionally, the cost of ISDN services varies among providers. This results from a lack of a central control of these services or any standardization of rates. Implementation of ISDN throughout the United

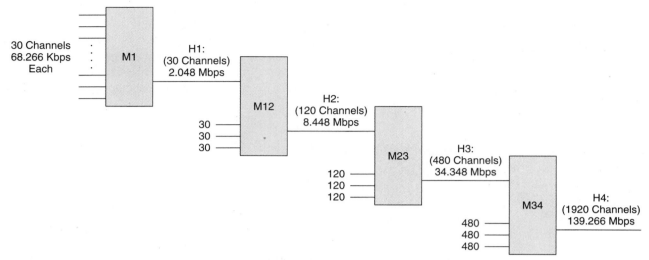

**FIGURE 9-2** Multiplex Hierarchy Using Synchronous Time Division Multiplexing (STDM)

States has been hampered by these problems. The rise in demand for ISDN services to facilitate Internet access has begun to overcome some of the previous roadblocks toward full ISDN implementation.

### ISDN Frame Formats

ISDN specifications define a network architecture modeled according to the Open Systems Interconnection (OSI) seven-layer network model. The network termination units address the physical layer by describing the interfaces required by users to access the system. Devices included as part of a network termination 1 (NT1) are transceivers, modems, interfaces (such as RS232C), and network media (such as twisted-pair cabling).

Figure 9-3 illustrates the frame format for the basic rate interface (BRI) as it applies to the physical entry and exit (S) points to the network. There are two frame formats, Terminal Equipment (TE), defining those sent by terminal equipment to the network (entry); and Network Termination (NT) frames, defining frames sent the opposite way (exit). The first two bits of these frames (F: framing, L: DC balancing) are used for synchronization. Each B channel information byte following the F and L bits is followed, in turn, by a D channel signaling information bit. The A bit in the network termination frame activates or deactivates terminal equipment, putting this equipment online or removing it from the network.

Line arbitration for the sending of D channel control information by a terminal is achieved through use of the E bit in the network termination frame, which is a copy of the previous terminal equipment frame D bit information. S bits are undefined at this time, being reserved for future applications. When a terminal has D information to send, it first observes the E bits of the network termination frames. A specific consecutive number of E bits must be high before the station can send its D information. The consecutive high

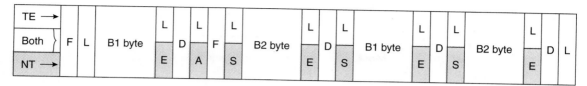

**FIGURE 9-3**  Basic Rate Interface (BRI) Frames

1.544 Mbps U.S. PRI Frame

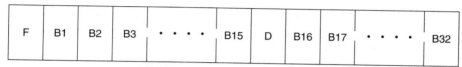

2.048 Mbps European PRI Frame

**FIGURE 9-4**  Primary Rate Interface (PRI) Frames

states of the E bits indicate that no D control information has been sent in any previous terminal equipment frames and the station can now send a terminal equipment frame with D control information.

Network termination and terminal equipment frames are transmitted at a rate of 4,000 frames per second. Each frame contains 48 bits, thus making the data rate 192 Kbps. In contrast, the primary rate interfaces transfer data at 1.544 Mbps (North America) and 2.048 Mbps (Europe), which are rates specified by the T1 channel specification. The North American version of the primary rate interface (PRI) frame (Figure 9-4) contains a framing bit followed by twenty-three or twenty-four B-channel fields and a D-channel bit.

The European PRI frame, also shown in Figure 9-4, begins with a framing bit followed by sixteen B-channel fields, a D-channel bit, and fifteen more B-channel fields. Frames are transmitted at a rate of 8,000 frames per second.

## ISDN Data-Link Layer

**link access protocol for D channel (LAPD)** and **LAPB for B channels**
ISDN data-link protocol related to HDLC.

The data-link protocol used by ISDN closely resembles the high-level data-link control (HDLC) protocol described in Chapter 7 and is called **link access protocol for D channel (LAPD)** and **LAPB for B channels.** Details of the protocol specifications are found in ITU Q920 and Q921. The basic LAPD frame (Figure 9-5) is identical to the HDLC data-link protocol frame. It starts with the 7E start delimiter flag followed by an address field. The use of the address field differs from the HDLC basic form by including

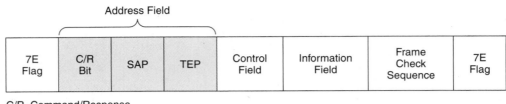

C/R–Command/Response
SAP–Service Access Point
TEP–Terminal End Point

**FIGURE 9-5** LAPD Frame

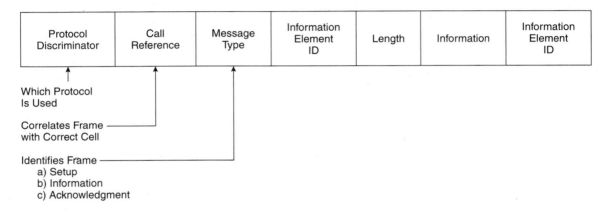

**FIGURE 9-6** ISDN Information Frame

**service access point (SAP)**
source address.

**terminal end point (TEP)**
destination address.

**transfer ID (XID)**
eXchange IDentification.

a **service access point (SAP)** identifier and a **terminal end point (TEP)** identifier instead of station source and destination addresses. The purpose is still the same, to identify a source entry (SAP) and destination exit (TEP) point in the network. An additional bit in the address field, called the COMMAND/RESPONSE (C/R) bit, is used to identify the frame as a command (0 for traffic from terminals, 1 for traffic from the network) or a response (opposite states) message.

The control fields identify the type of frame and keep track of the frame sequence as described in Chapter 7 for SDLC/HDLC protocols. Information, supervisory, and unnumbered control fields are the same as their HDLC counterparts with the following exceptions. The set normal response mode unnumbered frame is replaced with set asynchronous balanced mode extended. It functions similarly to the set normal response mode command by establishing a data-link for acknowledged data transfers (information frames). An additional unnumbered frame, **transfer ID (XID),** is included to allow stations to identify themselves for line management purposes.

The format for the information field consists of the subfields shown in Figure 9-6. It begins with a protocol discriminator, which identifies the type of protocol format the actual information payload is formatted in. The cell reference field correlates the current frame (cell) within the sequence of frames that comprise the entire message. The message type field following the cell reference identifies the message as a setup, pure data, or acknowledgement message. Information element IDs and length field aid in sequencing and determining how large the payload field is.

Frame check sequence and ending flag (7E) fields complete the LAPD frame. As with HDLC, supervisory frames cannot include a data field.

LAPB frames are similar to those of LAPD. XID and unnumbered information (UI) frames are not used with LAPB. Also, the frame numbers are restricted to modulo 8 (3-bit frame numbers) while LAPD uses modulo 128 (7-bit frame numbers).

Network layer specifications are included in the LAPD protocol. Call setup, flow control, call teardown, routing, error detection and recovery, and other network functions are described in ITU Q930 and Q931 sections of the LAPD protocol.

## ISDN Applications

In 1984 the hamburger chain giant, McDonalds Corporation, headquartered in Oak Brook, Illinois, decided that there was a requirement to integrate their numerous network and communications needs into a more efficient and cost effective system. Contracts between McDonalds, Illinois Bell Telephone, and AT&T were signed for the telephone companies to provide an integrated network based on the ISDN standard as specified by the ITU-ISDN standard.

Locally, McDonalds was divided between the home office (known as the Plaza), the Lodge, and Hamburger University. These facilities were up to 1.5 miles apart, with most interconnecting communications being provided on leased lines or the public telephone system. Networks at the sites and other services included an IBM Systems Network Architecture (SNA) network, office automation (OA), UNIX-based network supporting the product development area, a CENTREX system handling regular telephone and Telex communications, and various leased lines and packet-switch facilities to interconnect the networks.

Illinois Bell offered to establish a centralized system based on ISDN to integrate the functions of the various networks into a single information-transport network. Employees at the home office were equipped with ISDN telephone sets that included a ISDN data module that serves as an NT1 interface directly into the ISDN network. *Automatic number identification (ANI)* is included to display the caller's number before the call is answered. This is known as *Caller ID* today. Employees can use this ability to screen calls or to tap into a caller's database before actual dialogue is begun with the caller.

To manage the network, AT&T provided a package called NETPARTNER. This management software took the responsibility of maintaining the network from McDonalds and placed the maintenance and monitoring tasks onto AT&T. The network can still be monitored from a local terminal, but data are directly available to AT&T support teams. NETPARTNER tests ISDN lines for integrity and reports any problems that are detected. Translations between ISDN and analog lines are also checked and/or changed as required.

Office Automation (OA) users access the OA system by dialing a four-digit access number into either a X.25 packet-switched D channel or a B channel circuit-switch connection. The McDonalds office automation system makes the following services available to their employees:

1. Word processing
2. Database management
3. Graphics
4. Spreadsheets
5. Electronic mail
6. Time management
7. Access to other ISDN services
8. Printer access to all areas to obtain hard copies quickly

Printer interfaces throughout the system are served by a 9.6-kbps B channel circuit switch connection, allowing high-speed (eight to ten pages per minute) laser printer capability. No printer is more than 25 feet from employees. The quantity of printers coupled with the speed of printing results in efficient hard copy service. Other benefits of office automation via ISDN realized by McDonalds include:

1. Easy accessing of multiple hosts and applications.
2. Large and quick file transfer.
3. Easy implementation of employee situation change—office location change, new hire, termination, etc.

Use of IBM PCs and facsimile services is also incorporated into the McDonalds' ISDN network through terminal adapters (TA) that do the necessary conversion between IBM, facsimile, and ISDN formats and protocols.

McDonalds is noteworthy because it was the first nationwide business to incorporate the use of ISDN. The integrated nature of ISDN was used to consolidate existing services and to add many of the newer features described earlier. McDonalds is not the only large company willing to invest money and time to reap the benefits provided by ISDN.

The Mellon Bank of Pittsburgh, Pennsylvania, uses ISDN to handle many services between its Pittsburgh and Philadelphia banks, using Bell Pennsylvania as a local carrier and MCI for long distance connections. The main application is for video teleconferencing, transferring digitized video data at a rate of 112 kbps. Other uses by Mellon are customer-related. Using a system called Global Cash Management Services, Mellon handles cash disbursements, collections, and electronic funds transfers using ISDN as the prime network.

Other users of ISDN for services other than connecting to the Internet span the globe and include such diverse subscribers as Australia Telecom, Shell Oil, and Tenneco Corporation. In the United States, local Bell Telephone companies in conjunction with long distance carriers such as AT&T, MCI, and Sprint provide commercial ISDN services to many companies like Hayes Modem, Digital Equipment Corporation, and Southwest Banks.

A large number of subscribers employ ISDN for teleconferencing, telemarketing, and to expand upon or build wide area networks (WAN). Eastman Kodak implemented

ISDN as early as 1988 for voice conferencing. The list of ISDN users keeps growing, and now, with the benefits of direct digital access to the Internet, the future growth of ISDN usage is assured.

---

### Section 9.2  Review Questions

1. What does the I series of standards specify for ISDN?
2. What does the Q series specify?
3. What are the purposes of the B and D ISDN channels?
4. What are the two main types of ISDN services?
5. What is the purpose of a terminal adapter in the ISDN topology?
6. How do the two terminal equipment blocks differ? What function do they provide?
7. Packets in ISDN format enter the network at which reference point?
8. List the services provided by ISDN channels other than the BRI or PRI channels.
9. Describe a BRI data frame.
10. Describe a North American PRI data frame.
11. What is the purpose of F and L bits?
12. Which data-link protocol is used as the basis for LAPD and LAPB specifications?
13. How does the LAPD address field differ from an HDLC address field?
14. What items are specified in the ISDN network layer?

---

## 9.3  BROADBAND ISDN

**narrowband (N-ISDN)**
narrowband ISDN service.

**broadband ISDN (BISDN)**
broadband integrated services
digital network.

**subscriber's premises
network (SPN)**
multiplex broadband ISDN
service.

**broadband network
termination (BNT)**

The specifications discussed for ISDN to this point have been reclassified by ITU as **narrowband (N-ISDN).** In contrast, a second set of specifications for handling increased data rates on the public network have been drafted under the heading of **broadband ISDN (BISDN).** This set of specifications detail requirements for ISDN networks to handle 100-, 155-, and 600-Mbps transmissions carried on 1-KM cable segments utilizing repeater interfaces to extend network distances.

In composing this specification, the authors were required to meet N-ISDN interface specifications as well as broadband network needs. A typical network access arrangement is shown in Figure 9-7. A standard N-ISDN terminal (or network interface) and a broadband terminal interface (BTI) are serviced by the **subscriber's premises network (SPN),** which multiplexes the incoming data and transfers them to the broadband node, called a **broadband network termination (BNT),** which, in turn, codes the information into packets used by the BISDN network.

Data transfers to and from the BISDN network may be asymmetric. That is, access on and off the BISDN network may be done at different rates depending on system

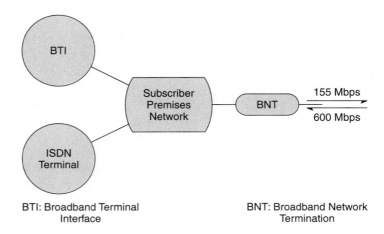

FIGURE 9-7 BISDN Configuration

requirements. The illustration shows data entry to the network to be at 155 Mbps and from the network, 600 Mbps.

Subscriber's premises networks can be a private branch exchange (PBX), token ring, or other local area network (LAN) configuration. The BNT interface with supporting software is capable of recognizing protocols and frame formats of the data supplied to the interfacing network.

## BISDN Services

N-ISDN services break down into three classes, *bearer service, supplementary service,* and *teleservice.* Bearer services include the user interface, setting transmission modes, and control signaling. Supplementary services include access to the public data network and user-to-user signaling. The responsibilities attributed to the teleservice class include:

1. defining type and category of information.
2. 4-kHz analog signal for telephone voice and a 7-kHz band for high quality audio.
3. transporting text, facsimile, mixed mode teletext, high definition video, and still or slow motion video telephony information.
4. handling video conferencing.

BISDN services expand user bearer services to include 155- and 622-Mbps subscriber access and additional teleservices, including LAN to WAN interconnections and high definition television (HDTV). These services are classified for BISDN as user network interfaces (UNI). Additional services classified as network-to-network interfaces (NNI) allow BISDN to be interfaced to any network node point.

Operation of BISDN networks is further distributed through three planes, the *user plane, control plane,* and *management plane.* User plane functions involve user information, flow control, and error recovery processes. Control planes are responsible for con-

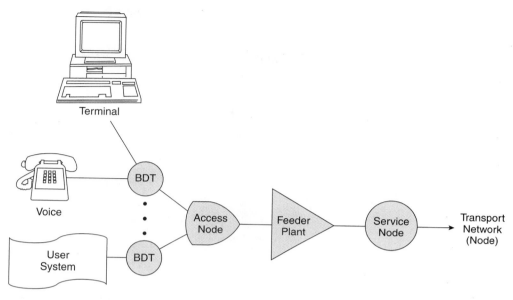

**FIGURE 9-8** BISDN Access

trol signaling, including establishing and breaking connections. Finally, the management plane manages network resources and provides maintenance and coordination between all functions of the BISDN network.

## BISDN Topology

A typical BISDN configuration (Figure 9-8) begins with data supplied to the system through various peripheral devices or networks. Each is interfaced to the *Access Node* of a BISDN network through a broadband distant terminal (BDT), which is responsible for electrical to optical conversions, multiplexing of peripherals, and maintenance of the local system. Access nodes concentrate several BDTs into a number of high-speed fiber lines directed through a feeder point into a **service node.**

**service node**
manages ISDN system access control functions.

The service node manages the majority of control functions for system access. It handles call processing, administrative functions, switching, and maintenance functions. The service node uses distributed management through functional modules within its area of influence. These functional modules are connected in a star configuration (Figure 9-9) and include switching, administrative, gateway, and maintenance modules. The central hub oversees the operation of the modules and acts as the enduser interface for control signaling and traffic management.

User terminals close to the central office bypass the use of access nodes and are connected directly to the BISDN network through a service node. The larger bandwidth of fiber cables allows for higher data rates and numerous channel handling capacity for the BISDN system.

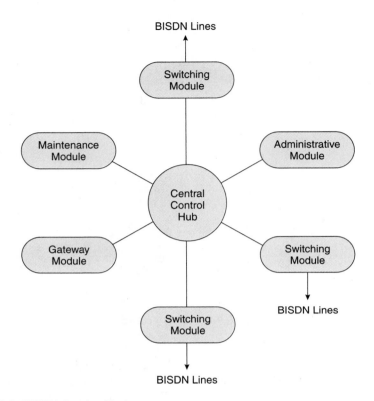

**FIGURE 9-9** BISDN Service Node

**Section 9.3 Review Questions**

1. Describe some of the additional benefits of using BISDN compared to narrowband ISDN (or N-ISDN).
2. What are the functions of the following BISDN units?
   a) Broadband distant terminal (BDT)
   b) Service node
   c) Broadband network termination (BNT)
3. What services are provided by a service node's central control hub?

## 9.4  IEEE 802.9: INTEGRATED VOICE AND DATA SERVICES

The 802 committee of the IEEE communications group has a section devoted to integrated services. The 802.9 specification details how ISDN and local area networks (LAN) are interfaced and used to transport voice, data, and video information. This specification

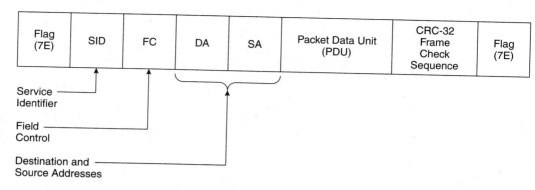

**FIGURE 9-10** MAC Frame

**access unit (AU)**
entry point into an ISDN network.

relies heavily on the ITU group of standards for ISDN as well as its own media access control (MAC) standards.

An **access unit (AU)** is designated as the interface between a user and the integrated network. The types of data handled by the network are classified as packet data and isochronous data. Isochronous data includes voice, image, video, facsimile, and text data. The format of the MAC frame for 802.9 is shown in Figure 9-10.

The MAC format uses the HDLC frame format as a model, beginning with a start flag sequence of six ones in between two zeros (7E in Hex). The control field begins with the service identifier (SID), which indicates the type of service provided, and the type of data-link frame used. The frame control (FC) field signifies the type of MAC frame (information or control) and the frame's priority.

**protocol data unit (PDU)**
field that holds data payloads.

Actual destination and source addresses fall between the control fields and the **protocol data unit (PDU),** which holds a higher-level payload packet. CRC-32 is the error detection scheme used in the frame check sequence (FCS) fields. This is followed by the ending 7E flag to complete the MAC frame.

---

**Section 9.4  Review Questions**

1. What does IEEE 802.9 specifically address?
2. What is isochronous data?

---

## 9.5 DIGITAL SUBSCRIBER LINE (DSL)

**digital subscriber line (DSL)**
high-speed digital service for transporting voice and data.

ISDN is a digital service supplying integrated voice and data traffic across the telephone system. An alternate digital service finding increasing use is the **digital subscriber line (DSL)** which is not a line actually. Instead, DSL describes modems that are used to carry digital information onto existing telephone lines. DSL is used to replace the local loop between telephone switching stations and the subscriber.

Unlike ISDN, which integrates voice and data onto a single channel, DSL uses narrowband filters called *splitters* to split the voice channel to carry voice and data on separate lines. This eliminates interference between the two signals and interference to the data signal from household appliances like light dimmers and vacuum cleaners. Since voice and data are using separate lines, an Internet connection can be maintained at the same time a voice communication is in progress over the same channel. DSL comes in many flavors, each with specific features and applications:

**HDSL high bit rate**

1. **HDSL** and HDSL2—**high bit rate DSL** is used as an alternate to T1 lines, carrying 1.544 Mbps full duplex over fiber optic lines up to 12,000 feet in length.

**VDSL very high data rate**

2. **VDSL—very high data rate DSL** is an asymmetrical service that carries up to 52 Mbps of data upstream from server to user and 2.3 Mbps downstream. It can be used on unshielded twisted-pair (UTP) cable up to 1,000 feet in length. VDSL is also specified for 45,000 feet of fiber optic cable. It is used for video conferencing, broadcasting, and bulk file transfers.

**SDSL symmetric DSL**

3. **SDSL**—symmetric DSL is used for lines that carry from 160 Kbps to 2.048 Mbps in both directions on cables up to 1,000 feet in length.

**IDSL integrated DSL**

**RADSL (rate adaptive DSL)**

4. **IDSL—Integrated DSL** is an ISDN-like symmetrical service that transports data at 128 Kbps on UTP cable up to 18,000 feet in distance. As mentioned earlier, the main difference between IDSL and ISDN is that IDSL splits data and voice into two lines, while ISDN truly integrates both onto a single line.

**MSDSL (multi-rate DSL)**

**CDSL**
customer DSL

5. Others include **RADSL** (rate adaptive DSL), **MSDSL** (multi-rate DSL), and **CDSL** (customer DSL).

## ADSL

**asymmetric DSL (ADSL)**

**discrete multitone technology (DMT)**

The form of DSL that is finding the largest acceptance is **asymmetric DSL (ADSL),** which uses a technique called **discrete multitone technology (DMT)** to increase the utilization of the system bandwidth. After voice and data are separated into different lines, the bandwidth is divided, using frequency division multiplexing (FDM) or echo cancellation techniques to create an upstream and a downstream transmission path. Each stream operates at a different data rate, thereby supplying the asymetric nature of ADSL. The frequency spectrum for an ADSL service is represented in Figure 9-11. The lower 4 KHz is set aside for **plain old telephone voice service (POTS).** For FDM-configured ADSL lines, the upstream and downstream signals are separated (Figure 9-11a), while echo cancellation techniques allow for overlapping (Figure 9-11b) to occur.

**plain old telephone voice service (POTS)**
original telephone system.

In addition, time division multiplexing (TDM) is employed to take these bands and divide them into multiple channels to carry different types of data simultaneously. The upstream data rate is from 16 Kbps to 1 Mbps, while the downstream rate can be anywhere from 1.5 to 9 Mbps.

**ADSL terminal units— central office (ATU-C)**

The topology for an ADSL system is shown in Figure 9-12. It starts with any number of different types of service entering an access point at reference point $V_C$ on the figure. The access node provides interfacing of broadband services (point $V_A$) to **ADSL terminal units—central office (ATU-C),** where data is converted to ADSL format at point

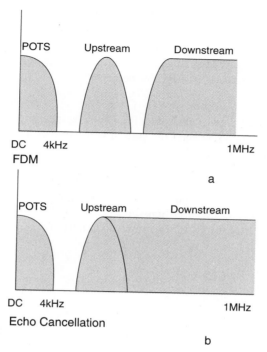

**FIGURE 9-11** ADSL Frequency Spectrum

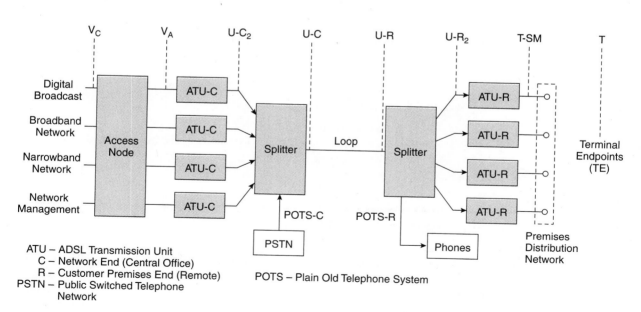

ATU – ADSL Transmission Unit
  C – Network End (Central Office)
  R – Customer Premises End (Remote)
PSTN – Public Switched Telephone
      Network

POTS – Plain Old Telephone System

**FIGURE 9-12** ADSL Topology

| 7E Flag | FF | 03 | PPP Protocol ID | PPP Protocol Data Unit (Payload) | FCS CRC-16 | 7E Flag |
|---------|----|----|-----------------|----------------------------------|------------|---------|

**FIGURE 9-13** ADSL PPP Frame Format

**public switched telephone network (PSTN)**
supplier telephone connections to ADSL lines.

U-C$_2$. Signals from the ATU-C units are fed to a splitter, which multiplexes them onto a single loop line (point U-C). Telephone connections from the **public switched telephone network (PSTN)** enter the system at the splitter level and are added in the POTS area of the ADSL spectrum.

At the remote end (point U-R), the ADSL signal enters another splitter to be demultiplexed. Telephone voice signals leave and go directly to PSTN. The remaining signals are fed to **ATU-R units (R for remote)** (point U-R$_2$) to be returned to their original format (point T-SM) and supplied to distribution networks at the remote premises and sent to their final destination terminal endpoints (TE at point T).

**ATU-R units**
ADSL terminal unit-remote.

### ADSL Frame

**point-to-point protocol (PPP)**
low-level data transport protocol.

An example frame format for data on an ADSL line between ATU-C and ATU-R units is shown in Figure 9-13. Notice that it is strongly based on the HDLC data-link format. This frame format is used to transport **point-to-point protocol (PPP)** data packets. PPP is a basic lower-level networking protocol whose payload (also known as a protocol data unit or PDU) can be used to encapsulate higher level protocols.

The frame begins with the standard HDLC flag (7E in hex) followed by a PPP address code field of FF03 hex. Two bytes of protocol ID identifies the payload type and possible protocol that has been encapsulated in it. The frame check sequence field uses CRC-16 for error detection and the frame ends with another 7E flag. There are similar frame formats to support other protocol services using ADSL lines.

---

**Section 9.5 Review Questions**

1. How does DSL differ from ISDN?
2. What is the difference between symmetrical and asymmetrical service?
3. What is the purpose of a splitter?

---

## 9.6 PRIVATE BRANCH EXCHANGE

**private branch exchange (PBX)**
office telephone switching unit.

Originally, the **private branch exchange (PBX)** was developed to facilitate switching of vast numbers of telephone sets in office and other buildings. A PBX station contained the switching system, interface, and control mechanism to relay calls from outside lines to numerous users within a locality. In the simplest form, a PBX switch station consisted of a switchboard and an operator who made the appropriate connections.

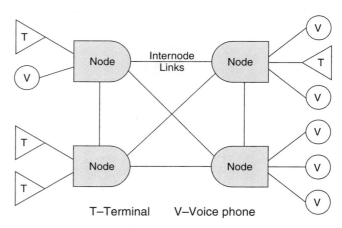

**FIGURE 9-14** Fully Connected PBX Nodes

By 1963, the switching and control mechanism came under computer control, but the circuits being switched were still analog voice lines. It wasn't until 1971 that the first data lines carrying digital information were installed using PBXs. Voice communications became digitized through the use of analog-to-digital conversions produced by CODECs operating into a pulse code modulated (PCM) encoded line.

Installation, control, and use of PBX systems remained in the domain of the telephone company until 1968, when the Federal Communications Commission (FCC) handed down a ruling permitting outside vendors to interconnect equipment to existing telephone lines. These vendors were required to adhere to rule 68 of the FCC code, which stipulated stringent electrical and interfacing requirements for utilizing telephone lines. The ruling opened the PBX networks to integrated data and voice communications.

## PBX Networks

PBX networks are interconnected using node concepts similar to those developed for the system network architecture (SNA) network. Entry and exit to the network, as well as interconnections within the networks, are managed through nodes. Figure 9-14 is an illustration of fully connected PBX nodes. Fully connected systems have a small number of nodes due to the limitation imposed by the number of interconnecting cables used by the network. The number of connections is computed using the formula developed in the discussion of switching methods and reproduced here for reference:

$$C = \frac{n(n-1)}{2} \qquad (9\text{-}1)$$

where *n* is the number of nodes and *C* the number of interconnections required to directly connect all the nodes.

Also shown in Figure 9-14 are data terminals (T) and voice phones (V) connected to each node. The disadvantage of a fully connected PBX is the same as that of directly connected networks. As the number of nodes increases, the number of connections increases by a factor close to one-half the square of the node count. For larger systems, PBXs utilize digital-switched networks.

## PBX Control Mechanism

The PBX control mechanism manages and coordinates activities of the digital-switch network, manages interfacing to the network, and handles miscellaneous network resources. The mechanism, a combination of hardware and software, is responsible for accepting input data, translating that data into a form recognized by the PBX network, and supplying output data to the destination node. The mechanism's functions are very similar to the SNA network systems service control points and logical units.

## PBX LAN

A PBX LAN using a digital-switch network is shown in Figure 9-15. PBX switching networks handle various types of information switching. Voice entering in analog form connects to the system via an *integrated interface* (I), where it is converted into digital form and integrated into a data channel. The voice data are separated from the data at the destination integrated interface. At this point, the voice data are returned to analog form. Voice information can also be admitted into the system via an analog interface, which also converts it into digital but does not integrate it into a data channel. In this instance, the digitized voice information is routed separately to a destination node.

Computer terminals supply digital data directly to the system using a T1 data line. Use of this line demands that the data already be placed into the packet form used by the PBX network. Parallel data are converted to serial and translated into the required form through PBX modems or CODECs connected between the network switch data interface (D) and the computer terminals.

Totally foreign networks can be interfaced to the PBX network by using a PBX gateway, which is the hardware and software element that translates the foreign network's protocols and data stream into PBX format. Translating between two differing protocols and packet formats within the gateway is the task of packet assemblers/disassemblers (PAD), which assemble segments of asynchronous data into X.25 formatted packets. Besides setting the data into the correct format, the PAD also translates addresses between PBX extensions and X.25 addresses.

## PBX Bridge

A system called a PBX bridge is employed to interconnect PBX networks. In PBX terminology, a bridge interconnects two adjacent networks, be they PBX or other types, as shown in

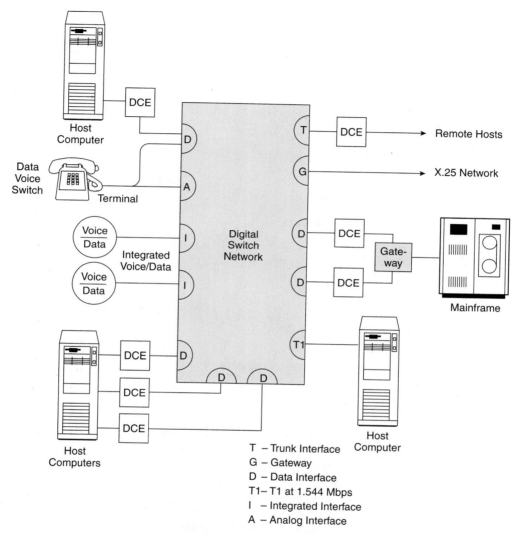

**FIGURE 9-15** PBX LAN Example

Figure 9-16a. Half bridges connect a network segment to an intermediate unit. The intermediate unit sends the data to another half bridge, which then routes them to another network segment (Figure 9-16b). It should be kept in mind that the definitions for PBX gateways and bridges differ from the general network definitions. For general use, bridges are used to interconnect similar networks and gateways to interconnect dissimilar networks.

## PBX Interfaces

Connection paths into the PBX system originate from many sources. They are digitized, if not already in digital form, at various data rates depending on the medium used to convey

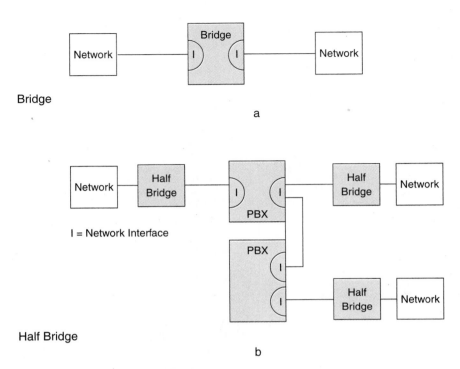

Bridge

a

Half Bridge

b

**FIGURE 9-16** PBX Network Bridges

**voice grade medium (VGM)**
cable that transports digitized voice.

the data. There is a **voice grade medium (VGM)** cable that transports digitized voice data at 1 Mbps at distances up to 2,000 feet from source node to destination node.

On the other hand, integrated voice and data are composed of digitized voice, digital data, and control bit streams, which are multiplexed into a single synchronous stream of data using a link interface control, as seen in Figure 9-17. This channel is built from a 64-Kbps voice channel, one or more 64-Kbps data channels, and an 8- to 64-Kbps control channel.

A 24-channel T1 line running at 1.544 Mbps can be used to interface computer terminals directly into a PBX network using the network's protocols and formats. Much of the transfer speed is easily gained since the need to translate between protocols and formats is unnecessary.

Connections into a PBX network as classed under switch-network functions based on a connection table that is updated to include added users and delete nodes no longer connected to the network. Keeping track of nodes that are moved from one entry point to another in the network is another function of the connection table. This table also includes classes of service for each node.

## PBX Session

An example of a communication session on a PBX begins with a signal from a user node requesting access to the network. This user's address is checked against the connection

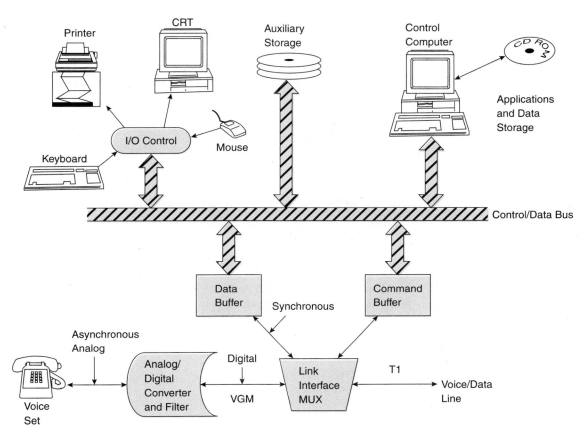

**FIGURE 9-17** PBX Integrated Voice/Data Unit

table for purposes of identification, access authorization, and to determine what kind of service the user requires and is entitled to. Once it is determined that the user can have access, it is assigned a time slot that is used by the central control of the PBX switch to acknowledge the user's request.

In response to the acknowledgement, the user issues a call command identifying the intended destination point. Once again, the connection table is checked, this time to determine the validity and availability of the destination station. If the destination station is in use, a busy signal is returned to the originating user.

If the destination station is not currently in use, a time slot is set aside to be used by the destination station. The connection table is updated to show that both stations are in use. The destination uses the assigned slot to issue a ready signal to the PBX controller. The controller then completes the connection to the originating station, and traffic can now flow back and forth between them.

At the end of the dialog between the two stations, one or the other issues a disconnect command and the connection is broken. The time slots used by the stations are freed for other use and the connection table is upgraded to reflect the nonbusy condition of the two nodes.

**Section 9.6 Review Questions**

1. What was the original use for PBX systems?
2. What type of services can today's PBXs handle?
3. What is the purpose of a time slot?

# 9.7 ASYNCHRONOUS TRANSFER MODE (ATM)

ISDN and other high-level protocol data units are routed through complex multiple networks using a routing protocol such as asynchronous transfer mode (ATM), sometimes referred to as *cell relay*. Another widely-used routing protocol is frame relay, which does the same job as ATM but uses a different format. Other applications for the use of these routing protocols includes internetworking transfers of the following data types:

- Imaging.
- Teleconferencing.
- Multimedia.
- Video on demand (movies).
- Distance learning online.
- Interactive video conferencing.
- Video and text libraries.
- Network interconnectivity.
- Internet access.
- File transfers.
- Distributed data bases.
- Campus backbones.

The benefits of these routing protocols include, but are not limited to:

- Integration of different types of data traffic onto a single channel.
- Guaranteed bandwidth and resource allocations.
- Bandwidth on demand—connections made as needed.
- Supports several levels of quality of service (QoS).
- Autoconfiguration.
- Recovery from system failures.

## ATM Cells

**cell**
ATM data packet.

ATM uses a fixed packet size of 53 bytes called a **cell** to encapsulate upper-level payloads. The first 5 bytes of the cell form a header for the ATM cell, while the remaining

48 bytes are used to carry the actual payload information. The ATM protocol is not influenced by the actual configuration of the payload. From its standpoint, the payload is data. ATM leaves the interpretation of that payload to the protocol that generated it. The payload is segmented into 48-byte sizes and packaged into ATM cells. This makes the ATM routing protocol predictable since headers and payloads are always the same size (5 and 48 bytes each). Switches located in the routing path know where to find header information as each cell passes through them.

**latency**
time it takes for a packet to migrate through a network.

Cell size is small enough to reduce **latency** which is the time it takes a packet to migrate through a network, and large enough to minimize the ratio of payload to header (almost 10 to 1). These cells can be multiplexed with other payload cells and transported through ATM routers and delivered to an endpoint network where they are demultiplexed and reassembled into the original payload format and placed onto the endpoint network to be eventually routed to the user terminal or node.

**user-network interface (UNI)**
handles ATM transfers between user equipment.

**network-tonetwork interface (NNI)**
handles ATM transfers between networks.

Two header formats are depicted in Figure 9-18, one of which is for the **user-network interface (UNI),** which deals with transfers between user equipment, ATM end systems (ES), or ISDN broadband terminal equipment (B-TE) and terminal adapters (TA), network termination (NT), or ATM intermediate systems (IS). The second format differs slightly from the first in that it lacks a generic control field and has an extended virtual path identifier. This header format is used for **network-to-network interface (NNI)** services between networks or between nodes and networks.

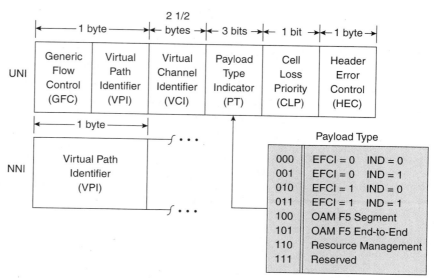

EFCI – Explicit Forward Congestion Indication
IND  – Last Cell Indicator (AAL5)
OAM – Operations, Administration, and Maintenance

**FIGURE 9-18** ATM Headers

**generic flow control field (GFC)**
identifies level of congestion control and cell priority.

**virtual path identifier (VPI)**
ATM network address.

**virtual channel identifier (VCI)**
ATM end point address.

**cell loss priority (CLP)**
determines which cells be discarded to avoid congestion.

**header error control (HEC)**
checks errors in headers.

The five-byte ATM header shown in Figure 9-18 begins with a **generic flow control field (GFC),** which identifies the level of congestion control and cell priority. This field helps determine if a transmitted cell can be discarded in the event that traffic on the network becomes highly congested. Two address fields follow the generic flow control field, the **virtual path identifier (VPI)** and the **virtual channel identifier (VCI).** VCIs contain a 16-bit value that identifies a single virtual channel across an ATM network. VPIs are 4- or 8-bit values that identify a group of VCIs and provides a logical link between networks or nodes where the VCIs are located.

The payload-type (PT) field uses three bits to indicate which ATM adaption layer is used to format the data in the payload, an explicit forward congestion indication, and whether the cell contains management information. ATM adaption layers are discussed in detail following this paragraph. The forward congestion indication alerts the application to possible delays that may occur because of following congestion. The **cell loss priority (CLP)** bit determines whether or not the current cell can be dropped when congestion becomes a problem. Dropped cells will have to be retransmitted by the sending station.

The final field in the ATM header is a one-byte **header error control (HEC)** field, which is used to tell upper layers of the ATM stack to perform error detection on the cell header and to drop a cell with a bad header before the payload is sent to the routing layers of the ATM stack. HEC can perform single bit error correction or multiple error detection of errors in the header. The ATM protocol suite also uses HEC information to aid in locating cells that are directly mapped into a time division multiplexed (TDM) payload using a system called *HEC-Based Cell Delineation.* When several headers are located in the received data stream, ATM knows where to find the 53-byte cells.

## ATM Layers

The ATM stack shown in Figure 9-19 is loosely comparable to the lower three layers of the open systems integration (OSI) model. The lowest layer is the ATM physical (PHY) layer, which is further subdivided into two sublayers, transmission convergence (TC) layer and the physical medium dependent (PMD) layer. The responsibilities of the convergence layer include:

- Regulate cell rate.
- Establish the header error control sequence and method.
- Generate and recover frames (cells).
- Convert between bitstream rates to ATM cell rates.
- Generate HEC.
- Cell rate decoupling.
- Uses the HEC-based call delineation process to locate cells in a directly mapped time division multiplexed stream of data.

Cell rate decoupling deals with inserting idle or unassigned cells into the ATM stream when a cell is expected but not yet present. This aids in maintaining a desired flow rate through the system. When a time division multiplexed slot is ready to be sent and there is no cell available, the transmission convergence function inserts an idle or unas-

| OSI* Layer | ATM Layer | Sublayer Responsibilities | |
|---|---|---|---|
| Network Layer | ATM Adaption Layer (AAL) | Convergence Sublayer (CS) | Common Part (CP) |
| | | | Service Specific (SS) |
| Data Link Layer | | Segmentation and Reassembly (SAR) | |
| | ATM Cell Layer | Flow Control Cell Header Generation and Extraction Cell VPI/VCI Translation Cell Mux/Demux | |
| Physical Layer | ATM Physical Layer (PHY) | Transmission Convergence (TC) Sublayer Cell Rate Decoupling HEC Sequence Frame Adaption Frame Generation and Recovery | |
| | | Physical Medium Dependent (PMD) Bit Timing Physical Medium Standards | |

\* Loose Interpretation

**FIGURE 9-19** ATM Protocol Stack

signed cell to that slot. This cell is stripped at the received end, which sends assigned cells to their proper destination.

The physical medium layer does not specify parameters directly, but relies on existing physical layer protocols. Its main responsibility is establishing the means of accessing the medium used and to set up any bit timing required to allow the physical medium to accept ATM cell transmissions. Some physical protocols that are supported by ATM include various levels of SONET, ISDN, and fiber optic network services. Table 9-2 lists some of these services.

---

**TABLE 9-2**

## ATM Supported Services

| Data Rate | Service Supported |
| --- | --- |
| 1.544 Mbps | ITU DS1, ISDN PRI |
| 2.048 Mbps | ITU E1, ISDN PRI |
| 6.312 Mbps | ITU DS2 |
| 34.368 Mbps | ITU E3 |
| 44.736 Mbps | ITU DS3, ANSI DQDB |
| 51.84 to 84 Mbps | SONET STS-1 |
| 100 Mbps | FDDI |
| 139.264 Mbps | ITU E4 |
| 155.5 Mbps | SONET STS-3c, ITU STM-1, Fibre Channel |
| 622.08 Mbps | SONET STS-12c, ITU STM-4 |

ITU—International Telecommunications Union.
ANSI—American National Standards Institute.
DQDB—Distributed Queue Dual Bus.
STS—SONET—Synchronous Optical Network.
SMT—Synchronous Digital Hierarchy (SDH).
FDDI—Fiber Distributed Data Interface.

OSI's data-link layer is represented by ATM as the ATM cell layer whose duties include:

- Flow control.
- Cell header generation and extraction.
- Translation of VPI and VCI addresses.
- Multiplexing and demultiplexing of ATM cells.
- Validation of ATM headers.
- Cell loss priority processing.
- Explicit forward congestion control.

Cell loss priority and explicit forward congestion control are methods to reduce the traffic load when heavy traffic leads to an overload and causes congestion with cells backing up and waiting to be transferred. A bit in the header identifies whether a cell may be dropped from the transmission queue when congestion occurs. This bit assigns a lower priority to that cell and if the traffic on the system becomes congested, the cell is dropped and will have to be retransmitted at a later time.

**ATM adaption layer (AAL)**
ATM network layer.

The larger work of the ATM protocol suite falls into the **ATM adaption layer (AAL),** which is similar in nature to the OSI network layer, although some of the issues addressed by the AAL can apply to both the data-link and transport layers.

**quality of service (QoS)**
level of data delivery service for reliability and speed.

Before we can explore the different ATM adaption layers, there a few additional facilities and parameters associated with the ATM protocol that need to be discussed. These are ATM switching methods and **quality of service (QoS).** Quality of service includes the type of data being transported, the ability to assure delivery of that data, recovery from faults and data errors, and other factors that effect the quality level of the transfer service provided by the ATM network.

QoS concerns can be summarized in these four areas:

1. Negotiation—resolving differences between users over requests and what can be provided.
2. Mapping—between applications, system, and network QoS.
3. Resource reservation—local end operating systems and intermediate network systems.
4. Delivery—carrying out activity to support QoS.

## Bit Rate Types Used by ATM

**constant bit rate (CBR), variable bit rate (VBR), available bit rate (ABR), unspecified bit rate (UBR)** defines amount and type of traffic a channel will handle.

Traffic routed by ATM networks falls into three main bit-rate categories, **constant bit rate (CBR), variable bit rate (VBR), available bit rate (ABR),** and **unspecified bit rate (UBR).** Each has its own characteristics and applications as shown in Figure 9-20.

Constant bit-rate traffic is data sent at a constant level of use. Peak periods of traffic carry the same volume as nonpeak periods. This type of traffic has a low latency since the communication link remains constantly on and traffic is transferred on a dedicated communication channel. This type of channel is called a **permanent virtual channel (PVC)** and is established before any data is transmitted onto it. PVC links require a portion of the system bandwidth to be constantly dedicated to that channel even if no data is being sent on it. Data cells or packets transferred using a permanent virtual circuit arrive at the destination in the order they are sent since the same path is used to transfer all of the cells and no delays are experienced.

**permanent virtual channel (PVC)** constantly established communication link.

Constant bit-rate traffic has the highest quality of service since the channel is always open and guaranteeing delivery of data packets. Applications for CBR-type traffic include time-sensitive information, like real time voice and video. When these cells arrive slightly out of sync, they would cause disruption in the audio or jitter in the video presentation. With constant bit rate data flow, delays between cells are minimized.

**switched virtual circuit (SVC)** communication links that are established when needed.

Variable bit-rate traffic is a constant flow of traffic whose volume varies at different times of the day. This traffic is usually sent using a PVC, but could also be transmitted using a **switched virtual circuit (SVC).** SVCs are circuits that are established when they are needed to transmit data. The routing for the circuit is selected by routers in the network as the first frame or cell is transmitted. Another name for SVCs is *Bandwidth on demand,* since the bandwidth needed for the channel is allocated only when it is needed. For variable bit-rate traffic to take advantage of a SVC, there must be periods of the day when the traffic level is zero and the transmission channel is not needed at all. Traffic sent part of the time and not sent at other times is referred to as **burst** traffic, since the level of traffic literally bursts onto the line for a fixed duration of time.

**burst** heavy data traffic for a short duration.

Highly bursty traffic, which experiences high-level peak periods and extended "quiet" times falls into the category of the available bit-rate traffic. This traffic always uses SVC channels and the lowest quality of service. Highly bursty traffic can experience congestion if the peak level becomes too high for the circuit to handle. In this case, cells may be dropped to be transmitted at a later time. The parameter that specifies the percentage of high priority cells to those that can be dropped is called the *acceptable cell loss ratio.*

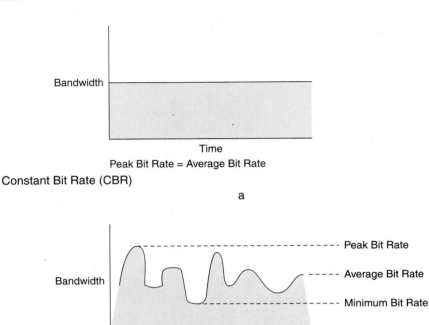

Bandwidth

Time

Peak Bit Rate = Average Bit Rate

Constant Bit Rate (CBR)

a

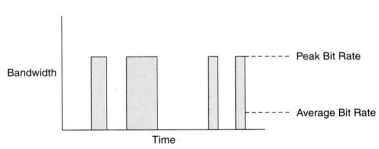

Bandwidth

- - - - - - - - - - - - - - - - - - - - - - - - - - Peak Bit Rate

- - - - - Average Bit Rate

- - - - - Minimum Bit Rate

Time

Peak Bit Rate > Average Bit Rate

Variable Bit Rate (VBR)

b

Bandwidth

- - - - - - Peak Bit Rate

- - - - - - Average Bit Rate

Time

Peak Bit Rate (Bursts) >> Average Bit Rate

Available Bit Rate (ABR)

c

**FIGURE 9-20** ATM Traffic Types

The last category of bit-rate traffic is a catchall category, aptly designated as the un-specified bit rate. Bandwidth is not reserved for this category, leaving all use of it to bandwidth-on-demand switched virtual circuits. There is no guarantee of delivery of data or quality of service and it experiences a high cell loss in the event of congestion. An-other name for unspecified bit rate traffic is *best effort data delivery* and is available to users that can tolerate possible data loss.

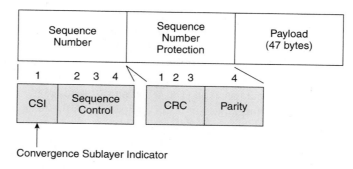

**FIGURE 9-21** AAL-1 Format

Other ATM connection parameters besides the type of traffic include *peak cell rate* and *sustainable cell rate (SCR),* which define the maximum traffic capacity a channel can carry. SCR is found by dividing the number of burst cells allowable for a specific time (committed burst, Bc) by that time period (committed time, Tc): SCR = Bc/Tc. If burst cells are sent in excess of the sustainable cell rate, the system can do several things:

1.  Send the cells if bandwidth is available.
2.  Store the cells in a buffer until bandwidth is available.
3.  Discard the excess cells if cell loss priority bit is set.

Time constraints include *cell delay variation tolerance,* which is a measure of how much variation in milliseconds data traffic may experience. This is crucial in time sensitive transmissions like video, where intolerable variances would cause the image to "jitter" on the screen. Burst size in cells specifies the maximum time burst traffic can be tolerated on a given system.

## ATM Adaption Layer (AAL)

**convergence layer**
ATM layer responsible for placing data into a common format.

**segmentation and reassembly (SAR)**
ATM process for dividing messages into cells and reassembling them correctly at the destination point.

The ATM adaption layer (AAL) details the classes of service for the ATM network with each layer being divided into two sublayers, the **convergence layer,** which is responsible for placing data into a common format and mapping cells onto the transmission medium and the **segmentation and reassembly (SAR)** layer, which is actually responsible for segmenting upper level protocol data units into ATM cells and reversing the process once the cells are delivered to the endpoint network. The convergence layer is also responsible for assuring necessary error control and the sequencing of cells. The adaption layer is also responsible for timing and flow control and the detection and handling of out-of-sequence cells. This layer has six different classifications of service designated AAL-0 through AAL-5.

The AAL-0 service is used for the transport and relay service of traffic already in ATM cell format. It does not require any protocol or data-rate translations. Class A or AAL-1 service handles constant bit-rate traffic over existing telephone lines. It is a connection-oriented service using permanent virtual circuits to carry voice and video information. Applications for AAL-1 service are primarily for voice traffic. The AAL-1 frame format is shown in Figure 9-21. It begins with a 4-bit sequence number to aid in sequencing of cells and a 4-bit sequence number protection to assure that the sequence number value is correct. The sequence number has a one-bit convergence sublayer indicator to indicate whether this cell refers to the convergence or SAR sublayers. The remaining three bits indicate the sequence number. In the protection field, a three-bit CRC (cyclic redundancy check) and one-bit parity are used for error detection of the sequence number field. The remaining 47 bytes of the ATM cell payload field are used for data of upper-level protocol payloads. This frame then becomes the 48-byte payload for an ATM cell, preceded by a 5-byte header.

Class B or AAL-2 service is the same as class A except that it uses variable bit-rate traffic instead of constant bit-rate. It used to carry compressed voice and video data and some LAN-type traffic.

Class C or AAL-3/4 service remains connection-oriented, delivering variable bit-rate data used to interconnect or emulate local area networks (LANs). A class D service also utilizing AAL-3/4 is for a connectionless, switched virtual circuit service that handles packet-type datagram traffic. The format of the cells for AAL-3/4 appear similar to that of AAL-0 and AAL-1 with the exception that header information identifies the type of cell and the length of the payload in use. An additional field for use when multiplexing ATM cells is included, called the Message ID (MID) field.

The simplest and most efficient class of adaption service is AAL-5, known as SEAL (*simple and efficient adaption layer*). It is used primarily for bursty LAN traffic. AAL-5 does not support any multiplexing and uses available bit-rate type traffic. The frame format for AAL-5, shown in Figure 9-22, includes multiples of 48-byte payload data and a 12-byte trailer, which are all segmented into ATM cells. The trailer includes padding to assure that the payload resides on 48-byte bounds, a 2-byte control field (CF), a 4-byte payload length field (LF) and a 4-byte CRC-32 error detection character. AAL-5 does not include sequencing numbers or any way to detect out of sequencing problems.

## ATM Address Resolution Protocol (ARP)

**address resolution protocol (ARP)**
translates between non-ATM and ATM addresses.

Translation of addresses from upper-level protocols into ATM virtual channel and virtual path identifiers is accomplished through the use of an **address resolution protocol (ARP),** which does as its name implies, resolves address translations. The process of establishing an ATM address begins with an ARP request to an ATM server to establish connection. The server responds with an ATM address from an internal table for the node that deals with the upper-level protocol address. The terminal, after receiving its ATM address, begins signaling procedures and requests a connection to the destination. ARP checks ATM switches for

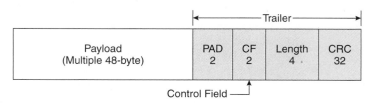

**FIGURE 9-22** AAL-5 Format

routing through the network to the requested destination. Once the destination is reached, a connect message is returned along the now established path between the two node points.

ATM addresses are required to identify *network service access points (NSAP)*, including networks, subnetworks, and end system nodes. They may not necessarily identify end users. Keep in mind that ATM is a routing protocol responsible for transporting data from an entry point that is more often a network server or hub to a destination node that may also be a network node or server. The beginning of the address, Figure 9-23, is called an *Initial Domain Part* and specifies the administrative authority that allocates the address, or basically, which routing tables to use. The last seven bytes of the address comprise the actual node address, the end system identifier, and a selector used to identify different logical applications at the end system. The middle portion of the address is dedicated to the company hierarchy used to identify switches and subnetworks within the same address space. These include divisions within the company, campus buildings, offices, switch locations within a site, etc.

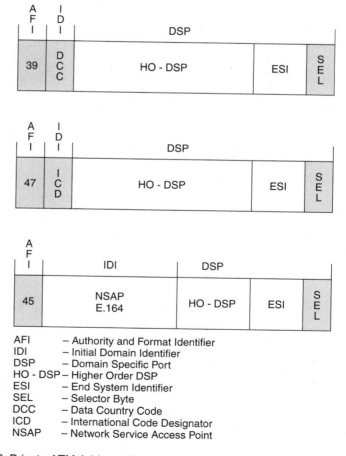

AFI       — Authority and Format Identifier
IDI       — Initial Domain Identifier
DSP       — Domain Specific Port
HO - DSP  — Higher Order DSP
ESI       — End System Identifier
SEL       — Selector Byte
DCC       — Data Country Code
ICD       — International Code Designator
NSAP      — Network Service Access Point

**FIGURE 9-23** Private ATM Address Formats

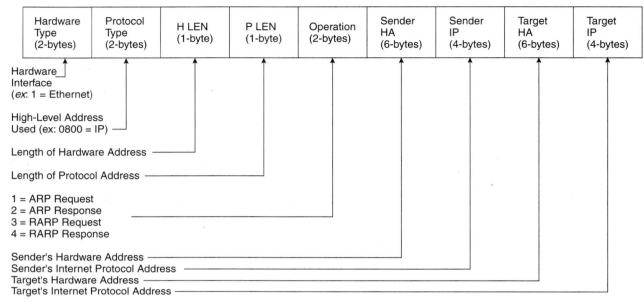

**FIGURE 9-24** Address Resolution Protocol Format (ARP)

The format for an ARP frame is shown in Figure 9-24. The information included in the frame is:

- Definition about hardware and protocol types used.
- The length of the hardware and protocol addresses to be translated.
- ARP operation to be performed.
- Sender's hardware and Internet protocol addresses.
- Target's hardware and Internet protocol address.

With this information, the address resolution protocol can translate several layers of addressing so the ATM cells formed know where to route the payload encapsulated into ATM cells.

## Internetworking with ATM

The ability to transport different protocol payloads across many platforms and networks is ATM's strongest value. One protocol in the ATM suite used to allow programs from different computers to communicate with each other is called the **advanced program-to-program communication (APPC)** protocol. A companion protocol called the **advanced peer-to-peer network (APPN)** protocol is plug-and-use networking software that keeps

**advanced program-to-program communication (APPC)**
allows programs from different computers to communicate with each other.

**advanced peer-to-peer network (APPN)**
application that monitors computers on an ATM network.

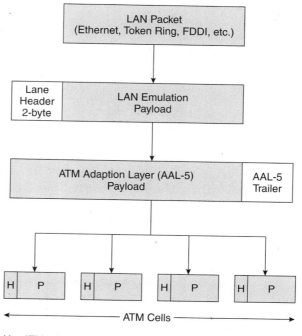

**FIGURE 9-25** ATM Lane Format

track of every computer on the network. It selects the best path to route data among nodes, including intermediate network nodes and end user nodes.

Classical Internet protocol (IP) can be carried over ATM using AAL-5 service. An ARP server maintains address tables and facilitates the translation between IP and ATM addresses within a single virtual subnetwork. Other applications for ATM include the emulation of local area networks using a protocol suite called Local Area Network Emulation (LANE).

LANE uses ATM as a backbone for high speed LAN traffic. LANE clients (LEC) may be Ethernet packets, token ring packets, fiber distributed data interconnection (FDDI) frames, or any other LAN-type traffic. Other clients could be servers, work stations, bridges, or switches on these networks. Each has its own ATM address and at least one media access control (MAC) address. LANE emulation servers and configuration servers maintain tables of LANE end stations, encapsulate ATM LEC packets into LAN formats, map between ATM and MAC addresses, configure emulated networks and maintain and manage switches and routers throughout the emulated network. Figure 9-25 shows how LAN payloads are encapsulated into ATM cells for transport, first by being placed into an emulation frame, loaded through AAL-5 format and segmented into ATM cells. The process is reversed at the destination end station.

## ATM Switches

ATM switches have been developed to operate as a fault recovery backbone for networks interconnected through the ATM protocol. These devices/software contain embedded segmentation and reassembly (SAR) functions so that the switches directly support Ethernet, token ring, fast Ethernet, and FDDI traffic. They are used to dynamically distribute traffic loads throughout the network. Like any switch, ATM switches are concerned with blocking, buffering, and traffic management, particular during congested periods.

ATM switches experience the possibility of two types of blocking. *Fabric blocking* occurs when the capacity of the ATM switch is less than the aggregate sum of all its inputs. *Head of line blocking* results from a simplistic approach to a queuing structure. A single queue services each input to the switch. Head of line blocking occurs when a cell is waiting to be sent through the switch. Additional cells applied to that input begin to back up because of the blocked queue.

Buffering in the ATM switch is required to handle multiple inputs that are to be switched to a single output and when an input wants to supply cells to a slower output. Two types of buffering are found in ATM switches, output buffering, and distributed input buffering. *Output buffering* is usually used in small ATM switches that handle a small traffic load. **Distributed input buffering with output control (DIBOC)** places buffers at each input port. Cells are sent across the switch fabric only when output bandwidth is available.

*Traffic management* is concerned with controlling the movement of data between ATM switch and an attached device. One type of preventative traffic management is known as *generic cell rate algorithm*, which enforces a number of traffic rate parameters including peak cell rate, maximum burst size, and sustained cell rate. Static-based enforcement of the algorithm throws away any cell that exceeds any of the parameters. In contrast, congestion-based enforcement will base excessive cells if there is bandwidth to handle them. When the system becomes congested, the cells with out-of-limit parameters are removed.

Reactive traffic management is used with available bit-rate traffic. User and switch can regulate the rate at which cells are received to maintain a minimum cell rate for guaranteed throughput and avoid congestion.

One additional method of managing traffic uses an explicit rate processing scheme that monitors the traffic for instantaneous information about the availability of bandwidth on a given channel in the context of the maximum current rate at which cells can be sent on that channel.

**distributed input buffering with output control (DIBOC)**
small ATM switch output buffer.

### Section 9.7 Review Questions

1. What types of data are routed using ATM?
2. What are the benefits of using ATM to route data?
3. How large is an ATM cell?
4. What is the difference between a VCI and a VPI?
5. What OSI model layers do ATM stacks address?
6. What are the four types of ATM traffic rates?
7. What are the benefits of PVCs?

8. What are the benefits of SVCs?
9. What is meant by bursty traffic?
10. What sublayers make up an AAL?
11. Which AAL is used for:
    a) data already in ATM format?
    b) highly bursty traffic?
12. What is the purpose of the address resolution protocol?
13. What types of nodes and networks are referenced by the ATM network services access points?
14. Which AAL service is used to transport Internet data?
15. What does LANE stand for?
16. What is meant by fabric blocking in ATM switches?
17. Why is buffering required in ATM switches?
18. What types of traffic management are available in ATM switches?

## 9.8 FRAME RELAY

An alternate protocol for routing information is called *frame relay,* which has existed longer than ATM, but does not find as wide an acceptance or use. Frame relay, unlike the fixed cell format of ATM, segments payloads into frames that do not have a fixed size. It uses packet switching and routing tables to guide these frames through multiple networks similar in method to ATM routing. Frame relay is subdivided into two operating planes, *control* and *user.* The control plane establishes and terminates logical connections and translates protocols between user and the networks. Error and flow control are also addressed under this plane. The user plane controls end-to-end functions including delimiting frames, aligning packets, and multiplexing frames. Size control to assure that a particular frame is neither too big or too small for the designated system falls under this plane as well.

The format of a frame relay frame, shown in Figure 9-26, is similar to an HDLC data-link frame. It begins and ends with delimiting flags. Following the beginning flag is a frame relay header field, which describes the frame much like the HDLC control field describes an HDLC frame. The header is composed of these fields:

- 10-bit data-link connection identifier (DLCI) plus a 2 (3–4 byte extended) address field.
- The command/response (C/R) indication, which is application specific indication of flow direction.
- The extended address (E/A) bit indicates if the address field has been extended beyond the default 2-byte size.

**forward** and **backward explicit congestion notification (FECN and BECN)**

**discard eligibility (DE)** determines which frames can be discarded to avoid congestion.

- **forward** and **backward explicit congestion notification** (**FECN** and **BECN**) deals with whether or not a frame is discarded due to congestion.
- **discard eligibility (DE)** is similar to ATM's cell loss priority bit, in that, when set, this field makes the frame eligible for discarding in case of congestion.

The payload data field follows the header field. It includes the encapsulated packets from the protocol the frame relay frame is transporting. This field is variable in length and depends on how the payload was segmented and encapsulated into frame relay frames. A frame check sequence (FCS) field, which uses cyclic redundancy checking (CRC-16) for error detection, completes the frame relay format fields.

Frame relay is good for data traffic because of its low overhead (less than 10 header bytes for hundreds or thousands of data bytes) and variable length for varying message sizes. It is a poor routing protocol for integrated services since it exhibits variable delays through the transmission path. These delays cause distortions in video and other type traffic.

**Section 9.8  Review Questions**

1. How is frame relay protocol similar to ATM?
2. How does frame relay differ from ATM?
3. What data-link protocol is the basis for a frame relay format?

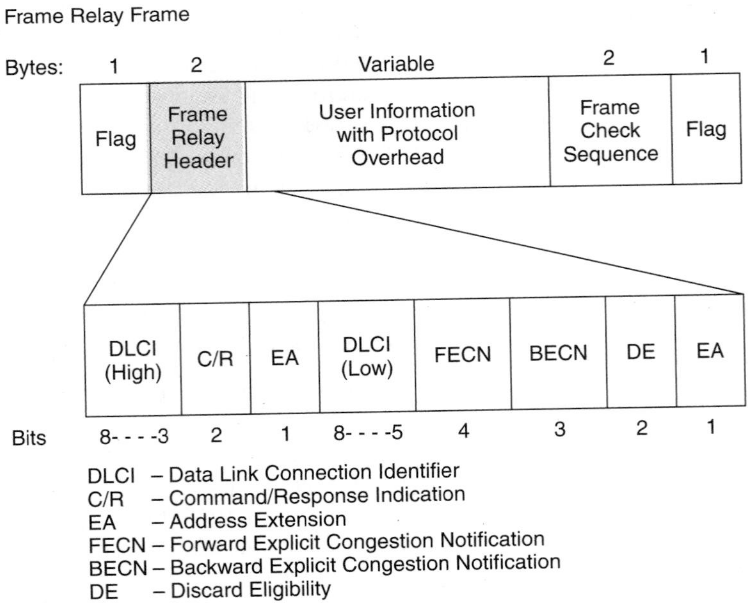

Frame Relay Frame

**FIGURE 9-26** Frame Relay Format

# SUMMARY

Integrated services of voice, data, teleconferencing, and other applications are receiving wider attention and acceptance with the data and digital communications community. The growth of the Internet is one area in which integrated services are in demand. Transporting text, graphics, and multimedia data across the "Net" efficiently has become a big concern of network designers. ISDN is supplying one method for accessing the Internet and delivering those integrated services. For applications that require higher data transfers or larger bandwidth allocations, a second form of ISDN, known as broadband ISDN, or BISDN, is available. Because of the increasing use of BISDN, the original specification is being renamed N-ISDN for narrowband ISDN. Within internal internetworks, routing protocols ATM and frame relay continue the job of routing these services through networks to their destination user. ATM is a cell-based service that encapsulates packet data into fixed units known as cells and transports them across numerous networks. Frame relay contrasts with ATM in the use of variable size frames to handle the traffic across the networks.

Table 9-3 shows a comparison of the routing protocols discussed in the text to this point.

## TABLE 9-3

### Protocol Comparisons

| Parameter | X.25 | ATM | ISDN | Frame Relay |
| --- | --- | --- | --- | --- |
| *Bandwidth* | 56 Kbps | 1.544 Mbps | 64-128 Kbps BRI 1.544 Mbps 2.048 Mbps PRI | 64 Kbps—1.544 Mbps |
| *Type* | Packet Switch | Cell Switch | Packet/ Circuit Switch | Packet-Switched Frames |
| *Features* | Bursty Traffic; Secure; Any-to-Any Connection | Integrated services; Qos; Reliable; Secure; All Types of Traffic | Integrated Services; Secure; Reliable; Bursty Traffic | Variable Packets; Bandwidth on Demand (SVC); Point-to-point Connectivity |
| *Weak Points* | Limited Bandwidth; Marginal for LANs | Lack of Standards | Variable Tariff Rate; Difficult to Install and Configure | Expensive; Requires Dedicated Lines; Complicated to Make Moves and Changes |

# QUESTIONS

**Section 9.2**

1. What does the acronym ISDN mean?
2. What does the acronym BISDN stand for?
3. How does the ISDN system differ from the regular telephone system?
4. What is meant by integrated services?
5. What type of information is conveyed on the B channel?
6. What is the D channel used for?
7. What is the composition of the basic rate interface (BRI)?
8. What is the data rate for a BRI service?
9. What is the composition of the North American primary rate interface (PRI)? At what data rate is this service used?
10. What is the composition of the European primary rate interface (PRI)? At what data rate is this service used?
11. What function does the terminal adapter (TA) provide?
12. What is the difference between terminal equipment 1 (TE1) and TE2?
13. Describe the R, S, and T reference points of the ISDN model.
14. Describe the U and V reference points of the ISDN model.
15. What is the purpose of the network termination (NT) units of the ISDN model?
16. In what format is the data presented to the NT units? What is the difference between the NT1 and NT2 units?
17. Which channel-type is used for each of the following purposes?
    a) Conventional analog voice.
    b) Carry broadband ISDN at 139 Mbps.
    c) Packet-switched data at 64 Kbps.
    d) Hybrid North American PRI.
18. Which channel type is used to carry common T1 data?
19. Which ISDN functional unit applies to the physical layer of the OSI model?
20. List some of the devices interfaced to the ISDN system by the network termination unit.
21. In the BRI frame format, what is the purpose of the F bit?
22. In the BRI frame format, what is the purpose of the L bit?
23. In the BRI frame format, what is the purpose of the S bit?
24. In the BRI frame format, what is the purpose of the E bit?
25. What is the name for the ISDN data-link protocol?
26. Which-data link protocol does the ISDN protocol closely follow?
27. What does the service access point identify?
28. What does the terminal end point identify?
29. What does a 1 in the command/response bit signify about the message?
30. What does a 0 in the command/response bit signify about the message?
31. Which HDLC unnumbered frame does the asynchronous balanced mode extended function like?
32. What is the name of the added unnumbered frame format? What is its purpose?
33. What is the purpose of the protocol discriminator in the LAPD information field?

34. What is the purpose of the cell reference field in the LAPD information field?
35. What types of LAPD messages are identified by the message type field?
36. Which fields of the LAPD frame are used for sequencing and payload size?
37. What does ANI stand for?
38. What is the function of ANI?
39. What is noteworthy about the McDonalds' ISDN implementation?

### Section 9.3

40. With the advent of BISDN applications, regular ISDN services has been re-named. What is the new name?
41. What is the function of a BISDN subscriber premises network and broadband terminal interface?
42. What function does the broadband network termination unit provide in the BISDN system?
43. What services are provided by the N-ISDN system?
44. What services do BISDN systems add to N-ISDN systems?
45. What functions are provided by the BISDN user plane?
46. What functions are provided by the BISDN control plane?
47. What functions are provided by the BISDN management plane?
48. What is the term for the connection point of peripherals to a BISDN system?
49. What is the function of the BISDN service node?
50. What type configuration are service node modules placed in?

### Section 9.4

51. Which IEEE 802 specification addresses integrated services?
52. What is the IEEE 802.9 interface unit between user and integrated network called?
53. Which data-link protocol is used as the basis for IEEE 802.9 MAC frame?
54. What is the purpose of the MAC frame's SID field?
55. Which field indicates the type of MAC frame in use?
56. Which MAC frame holds the higher level payload?

### Section 9.5

57. What type of line is a DSL?
58. Why is ADSL asymetric?

### Section 9.6

59. What type of switching was employed by early PBXs?
60. Define PBX bridge and half bridge use.
61. What is returned to an originating station by a PBX control mechanism if the station's access is verified?
62. What types of information are contained in PBX connection tables?

### Section 9.7

63. What is the size of an ATM cell?
64. What fields make up an ATM cell? How many bytes are in each field?
65. What does a virtual channel identifier identify?

66. What does a virtual path identifier identify?
67. Describe the differences between a permanent and switched virtual channel.
68. What is the prime purpose of the ATM and frame relay protocols?
69. What service connections are categorized as UNI?
70. What service connections are categorized as NNI?
71. Describe the difference between the constant bit- and variable bit-rates used by ATM.
72. What type of data is described by the available bit rate?
73. What does the header error control (HEC) do?
74. What does the ATM protocol use to aid in locating cells that are directly mapped into a TDM payload?
75. What are the responsibilities of the ATM physical (PHY) layer?
76. What are the two sublayers of the PHY layer? What are their specific duties?
77. Explain the purpose of cell rate decoupling.
78. Which ATM layer is involved with header generation and cell multiplexing?
79. What functions are used to reduce traffic congestion at peak periods?
80. What does the cell loss priority bit signify?
81. Define bursty traffic.
82. What is included under the heading of Quality of Service?
83. Define bandwidth on demand. What is the advantage of a system that uses bandwidth on demand?
84. What is the main disadvantage of using bandwidth on demand?
85. Define acceptable cell loss.
86. Define best effort data delivery.
87. What is the purpose of the ATM adaption layer (AAL) convergence sublayer?
88. What does segmentation and reassembly perform?
89. Which AAL layer applies to traffic already in ATM cell format?
90. Which AAL layer applies to bursty LAN traffic and is the simplest of the layers to implement?
91. What is the difference between class A and class B AAL services?
92. Which class of AAL service is used when emulating LANs?
93. Which protocol is responsible for translating ATM VCI and VPI addresses?
94. What is supplied by the initial domain part of the address resolution protocol (ARP)?
95. What type of nodes are included as network service access points?
96. What two ATM protocols are used to interconnect computers and networks?
97. Which AAL layer is used for Internet traffic?
98. Define LANE.
99. Define fabric blocking.
100. Define head or line blocking.
101. How do output buffering and distributed input buffering with output control differ?
102. What parameters are addressed using generic cell rate algorithm for management?
103. What type of traffic is reactive traffic management used with?
104. What does explicit rate processing monitor?

**Section 9.7**

105. How does frame relay differ from ATM?
106. What are the purposes of the frame relay control and user planes?
107. Which data-link protocol is used as the basis for a frame relay format?
108. Which frame relay format field resembles ATM's cell loss priority field?

# RESEARCH ASSIGNMENTS

1. You are considering an ISDN-type network for your office. Ethernet and token-ring LANs are already in place along with a PBX. Consider and list the hardware and software requirements to interface existing networks into one cohesive network that also interfaces with ISDN. You will have to conduct research in the subject areas of previous chapters in addition to this chapter's material to complete your list. Give a reason for each selection on your list.

2. Research the current applications, including Internet access, for ISDN. Include why the selection to use ISDN was made.

3. Research BISDN applications. Include why BISDN was selected over ISDN.

4. ATM and ISDN are platforms used for accessing the Internet. Research how each is applied to the handling of all types of Internet traffic.

5. Frame relay is a viable routing protocol and is used for numerous applications today. Research the use of frame relay, explaining why it is preferred over ATM for these applications.

# ANSWERS TO REVIEW QUESTIONS

**Section 9.1**

1. Routing of data through interconnected networks.
2. Direct digital interface allows for quicker Internet connection and faster downloads.

**Section 9.2**

1. ISDN principals, service capabilities, network characteristics, network interfaces, internetwork interfaces, maintenance principals.
2. Actual protocols used.
3. B channel: data traffic.
   D channel: signaling information.
4. Basic rate interface and primary rate interface.
5. Convert non-ISDN frames to ISDN format.
6. Both supply user interface to the ISDN system. One is used for ISDN type data and the other for non-ISDN type data.
7. S reference point.
8. Analog voice, packet switch data, circuit switch data, facsimile, video, and imaging.

9. Two B channel + one D channel.
10. North America: 23 B channels + 1 D channel.
    European: 30 B channels + 1 D channel.
11. Framing and synchronization.
12. HDLC
13. Uses a service access point and terminal end point instead of source and destination addresses.
14. Call set up and teardown, flow control, routing, error detection, and recovery.

### Section 9.3

1. Increased data rates; LAN to WAN interconnection; high definition TV (HDTV).
2. a) BDT—interface access nodes to peripherals.
   b) Service node—manages majority of control functions for system access.
   c) BNT—codes data into BISDN format.
3. Oversees operation of BISDN modules. Acts as an end user interface for control signaling and management.

### Section 9.4

1. Integrated data and voice services. ISDN and LANs are integrated and used to transport voice, data, and video information.
2. Isochronous data is data that is integrated (voice, imaging, video, fax, text, etc.).

### Section 9.5

1. ISDN integrates services onto a single line. DSL splits data and voice onto separate lines.
2. Symmetrical services use the same data rate for upstream and downstream traffic. Asymmetric doesn't.
3. A splitter separates data and voice into two lines.

### Section 9.6

1. Connect telephones in an office building.
2. Integrated voice and data services.
3. Acknowledge user requests and send ready signals.

### Section 9.7

1. Imaging, teleconferencing, multimedia, video, text.
2. Integration of data into a single channel, allocation of resources.
3. 53 bytes.
4. VCI identifies a destination node.
   VPI identifies the intermediate network path through interconnected networks.
5. Physical, data link, and network.
6. Constant bit rate (CBR).
   Variable bit rate (VBR).
   Available bit rate (ABR).
   Unspecified bit rate (UBR).
7. With PVC, a guaranteed virtual connection is always available.

8. SVC provides for the conservation of resources by supplying a connection only when it is used.
9. Bursty traffic is intermittent messaging—there are times when there are no messages to send.
10. Convergence and segmentation and reassembly (SAR).
11. a) AAL-0   b) AAL-5
12. Translates addresses between ATM and other protocols.
13. Networks, subnetworks, end system users.
14. AAL-5.
15. Local area network emulation.
16. Capacity of switch is insufficient to handle the amount of incoming data from multiple ports.
17. Handle multiple inputs that are switched to a single output (store and forward).
18. Preventative, reactive, and explicit rate processing.

### Section 9.8

1. Both are routing protocols.
2. Frames use variable size packets while ATM uses a fixed length packet size.
3. HDLC.

# CHAPTER 10

# The Internet and TCP/IP

## OBJECTIVES

After reading this chapter, the student will be able to:
* Have a background on the historical development of the Internet.
* Recognize the different functions and services available on the Internet.
* Decipher Internet and domain name service addresses.
* Describe the TCP/IP Protocol used for transporting Internet messages.
* Define the different types of firewalls and describe how they limit access to the Internet.
* Understand the security issues and methods to address those concerns.
* Tell the difference between the Internet, intranets, and extranets.

## OUTLINE

| | |
|---|---|
| 10.1 | Introduction |
| 10.2 | Internet History |
| 10.3 | Uses for the Internet |
| 10.4 | Accessing the Internet |
| 10.5 | Internet Addresses |
| 10.6 | Security on the Internet |
| 10.7 | Authentication |
| 10.8 | Firewalls |
| 10.9 | Intranets and Extranets |
| 10.10 | TCP/IP, the Technology Behind the Internet |

## 10.1 INTRODUCTION

**internet**
worldwide communications network.

The use of the **Internet** is the fastest growing application of a technology known to man. Uses have gone way beyond the initial intent to provide a means for scientists to share research information. The fastest growing sector of Internet use for the foreseeable future is commerce—the selling and buying of products and services. Because of the immense growth of use by the public, a new Internet system known as Internet 2 has been proposed and is in the design stages. Its purpose is, as you may have guessed, for scientists to share research information!

"Surfing the Net," "browsers," "e-mail," or "Internet"—a new language used for a new technology. Well, not exactly new. Actually, the Internet as an information-sharing collection of networks began in 1969 as a Department of Defense network for allowing research institutions to quickly share data. The Advanced Research Projects Agency (ARPA) contracted with Bolt, Baraulk, and Newman (BBN) to develop a packet-switched network, which was designated as the Advanced Research Projects Agency Network (ARPANET). It was a packet-switched network, and employed a basic networking protocol known as the **internet protocol (IP)** to initially link three nodes—University of California at Los Angeles (UCLA), Stanford Research Institute, and the University of Utah.

**internet protocol**
Internet network layer protocol.

The only services provided by **ARPANET** at the onset was remote log in (TELNET), file transfer, and remote printing. In 1972, ARPANET connected 37 sites to transfer text messages only between the sites. Electronic mail in the form of **e-mail datagrams** was included that year. Datagrams are messages that are sent without any anticipation of a verification of their reception. The purpose is to transfer large amounts of data quickly, and if there was any problem in getting the information, it had to be resent at a later time. However, ARPANET proved to be a very reliable system for transferring these datagrams.

**ARPANET**
advanced research projects agency network.

**e-mail**
system that allows users to communicate directly with one another through electronic messaging.

**datagrams**
unacknowledged packets sent on the Internet.

## 10.2 INTERNET HISTORY

By the mid-seventies, a full suite of protocols, called the Transmission Control Protocol and Internet Protocol (TCP/IP) were developed and proposed as the standard underlying technology for the expanded use of the ARPANET system. In 1975, the agency controlling ARPANET became the **defense advanced research projects agency (DARPA),** which immediately limited access to ARPANET to defense-related organizations. **USENET** was created in 1979 to add news and mail servers to the ARPANET system. USENET continues today as NetNews. TCP/IP was finally adopted as the standard protocol suite for the system in 1983. At that time, the network interconnected 500 defense-related research sites. However, purely military-related sites splintered away and formed their own network, **MILNET.** This eventually opened the door to allowing commercial and nondefense organizations to join the network.

**defense advanced research projects agency (DARPA)**

**USENET**
maintains newsgroup servers.

**MILNET**
MILitary NETwork.

Addressing used by ARPANET users became easier to handle in 1984 with the development of **domain name service (DNS).** This service took on the responsibility of putting ARPANET numerical addresses into readable text form. The service is still used today to continue this process.

**domain name service (DNS)**
function used to assign Internet addresses to users.

In 1987, the National Science Foundation scientists improved on the TCP/IP protocol to produce a faster backbone to interconnect college and university research centers. A program named **ENQUIRE** was developed to improve the ease of access to the Internet. Users could now gain access to the network without meeting defense-oriented requirements set by DARPA. By the late 1980s, the ranks of Internet users swelled to 100,000.

**ENQUIRE**
early application that facilitated access to the Internet.

**European center for nuclear research (CERN)**

**world wide web (WWW)**
network for exchanging data.

As the 1990s dawned, an international consortium under the blanket name of **European Center for Nuclear Research (CERN)** created the concept of web pages and the **WORLD WIDE WEB (WWW)** that allowed newcomers to exchange data on the Internet with ease. These early web sites contained text only. Graphics and multimedia were to come later on. To locate points of interest on the web, ARCHIE, the first of many search engines was introduced at McGill University. To further extend the ability to quickly find site locations on the network, the University of Michigan developed *GOPHER* in 1991, and McGill added VERONICA (very easy rodent-oriented netwide index to computerized archives) in 1993.

**hypertext markup language (HTML)**
programming language used for creating web pages.

**standard generalized markup language (SGML)**

To aid in creating web sites for distribution on the web, **hypertext markup language (HTML)** was developed as a subset of **standard generalized markup language (SGML)** in 1992. The concept behind HTML was to supply an easy-to-use, less sophisticated programming tool for creating web pages. A example of a HTML file and a web page are shown in Figure 10-1. The HTML code directs how the page looks as we will see in more detail later in the chapter.

This is a simple example of a web home page to keep the code minimized. Above is an image below a larger font heading. After this short note there will be a link to the author's actual home page. It is encased

in anchor tags and appears on the web page in a different color and underlined.

Go to my home page.

Of course actual web pages have a lot more to them, but this is being kept short and simple for illustrative purposes.

You can explore all about web page design, HTML, JavaScript and cascading style sheets at: http://www.devry-phx.edu/fac/miller/web_less.htm

**FIGURE 10-1a** HTML Example

```
<HTML>
<HEAD>
       <TITLE>HTML Example</TITLE>
</HEAD>
<BODY BACKGROUND = "pastel~1.gif">
<center><H3><FONT COLOR="#800080"><B>Web Home Page
Example </B></FONT></H3><BR>
<P>
<IMG SRC="Mprof.jpg" WIDTH=120 HEIGHT=126 BORDER=0
ALT="Professor Miller"></center><BR>
<P>
This is a simple example of a web home page to keep the code
minimized. Above is an image below<BR>
a larger font heading. After this short note there will be a link to the author's actual home
page. It is encased<BR>
in anchor tags and appears on the web page in a different color and underlined.<BR>
<P>
<A HREF="http://www.devry-phx.edu/fac/miller "TARGET="top"
NAME= "Millers Home Page"><IMG SRC="Mprof.jpg" WIDTH=120
HEIGHT=126 BORDER=0 ALT="Professor Miller">Go to my home
page.</A><BR>
<HR>
Of course actual web pages have a lot more to them, but this
is being kept short and simple for illustrative
purposes.<BR>
<P>
You can explore all about web page design, HTML, JavaScript
and cascading style sheets at:<BR>
<A HREF="http://www.devry-phx.edu/fac/miller/web_less.htm"
TARGET= "top" NAME="Web Design">http://www.devry-
phx.edu/fac/miller/web_less.htm</A>
</BODY>
</HTML>
```

**FIGURE 10-1b** HTML Code for Example

From these beginnings, other software developers created still easier tools to achieve the same ends. These tools boast that anyone can create web pages without being a programmer or even knowing HTML. Additionally, web pages have come a long way from basic text to graphics to animation, and finally, to multimedia effects.

**Mosaic**
first graphical web browser.

In 1993, the first graphical web browser, **Mosaic,** shown in Figure 10-2, was developed by students at the National Center for Supercomputing at the University of Illinois. A web browser is a program that allows you to access web pages and information on the Internet by using a mouse and point-and-click process. Web pages were now being stored on Internet servers, which acted as central storage depositories for Internet network information. By now, the number of users accessing the "Net" or the "Web" had grown to 1,000,000.

Web browsers do facilitate access to Internet sites, but they cannot present all of the functions that might be available at a particular site. There are sound files, movie-type

The different parts of the Mosaic window.

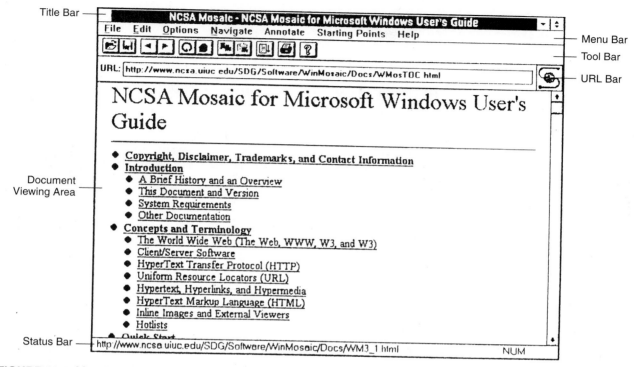

FIGURE 10-2 Mosaic Browser

graphics and a host of advanced features that are not supported directly by many browsers. To allow users to take advantage of these features, there are a number of plug-in or helper application programs. An example of a helper application is ADOBE Acrobat Reader (Figure 10-3), which allows you to view and hear multimedia applications that are in the **multimedia internet mail extension (MIME)** format. Other examples and the features associated with them are:

**multimedia internet mail extension (MIME)**
multimedia format.

1. COSMO Player to view three-dimensional graphics and virtual reality files.
2. Real Player for streaming video and audio files, which are those that are fed continuously for a fixed period of time.
3. QuickTime Player for large video files.
4. Others that play .MIDI extension audio files, movie videos, and other special features.

Today, we have many ways of accessing and using the Internet, some are direct, like the use of Web browser, and some indirect through online services like America OnLine, Prodigy, and CompuServe. The growth of users on the Internet has been, and continues to be, phenomenal given the short time of its awareness to the general public. Still, by 1997, only 11% of the American population actually access the Internet.

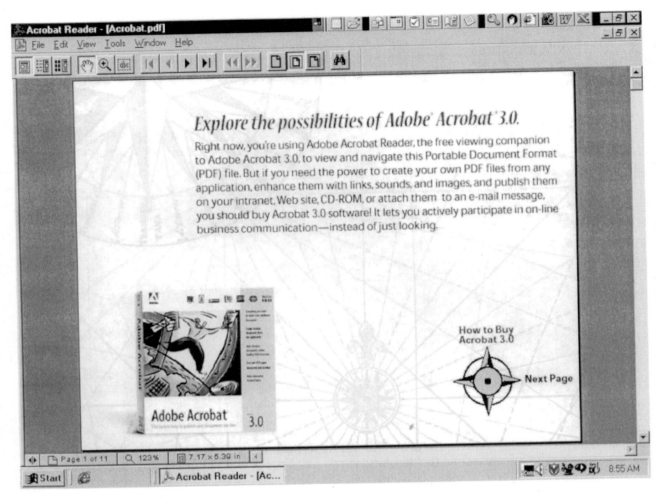

**FIGURE 10-3** Adobe Acrobat Reader

Essentially, the Internet is a collection of interconnected networks that no one single entity actually owns. The overall system is overseen by a group of volunteers dedicated to promoting the exchange of information using the Internet. This group is known as the *Internet Society,* which maintains the Internet through several subgroups. The *Internet Architecture Board* authorizes network and protocol standards used by the Internet and allocates and keeps track of Internet resources, including addresses. The *Internet Engineering Task Force* concerns itself with operational and technical difficulties that might arise in the daily operation of the system. Lastly, the *Internet Research Task Force* keeps its eyes on developments that may be made to improve the services provided by the Internet.

**Section 10.2  Review Questions**

1. What is the original name for the Internet network?
2. Which protocols are used to transport Internet messages?
3. What is the name of the first graphical browser?

## 10.3  USES FOR THE INTERNET

There are several basic uses made of the Internet network and they all fall into the general category of exchanging information. The promised Information Superhighway of the seventies has materialized as the Internet. The major uses for the Internet are:

1. E-mail.
2. Researching information.
3. Entertainment.
4. Electronic commerce.
5. Downloading and uploading files and software.
6. Education.
7. Chatting with people of like interest.
8. Game playing.

### Electronic Mail

High on the list of uses is electronic mail or **e-mail.** Sending messages between people on the Internet is cheaper and quicker than using the post office or other conventional mail services. A person can post a message in the morning knowing the recipient will have it within a few minutes.

Many e-mail program packages are available by downloading them from the Internet. Others are bundled with other software to become part of a *software suite*. Still others are commercially available for a price. One of the more popular e-mail programs is Eudora, which comes in two flavors. One is Eudora Lite, whose opening screen is seen in Figure 10-4, which can be downloaded from the Internet and the other is a fuller featured package available for purchase. Eudora Lite is more than adequate for everyday general use for sending and receiving messages. Messages can be sent to individuals or groups. Eudora Lite also allows users to attach files and send them along with their messages. For heavy users, sometimes called power users, the more fully featured Eudora version is required.

Besides sending and receiving text messages, e-mail can be used to attach and send files with the text messages. Messages can also be sent to multiple users at one time. One

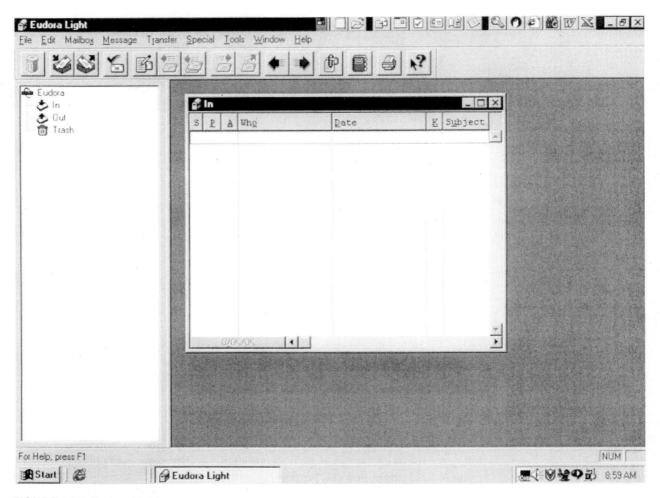

**FIGURE 10-4** Eudora Light

additional service allows a user to subscribe and use e-mail lists that link users with a common interest. Messages posted to a mail list are sent to all subscribers on that list.

Undeliverable e-mail messages are "bounced" back to the sender. Causes for undeliverable mail include:

- Unknown recipient (incorrect e-mail address).
- Unknown host system (something wrong with the domain address).
- Network is unreachable (possibly a breakdown in the system).
- A software problem at the destination causes a time-out to occur. Message could not be delivered within an allotted time frame.
- A problem exists with the destination server that caused the message to be refused.

**simple mail transfer protocol (SMTP), post office protocol (POP), Internet mail access protocol (IMAP)**

A protocol for e-mail service that is widely used is the **simple mail transfer protocol (SMTP),** which describes how e-mail moves between hosts and users using **post office protocol (POP)** servers. This system allows access to mailboxes remotely on TCP/IP networks. A subset of POP is the **internet mail access protocol (IMAP),** which supplies a standard way for clients to read and send e-mail messages. Mail is stored on the server so that the same message may be sent to different endusers. Mail can be downloaded based on selected criteria such as date, user, new, etc.

**common messaging calls (CMC)**

defines cells for sending and receiving e-mail messages.

An open standard for e-mail use, X.400 was developed for the International Telecommunications Union (ITU). It includes a standard called the **common messaging calls (CMC)** that specify eight basic cells for sending and receiving messages, calendering and scheduling processes, and specifications for electronic forms, voice, and fax over networks. X.400 is an open transport protocol for e-mail, which includes the following subsections:

- User agent (UA) defines message service you are authorized to use and transmits these messages to local message transfer agents (MTA).
- Message transfer agent (MTA) routes messages among UAs, distributes messages to dissimilar mail systems, notifies users when mail has been sent and received, and sets expiration time (time-out) for mail delivery.

X.400 was the first attempt at standardizing e-mail use and processes. A companion standard, X.500, is an enterprise-wide directory service of e-mail and other addresses. It uses a hierarchal information tree, starting with country of address as the top level down to individual users.

## Newsgroups

**newsgroups**

Internet sites used for sharing information on a particular topic.

Sharing knowledge with others about a common topic of interest using **newsgroups** is another wide use for the Internet. Essentially, newsgroups work similarly to the old idea of bulletin boards. You would join a newsgroup, and once a member, you could post messages called *articles* about the subject of the group, make inquiries, or read the current crop of information placed in the group's site. A set of articles with a common theme are known as a *topic,* and usually form a particular newsgroup. A patchwork of queries and responses within a topic form a **message thread.** Many newsgroups are accessed by connections between the Internet and a system known as USENET, which maintains newsgroup servers.

**message thread**

queries and responses about a spectic newsgroup topic.

## File Transfer

**file transfer protocol (FTP)**

protocol that allows the uploading and downloading of files to/from the Internet.

Downloading, uploading, and sharing programs and data files is a third use of the Internet, which falls under the category of *File Transfer.* Files may be transferred from within newsgroups, from web sites, or directly using the **file transfer protocol (FTP).**

While the file transfer protocol is the main protocol for supporting the uploading and downloading of files on the Internet, there are several other supporting protocols that assist in finding these files. One of the oldest of these is **ARCHIE,** which is a play on the word *archive.* Drop the "V" and you have ARCHIE. ARCHIE accesses a database of file

**ARCHIE**

database of FTP sites.

**wide area information server (WAIS)**
Internet location search engine.

**VERONICA**
very easy rodent-oriented net-wide index to computerized archives database of GOPHER documents.

transfer protocol (FTP) sites, listing the files at each of the sites. FTP is a protocol that allows Internet users to download and upload files to and from FTP servers. Other search engines used to uncover locations on the Internet include GOPHER and **wide area information server (WAIS).**

GOPHER was developed by the University of Minnesota, whose mascot is the Golden Gopher (hence the name), as a menu-driven index of databases on the Internet. A companion program called **VERONICA** for Very Easy Rodent-Oriented Net-wide Index to Computerized Archives searches for text that appears in a GOPHER menu. VERONICA is a self-updating database of GOPHER documents. A third search assistant application is known as JUGHEAD, which allows you to search for specific information at a GOPHER site.

It has been suggested that ARCHIE was selected as a term because of its originator's fondness for the comics of the same name. This may or may not be based in fact. The ensuing result was the selection of the use of VERONICA because she is another character in the same comics. Effort was then made to fit a meaning to that acronym, which resulted in the somewhat unwieldy expression "Very Easy Rodent-Oriented Net-wide Index to Computerized Archives." Rodent? Guess *mouse* wouldn't work! In any case, the madness does not stop there. A third database search engine, called JUGHEAD, has arrived. It is similar to VERONICA, but allows you to narrow your search to a specific group of GOPHER subjects. JUGHEAD is not an acronym for anything, but he is another character in the ARCHIE comics.

WAIS is a full-text indexing program for large text files, documents, and periodicals. The University of Minnesota GOPHER lists all WAIS servers. One additional search engine, called WHOIS, is a TCP/IP utility that allows users to query compatible servers for information about Internet users.

## Web Sites

Probably the largest use of the Internet is browsing through web sites. Anyone who has access to the world wide web (Internet) can create a web site. There are sites for just about every subject imaginable. Thus far, the Internet is unregulated, but growing concern over some of the material placed in web sites that are accessible to children, may eventually force the government to step in and place restrictions on creating and using web sites.

**home page**
opening page of a web site.

A web site starts with a **home page,** which like the cover of a book, describes, in general, what the web site is all about. Figure 10-5 is a copy of the author's home page. It takes several screens to view the entire page—only the first screen is shown in the figure. From the home page there are many *hyper-links* to other pages and/or other web sites that are related to the original home page. In the screen shot of the author's page, there are links to other sections on that page. It depends on what the site creator chose to place at the site. Each linked page could have links to other pages as well as return links to the home page. A hyper-link allows the user to move between pages by clicking on designated areas with a page. For instance, this page might, in abbreviated form, be a home page about the Internet. A description of the TCP/IP protocol might be placed on a separate page at this site. Within the home page, a line of text could say:

"The underlying protocol for the Internet is *TCP/IP,* developed in the middle seventies."

Frame with Links to Other Pages    URL    Home Page Frame

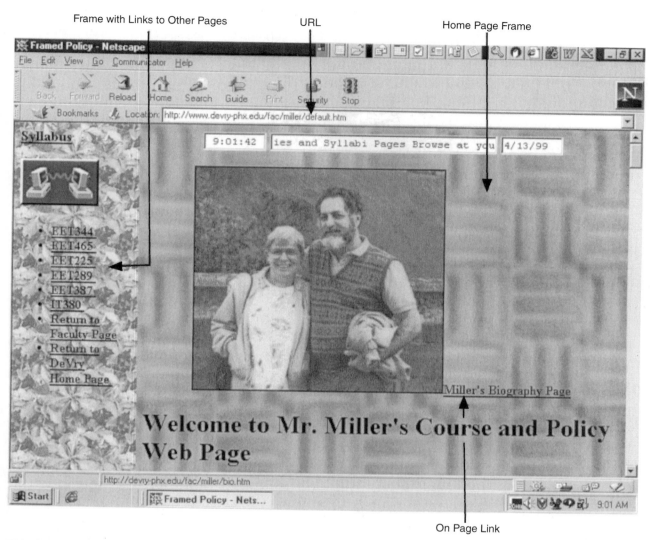

On Page Link

**FIGURE 10-5** Author's Home Page

By using the mouse and clicking on the underlined, highlighted area, the browser software would "jump" from this home page to the TCP/IP page.

## Other Internet Services

**request for commands (RFC)**
documents that define parts of the Internet.

Other services commonly found on the Internet include TELNET, which allows users to log in to remote computers on the Internet, and **request for commands (RFC),** which are documents that define specific parts of the Internet. These are usually highly technical articles that are not too useful to the casual user. They are accessed by requesting an RFC by its RFC number. An index of RFCs is available to assist the user in making the appropriate inquiry. An additional service is called **frequently asked questions (FAQ),** a collection of questions and answers most often asked by users. FAQs exist for most newsgroups, e-mail mailing lists and incorporated into web sites.

**frequently asked questions (FAQ)**
collection of commonly asked questions and answers on a particular subject.

### Section 10.3 Review Questions

1. List four uses of the Internet today.
2. List three e-mail protocols used on the Internet.
3. What programming language is used to create web pages?
4. Which Internet protocol is used to download files from the Internet?
5. List three search engines commonly used to find stuff on the Internet.

## 10.4 ACCESSING THE INTERNET

Before you can use the Internet (the Net), you have to gain access to it. This access is achieved in one of several ways, but before exploring these methods, a little understanding of the underlying network structure is required. Above all, the Internet is a collection of networks that are connected together through various protocols and hardware. The common base for all of these protocols is the TCP/IP suite of protocols discussed later in the book. The packets transported in TCP/IP format may be encapsulated in numerous other protocol frames for transport through intermediate networks. Some of the more common intermediate protocols used are ATM, ISDN, and FDDI (fiber distributed data interface).

### Internet Service Providers

As mentioned earlier, nobody truly owns the Internet, but it is maintained by a group of volunteers interested in supporting this mode of information interchange. Central to this control is the **internet service provider (ISP)** shown as unit B in Figure 10-6, which comprise a number of limited central locations. These locations are responsible for assigning Internet addresses to users and to make the connections between valid users. These central ISPs do not deal with a large number of individual users, but rather handle the distribution of addresses through intermediaries. Some of these intermediaries only serve the purpose of providing access to users. Others, like online services (unit C in Figure 10-6), provide additional services not related to the Internet.

**internet service provider (ISP)**
an entity which makes access to the Internet available to users.

### Intermediate Providers

The methods and technology behind Internet access begins with the ISPs. From there it can take many forms. The aforementioned online services is one of the better known.

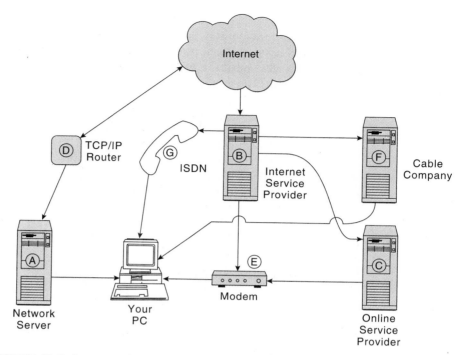

**FIGURE 10-6** Access to the Internet

America OnLine, Prodigy, and CompuServe are examples of online services that provide Internet access. With these, you must take care and check the extent of Internet service each will supply. Some limit access to e-mail service, while others include most services available on the Internet.

More direct connections can be used if you have a TCP/IP interface (unit D in Figure 10-6), a modem (unit E), or cable modem (unit F), and the appropriate software to connect directly to an ISP. Two protocols, **serial line interface protocol (SLIP)** and the **point-to-point protocol (PPP),** allow a user to dial into the Internet. They convert the normal telephone data stream into TCP/IP packets and send them to the network. With these, the user becomes a peer station on the Internet and has access to all of the Internet's facilities.

WINDOWS operating systems come with its own way of communicating with the Internet through the phone lines. WINDOWS uses WINSOCK.DLL to formulate the link to the Internet.

**serial line interface protocol (SLIP)** and **point-to-point protocol (PPP)** Internet dial up access protocols.

## ISDN Service

The widest availability for accessing the Internet incorporates the use of ISDN's basic rate interface. The use of ISDN for accessing the Internet has breathed new life into the

ISDN service. This was an application looking for a place to fit into. ISDN's slow acceptance was due mostly to a lack of a need for its capabilities. Being a digital interface, ISDN has provided a means for accessing web sites quickly and efficiently. In response to this new demand, telephone companies are rapidly adding ISDN services.

Of the two basic rate B channels, one is used to upload data to the Internet and one to download from the Internet. The D channel assists in setting up connections and maintaining flow control. There are three ways ISDN can be used to interface to the Internet, by using a modem, adaptor, or bridge/router. ISDN modems and adaptors limit access to a single user. Both terminate the line into an ISDN service. The difference between them is that the ISDN modem takes the Internet traffic and pushes it through the computer serial port, while, the faster ISDN adaptor connects directly to the computer's buses.

ISDN bridge/routers allow for local network connections to be made through ISDN to the Internet. The ISDN termination is made into an Ethernet-type LAN so that multiple users can achieve access to the Net through a single access address. Transfer rates between user and the Internet are between 56 and 128 Kbps.

### Direct ISP Service

**data service units (DSU) and channel service units (CSU)**
equipment to facilitate direct ISP service.

The most costly method of accessing the Internet is to use leased lines that connect directly to the ISP (unit C in Figure 10-6). This will increase access rate to anywhere between 64 K and 1.5 Mbps, depending on the system in use. Equipment called **data service units (DSU)** and **channel service units (CSU)** are set up in pairs, one pair at the customer site and the other at the ISP site. There is no phone dialing required since the connection is direct. Also, the only protocol needed to complete the access is TCP/IP, for much the same reason. Depending on the transfer rate required and the distance between the sites, cabling between them can be made with fiber optic cables or unshielded twisted-pair (UTP) copper wire.

### Cable Modem

One more way of accessing the Internet currently being developed is the use of cable modems. These require that you subscribe to a cable service and allows you two-way communication with the Internet at rates between 100 K and 30 Mbps. The cable modem performs modulation and demodulation like any other modem, but it also has a tuner and filters to isolate the Internet signal from other cable signals. Part of the concern for use of the cable modem is to formulate LAN adapters to allow multiple users to access the Internet. A medium access control (MAC) standard for sending data over cable is being formulated by the IEEE 802.14 committee.

---

**Section 10.4 Review Questions**

1. What is an ISP?
2. List four ways to access the Internet.
3. Which access method do you think provides the fastest and most complete service? Why?

# 10.5  INTERNET ADDRESSES

http://www.devry-phx.edu.us

What does it all mean? Actually to the ISP server, very little. The server wants to see something quite different. It wants to see a 32-bit number as an Internet address. Something like this equivalent decimal grouping:

198.168.45.249

**universal resource locators (URL)**
Internet address.

**hypertext transfer protocol (http)**
controls access to web sites.

**domain name**
part of an Internet address that is used to identify the user.

**domain top**
part of Internet address that designates the general group or organization a user belongs to.

The Internet addresses, known as **universal resource locators (URL),** are translated from one form to the other using an address resolution protocol. The first address is in the form we are most used to and that users use to access an Internet site. In this example, the address is for a web site, identified by the **hypertext transfer protocol (http),** which controls access to web pages. Following http is a delimiter sequence, ://, and identification for the world wide web (www).

The **domain name,** devry-phx, follows www and identifies the general site for the web. .edu is one example of a **domain top,** which is a broad classification of web users. Other common domain tops are:

* .com for commerce and businesses.
* .gov for government agencies.
* .mil for military sites.
* .org for all kinds of organizations.

Lastly, in this example is a country code, again preceded by a dot. Here we are using **us** for the United States, which is the default country. You do not have to specify the **us** for sites within the United States.

---

### EXAMPLE 10-1

What is the Internet address to access the White House on the world wide web?

### SOLUTION

Using the hypertext protocol, the address starts in the same way as the one used above, with http://www. This is followed by the domain name, whitehouse and the domain top, which, in this case, is .gov for government. Put together, the pieces form the White House address:

http://www.whitehouse.gov

This example will also illustrate what might happen if you are not careful with using the EXACT address. Mistakenly substituting .com for .gov will bring you some very unexpected results. Be forewarned, for those of you who do not like explicit sexual material, do not try this on the net—http://www.whitehouse.com is a sex site. This was by no means an accident!

Addresses may be followed by subdomains separated by dots or slashes (/) as needed. These addresses are translated into a 32-bit address shown as the second address example. When written, these addresses are placed in four decimal numeric groups separated by periods. The complete numerical address is deciphered by the system dependent on the size of the network and the number of host stations or nodes at each network and is summarized for the current Internet Protocol version 4 (IPv4) in Figure 10-7.

As you can see in the figure, the first part of the address defines a network, which contains a number of host nodes. Each node is identified in the second part of the address. Nodes may connect to a server, a terminal computer, or a subnetwork.

A number of addresses are reserved for network maintenance and multicasting capabilities (accessing multiusers at one time). These addresses are recognized by the use of a default mask, which is all logic ones in the network number field. The default mask for each class is:

Class A: 255.X.X.X
Class B: 255.255.X.X
Class C: 255.255.255.X

where X is a subnetwork value. Additional logic ones in the X field indicates the size of the subnetwork field. An example of a class C mask for a three-bit subnetwork field is:

255.255.255.224 = 11111110.1111110.11111110.111xxxxx

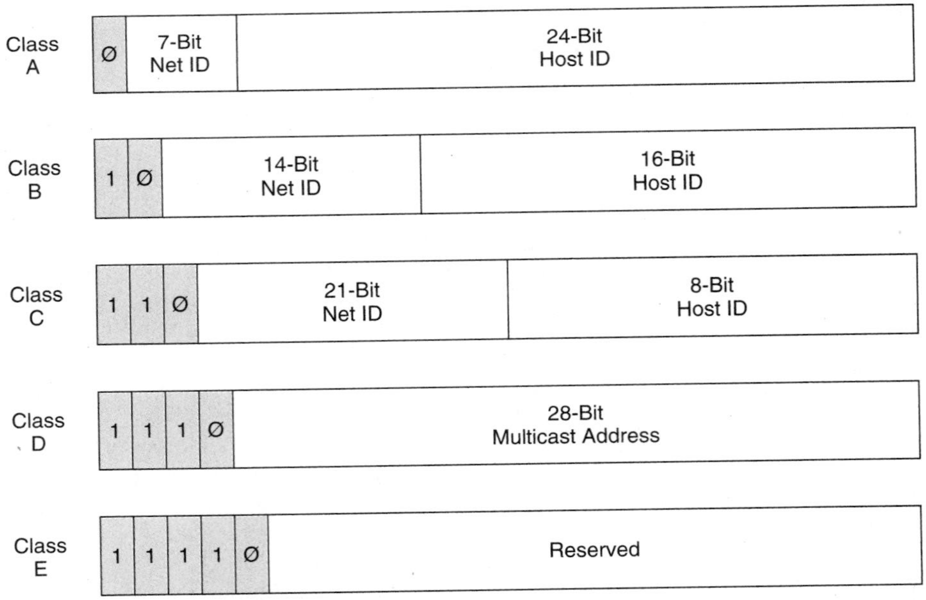

a) Directed Broadcast Address for a Net Has a Host Address of All Ones (1s)
b) General Broadcast Is Address of FFFFFFFF
c) Network Broadcast Has Host Address All Øs
d) Host Broadcast (to All Nets) Has Net Address All Øs

FIGURE 10-7 Internet Protocol (IP) Addresses

where xxxxx would designate the host access to be made to this subnetwork. This example uses a host number equal to zero, which is reserved to address the entire network or subnetwork. A message with a URL of all ones (255.255.255.255) is a broadcast message to the entire Internet.

Other sets of IP addresses are reserved for dedicated usage. One of these usages is for broadcasting a message to all the hosts on a network or system. By setting the host number equal to zero for any given network, the message using that URL would be sent, as a broadcast message, to all hosts on that network. Each class also has a broadcast address that allows a message to be sent to all hosts on a particular class of network. Those addresses includes logic ones in all bits of the host portion of the URL:

1. class A: <class A network>.255.255.255
2. class B: <class B network>.<class B network>.255.255
3. class C: <class C network>.<class C network>.<class C network>.255

One of the drawbacks to the Internet addressing scheme is that it is reaching its limit and ability to assign addresses to new users. A new internet addressing protocol, **IPv6,** has been developed to eliminate this problem. As part of the specifications within this protocol, address size has been increased to 128-bits from the current 32-bits. Other improvements include:

* More efficient routing.
* Autoconfiguration when adding new users.
* Improved security and performance.
* Interoperation with existing IPv4 protocol.
* Allows for growth on the Internet.

Internet addresses are used, for the most part, for a one-to-one communication link. However, there are times that a one-to-many transfer is desired. For instance, a member of a newsgroup may want to send the same message to other members of the group through an e-mail message. Doing it one-to-one would require sending a separate e-mail message to each member of the group.

**multicasting**
sending a message on the Internet to many destinations at one time.

An alternate method is to use **multicasting,** which allows a single message to be sent to many recipients. An IP-multicast is a single data stream that is sent to members of a multicast group. That is, users that have a specific multicast IP address. A router on the network will only pass the multicast message onto a subnetwork that has nodes that can respond to the multicast message. This reduces unnecessary multicast traffic on networks where there is no response to the multicast message. Multicast messages use the best effort delivery of the user datagram protocol to transfer the packets of data. The lack of acknowledgements for this service results in an unreliable delivery of multicast data, which is considered to be a major drawback to multicast use. The reason unacknowledged datagrams are used is to reduce the traffic load that would be created if each member of a multicast group had to acknowledge the reception of a multicast message. Another form of IP address is directed at supporting e-mail.

E-mail addresses, from a user standpoint, appear simpler than Internet web addresses. For instance, the break down of my e-mail address, MMiller@devry-phx.edu, as specified in the standard, domain name service (DNS), is shown in Figure 10-8.

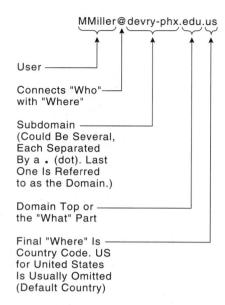

MMiller@devry-phx.edu.us

User ⎯⎯⎯⎯⎯⎯⎯⎯⎯⎯⎯

Connects "Who"⎯⎯⎯
with "Where"

Subdomain ⎯⎯⎯⎯⎯⎯⎯⎯
(Could Be Several,
Each Separated
By a . (dot). Last
One Is Referred
to as the Domain.)

Domain Top or ⎯⎯⎯⎯⎯⎯⎯
the "What" Part

Final "Where" Is ⎯⎯⎯⎯⎯⎯
Country Code. US
for United States
Is Usually Omitted
(Default Country)

**FIGURE 10-8** Domain Name Service (DNS) E-mail Address

The capitol Ms are optional and used for readability since addresses are not case sensitive. This address could have been written as mmiller@devry-phx.edu and would work just as well. MMiller is the user ID portion of the address and signifies the enduser at the domain following the @ symbol. For this address, the domain is devry-phx. As with Internet addresses, e-mail addresses end with a domain top preceded by a period—.edu in this address example.

**EXAMPLE 10-2**

What address is used to contact the president at the White House?

**SOLUTION**

One possible solution could be:

president@whitehouse.gov

Here president is used to identify the person and whitehouse.gov is the location specified by the domain name and domain top.

Other services on the Internet are accessed using similar addresses that are specific to the service used. These services are identified by the protocol used to handle them. For example, to access the file transfer function, addresses are preceded by ftp: (file transfer protocol):

ftp:// <ftp location address>

**Section 10.5  Review Questions**

1. What is the size, in bits, of current Internet addresses?
2. What do the sections of an Internet address specify?
3. What does the acronym URL stand for?
4. List four domain tops other than country code.
5. What is the term for the ability to send Internet messages to multiple destinations?

# 10.6  SECURITY ON THE INTERNET

When the Internet first came into being, security was not a concern. After all, the purpose was to share research and information freely and quickly. Now that the Internet is being used by a large and diverse population with differing applications, the problem of securing messages and information has become uppermost in many minds. In 1996, 54% of Internet users in the United States purchased merchandise online for sales totaling $500 million dollars. Projections for electronic commerce through the Internet expect that figure to rise to over $100 billion by the year 2000. Chief among the methods for paying for these purchases is the use of credit cards, which require the purchaser to send their credit card number over the Internet. Need for securing such a transmission is paramount to the success of electronic commerce.

## Cryptography Systems

Two driving forces have kept security concerns to the forefront of the data communications industry—privacy and theft. It is desirable to share certain information only with people who have a need to know that information. Some critical data in the wrong hands can have an adverse effect on the individuals involved. Theft of mobile phone numbers and network access codes can lead to all kinds of problems, starting with simple misuse of your account, which generates high bills, to trafficking in illegal material and contraband.

The two main types of security processes in use today are authentication and cryptography. Authentication is the process of verifying that you are the user of the account and that you have rights to particular services on a network. Cryptography involves coding messages so that only authorized users can decipher and read them. Many security systems include a combination of both types of processes.

Cryptography codes depend on the use of *keys* to encrypt and decrypt messages. There are two types of keys, symmetrical and asymmetrical. A *symmetrical key* is one that is used for both encrypting and decrypting data. In order for a cryptographic code to work using a symmetrical key, both the sender and receiver have to have a copy of the key. This means that key has to be stored at both sites or has to be sent, along with the message, to the receiver. Sending the key is the least secure method to use since the message may be intercepted and the key extracted from it.

*Asymmetrical keys* use two different keys, one for encryption and one for decryption. A *public key,* stored on a central server within the network, is used by any node to encrypt a message. A *private key* is held only by the receiving node and is used to decrypt the message. A reverse procedure can be used, where a message is encoded using the private key and decoded using the public key. This is less secure because anyone can have access to the public key, which could then be used to decrypt the message. However, the person decoding the message still needs to know which public key to use.

The value of key for decoding relies on how easy it is to compromise the key used. The more bits used for a key, the harder it is to "crack," and the more secure the code that uses it. One method for testing the validity of a key is to subject it to a *brute force attack.* A number of computers armed with several mathematical algorithms and random code generation are used to find a match to a key code. The longer it takes for the attack to be successful, the more reliable the key code. As an example, a 64-bit key code was subjected to a brute force attack. It cracked the key code in about three hours, suggesting that this length of code for a key was highly inadequate. Fortunately, the time it takes to crack a key code rises exponentially with the increase in the number of bits in the key. A key that is 128-bits long requires years to be deciphered.

Another type of test for the validity of a coding scheme is called a *plain text attack,* which uses specific blocks of text to find the most information about the key. Essentially, key combinations are tested and the text yielded by each test is compared to a coded text to find how close they resemble each other. If the text appears close enough, the key can be used to decipher most of the text sent.

## Data Encryption Standard

**data encryption standard (DES)**
64-bit key encryption method.

An older cryptographic code is based on the **data encryption standard (DES),** a 64-bit block cipher that uses a 56-bit key. Each 64-bit block of a message is coded separately using the 56-bit key. The key has a 8-bit checksum to assure that it is correct. A sixteen-step process is used to encode and decode the data blocks. The downside of this standard is the key length that was cracked in three hours using a brute force attack. However, this code was developed early on, before code cracking became more sophisticated. An improvement over DES, called *Triple des,* uses three separate DES algorithms, each with its own 112-bit key to encrypt and decrypt messages. This resulted in a highly secure, but cumbersome coding scheme.

## Pretty Good Privacy

**pretty good privacy (PGP)**
1,024-bit key encryption standard.

A scheme in popular use today is **pretty good privacy (PGP),** possibly because it is free and can be downloaded from the Internet. It uses asymmetrical 1024-bit keys to accomplish the coding and decoding of messages. A mathematical algorithm based on the public and private keys used assures another level of security. Developers of this method boast that it would take decades to crack this cipher.

## Clipper Chip

The government uses an *escrowed encryption system* as the basis for an encryption algorithm named SKIPJACK that is embedded in the *Clipper chip.* Laws and regulations involving the coding of data on networks or for wireless transmissions are insisting that the Clipper device be used with each system. This would allow law enforcement and other government agencies a method of "wire-tapping" into encrypted messages. Two sets of keys are provided with each Clipper chip, one for the user and one for the government. In order for a government agency to gain access to your data, they are required to supply an encrypted serial number to a central control. If they are authorized to tap into your transmissions, the central control will issue them the key to do so. There is a lot of controversy surrounding the use of the Clipper chip, which is yet to be resolved. Legislation to restrict and limit, if not totally eliminate this invasion of user privacy, is before Congress.

---

### Section 10.6  Review Questions

1. How does using a symmetrical private key system assure the privacy of your information?
2. What is the weakness of a private key system?
3. How does an asymmetrical key system work to secure data transmission?
4. What is meant by a brute force attack?

---

## 10.7  AUTHENTICATION

**authentication**
type of security that acknowledges users, through the use of passwords, as having authorized access.

Determining who the user is and what they have access to falls under the general heading of **authentication.** Part of the authenticating process lies in the use of the cryptographic keys described earlier. Additional methods of authentication use user names and passwords before the barest of accesses into a system are permitted. Sometimes this information is typed into a terminal when a user attempts to log on to a system. At other times, as in mobile communications, this information is supplied from a *smart card,* a credit card sized device that contains the necessary codes to allow access. The user inserts the card into a mobile phone or personal computer, which reads the codes and determines if the user is authorized to use the system.

### Smart Card

**subscriber identity module (SIM) and international mobile subscriber identity (IMSI)**
global mobile telephone smart card system.

One such smart card for authentication used for global mobile telephone service is called **subscriber identity module (SIM).** It uses an **international mobile subscriber identity (IMSI),** which is a numeric stream that identifies the user. The first three digits of the stream identifies the country of origin of the user. The operating facility that issued the

SIM card follows next and then the user ID number. Besides authenticating the user, the card is used for billing purposes by comparing the location of the user to the location of the called party. To prevent use by a stolen or lost SIM card, the user must supply their Personal Identification Number (PIN) in much the same manner as one does when using an automatic teller machine.

Another example of a smartcard is SGS-Thomson's ST19 series illustrated by the functional diagram in Figure 10-9. We will use this device to describe the operation of a smart card. As can be seen by the functional diagram, there are several operating units within the electronics of the smart card, but few connections to the outside world. Those connections include power ($V_{CC}$) and ground (GND), a central processing unit (CPU), clock (CLK) for internal synchronization, a system reset (RST) input, and two serial input/output ports ($IO_1$ and $IO_2$). This simple interface, defined by the International Standards Organization (ISO) ISO7816 document, makes for easy connection to a terminal or mobile unit.

The CPU controls the operation of the remaining units within the chip whose heart and sole resides in its memory sections. A read/write or RAM type memory of 128 to 960 bytes, depending on the ST19 version in use, supports the execution of instructions by the CPU. The read only memory (ROM) area is split into three distinct functions—system, user, and a cryptographic library. In the user ROM resides the chips operating system,

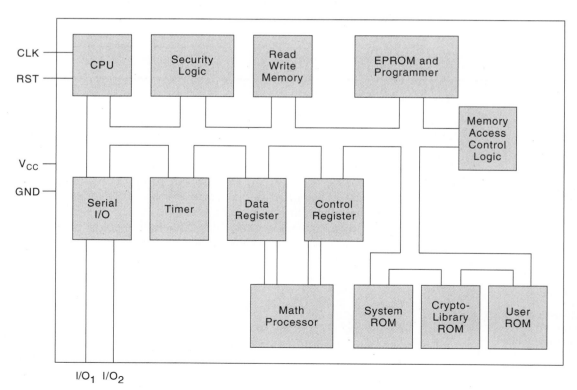

**FIGURE 10-9** ST19 Smart Card Microarchitecture

which holds the user application code for the chip. The system ROM retains basic I/O, testing, and security functions. An additional section of Electrically Reprogrammable ROM (EPROM) is used to hold key codes and personal data like social security number, selected user telephone numbers, and use history.

The security logic and math processor blocks combine to perform the actual encoding and decoding process, based on 512- or 1024-bit public keys for encoding and a unpredictable-number-generator-based secret key for decoding.

The cryptographic library ROM holds firmware code for the following functions:

* Basic math.
* Long random number generation.
* Cryptographic math routines.
* Digital key signature and authentication algorithms.

The remaining units of the ST19 provide timing and control processes for the other functional units. Data is entered and transferred out, serially, through the two serial I/O ports. The first data the card deals with is used for authenticating the user. After the user is allowed access, the ST19 then can be used for encrypting or decrypting user messages.

## Digital Signatures

*Digital signatures* is a form of authentication that encapsulates a key into a small amount of data and then scrambles the combination. The scrambled combination is recognized by an authenticating service that unscrambles the data and retrieves the key. There are several different methods of creating a digital signature. One is called a *hash-based signature,* which follows this basic procedure of concatenating a secret key to a file that is to be sent secured. The file plus the key is hashed to form a *hash value* that is sent with an encrypted form of the original file. The receiver of this file must have a copy of the secret key to evaluate the signature and decrypt the message. The weakness in this signature system is that both ends must have a copy of the secret key (symmetrical key) making its use susceptible to corruption by anyone at either end.

A second method of creating a signature uses public keys to verify signatures. It is the *Digital Signature Standard* developed by the U.S. National Institute of Standards and Technology (NITS) and the National Security Agency. The use of this signature only system is required only for companies that do business with the government. A private company called **RSA** has developed a similar public key signature algorithm that also includes encoding and decoding functions.

**RSA**
Rivest, Shamir, and Adleman.

## Certificate of Authority

**certificate of authority (CA)**
verifies validity of digital signature and public key owner.

The problem of issuing and verifying digital signatures presents the major stumbling block to the use of these tools. The concept of a *digital certificate,* based on the ITU X.509 standard, was developed to resolve that problem and add an additional level of authentication. A digital certificate, issued by a **certificate of authority (CA)** is a small block of data that contains a public key and an endorsement made by the CA, which includes name of holder, name of issuer, and a time limit for validation.

One such CA is VeriSign, which is a major provider of certificates for web servers that use the **secure sockets layer (SSL)** protocol for security. VeriSign acts as a root certificate, at the top of a possible hierarchy of certificates. The browser validates a VeriSign certificate by first checking VeriSign's signature in the root certificate. After VeriSign is authenticated, then the browser checks the user certificate issued by VeriSign.

VeriSign's certificates fall into three classes. Class 1 certificates are obtainable by identifying yourself and filling out a form on a web page. This type of certificate simply binds a public key to an e-mail address and is only used to verify your consistent presence. There is no actual verification of the user itself. A class 2 certificate bases the issuance on some relevant information about the user, found in a consumer database, such as a credit rating. There is no guarantee with this class that someone with access to enough data, couldn't "steal" an identity for a certificate. The class 3 certificate requires an individual to appear with a notarized application before being issued a certificate. This class provides the highest level of credibility to the use of certificates.

CAs must maintain two crucial databases to be effective. One is a list of current certificates that it has issued and the other contains certificates that have been revoked. Certificates that have been compromised or have passed their expiration date need to be disavowed and revoked.

The process to use a certificate begins by a user applying and getting a certificate from a CA. That user then begins a purchase over the Internet by filling out an order form or the like, signing the file containing the order, and sending this along with a copy of their certificate to the store offering the goods. The store checks the certificate with the CA and if the user is authenticated, the store and user exchange digital signatures setting up a secure trust relationship. The transaction can now be completed with critical data, such as credit card numbers, etc., securely transferred between user and the store.

## Secure Electronic Transactions

The credit card companies themselves, becoming very aware of the potential of electronic commerce, have devised an authenticating system using digital signatures called **secure electronic transactions (SET).** It uses two types of certificates in a two-level hierarchy scheme. They are the corporate certificate authority and the geopolitical certificate authority. These handle certificates based on company use and on the physical location of the transaction users. Like SSL, the process begins by a user deciding to make a purchase electronically. That user sends the order form and a signed encrypted authorization to a store that passes the authorization onto a local bank. The bank checks the digital signature and decrypts the credit card number. Next, the bank checks with the issuer of the credit card for validation. If the bank gets a valid return, it informs the store that it is alright to proceed with the transaction. The user gets the goods and a receipt while the store returns to the bank to verify the transaction and request the funds for the purchase. The bank pays the store and adds the transaction amount to the user's monthly statement. All of these communications are done using secure lines and authenticating digital signatures at every point of the transaction process.

---

**secure sockets layer (SSL)**
security protocol to verify certificate of authority issuer.

---

**secure electronic transactions (SET)**
authenticates credit cards.

### Web of Trust

Pretty Good Privacy (PGP) protocol uses a network of digital signatures known as a *web of trust* to easily expand the availability of access to authorized users. In this case the concept of an authorized user gets a little stretched. The system works by issuing a digital signature for an individual's public key to a select group of users. This group may be composed of relatives, friends, business associates, or whatever you decided to set up. Each member of the group can send secure messages to the individual whose public key is in the signature. An outside user who happens to be a friend, etc., to one member of the group can also gain access to the individual public key if that member grants that friend the use of the signature. Further, a friend of the friend, (you probably caught on already) can access through the friend who has gained access through the member. The thought behind this is to prevent limiting access to people who might have a legitimate reason for having it. It relies on the integrity of the group members and the outside people they allow to gain access to the individual's public key.

---

**Section 10.7  Review Questions**

1. How is a smart card used for authenticating a user?
2. Besides authentication, what else is a smart card used for?
3. What is a digital signature?
4. What is a certificate?
5. What is meant by a web of trust?

---

## 10.8 FIREWALLS

Another security concern is the limiting of access to certain web sites. The most basic way of limiting this access is the use of a user ID and password to authenticate a user. A more sophisticated method is the use of a *firewall*. Authentication is used by the destination site to limit access while firewalls are used by the source host. Essentially, a firewall checks a web access to determine if it is a secure site. It uses packet filtering or screening routers to determine if a user can have access to the desired web site. Each packet of data is scanned and passed on or rejected, based on some criteria set up within the firewall protocol. The firewall acts as a barrier to user access as shown in Figure 10-10. Users can still access other web sites that are not denied by the firewall as also shown in the figure.

Recently, this author had an experience that will illustrate the firewall process. I had registered for an online class through a college's distance learning program using the Internet. Deciding to use the Internet access provided by our school, I attempted to access my class. The home page for the college offering the class appeared without any problem. On this home page was a hyper-link to the current course listings. I clicked on this link and successfully got the list of courses. To this point, the web site was not secure.

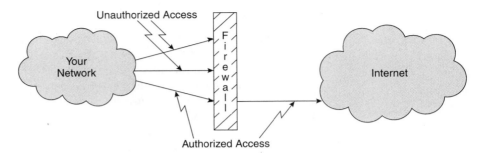

**FIGURE 10-10** Firewall

Clicking on my course was supposed to produce the authentication window asking me for my ID and password to allow access to the course I registered for. Unbeknownst to me, our Internet access had a firewall, which denied access to secure web sites, thus restricting student access to places they shouldn't be from our location. The firewall denied me access to the authentication window even though I knew I was an authorized user. It simply recognized that this was a secure site and denied access. The problem was resolved, later, by using a special access account on our system that bypassed the firewall and allowed me access to the class.

### Types of Firewalls

There are three basic types of firewalls, *screened host, screened subnet,* and *dual-homed gateway.* A screened host uses a screening router that directs all Internet traffic to a *bastion host,* which stands between the local network and the Internet. The bastion host's job is to decide which accesses to block and which to send onto the Internet.

A fairly simple firewall to implement is known as the bastion firewall. The bastion host is located between the Internet and the internal network in similar manner to the firewall shown in Figure 10-10. All Internet traffic must pass through the bastion host. To access the Internet, a network host must access the bastion host (this is usually transparent to the user). The bastion host checks the authorized accesses the requesting network host has and if its request is allowed, the connection to the Internet is made. The main drawback to using bastion host firewalls is that each type of access that is authorized for each network host must be set up in the bastion host. This includes web browsing, e-mail, file transfers, search engines, etc.

A screened subnet firewall (Figure 10-11) uses a network of screened hosts. On the secured side of each bastion host is a router connecting one or more networks to hosts. The screened subnet firewall operates like the screened host for larger networks.

Another firewall method uses an application gateway that inspects data before allowing connection in the first place. With this type of firewall, I would not have been allowed access to the college's home page, let alone any other pages. The dual-homed gateway is an applications layer firewall that checks all traffic using a bastion host between the network and the Internet. This firewall blocks applications from use rather than individual accesses within an established application.

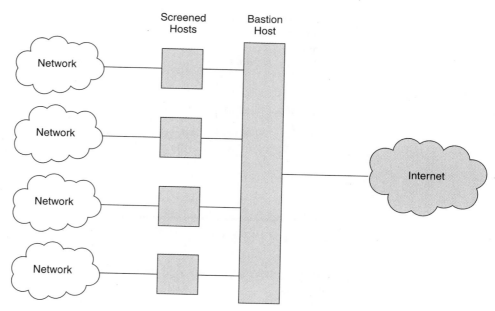

**FIGURE 10-11**  Screened-Hosts Firewall

Circuit level gateways are firewalls that authenticate TCP and user datagram sessions when setting up sessions between two hosts. This method maintains tables of authenticated users and terminates sessions when finished.

## Firewall Filtering

In addition to these types of firewalls, there are also three generations of firewalls, *packet-header filtering, application-level proxies,* and *stateful inspection.* Packet-header filtering is the first generation firewall that filtered out unauthorized user access based on the information in the header of an Internet packet. This type of firewall can be compromised by a method called *address spoofing,* which places false information into the header so that the firewall accepts the user as valid.

Application-level proxies are the second generation firewall. Application-level and circuit-level gateways protect networks by changing the names of IP addresses as well as checking the validity of incoming packets. Proxies work with proxy servers that provide two basic functions—the control of access to certain TCP/IP services and the management of network resources. Access to individual services such as FTP, HTTP, GOPHER, etc., can be restricted using a proxy server. This aids in allowing general access to, say, employees of a company for company business, while restricting access to noncompany functions.

The third generation, stateful inspections, check for certain bit patterns within a packet, regardless of the overall content of that packet. It uses an archive system that allows access to a user that has made a previous access to a currently selected location. Stateful inspection firewalls analyze all protocol layers, comparing past sessions to the current one. This type of security is not dependent on predefined application information, but uses rules established by the user in previous sessions.

---

**Section 10.8  Review Questions**

1.  What is the purpose of a firewall?
2.  What are the different types of firewalls?
3.  What is meant by filtering?

---

## 10.9  INTRANETS AND EXTRANETS

Using the Internet to exchange information easily and relatively quickly (try comparing an Internet access for a research assignment to searching for material in a library or by inquiring various sources for information!). However, many companies became concerned because employees began to spend too much work time surfing the net. Either it took a long time to gain access to a desired site or employees found many diversions to take their time away from the business at hand. Yet the Internet was an efficient way to handle information resource transfers. What became apparent was that the technology underlying the Internet network was providing the functionality that company executives were admiring. There had to be a way of separating the distracting features of the Internet and retaining the beneficial, from the company's standpoint, aspects of the system. From this combination of desires arose the **intranet.**

**intranet**
a local Internet type of network with no external access.

Intranets are basically "baby" Internets. They use the same network facilities that the Internet does, but access is restricted to a limited sphere. For instance, a company can set up an intranet within the confines of the company itself. Access can be tightly controlled and limited to authorized employees and staff. There is no connection to the Internet or any other outside network. Functions like web sites, file uploads and downloads, and e-mail are available on intranets within the confines of the network.

Since frivolous sites are no longer available, there is no employee time lost due to accessing them. There is, of course, the limitation of the networking area. The very benefit of restricting access to all of the facilities available on the Internet also restricts communication to other desirable locations. This is where the **extranet** steps in.

**extranet**
a restricted type of intranet network with external access.

An extranet is a network that connects a number of intranets into a truly mini-Internet. Access is extended to all the intranets connected through the extranet, but, again, not to the Internet. Extranets require a constant Internet connection and a hypertext transfer protocol (http) server.

Extranets can also be used to connect an intranet to the Internet so that remote offsite access can be made into a company's intranet by an authorized individual. This can facilitate multimedia conferencing and telecommuting. A system known as *distributed computing environment,* coupled with encryption techniques, are employed for access control through an extranet. Basically, it uses passwords and smart cards to log in to a gateway server that checks the requester's security credentials. If the user checks out, he or she is allowed access into the company's intranet structure.

A number of URL addresses are set aside for intranet and extranet use. Essentially because intranets are self-contained networks, the same set of addresses can be used by all intranets without conflict. Extranet addresses are designed to recognize the intranets they connect and correctly preface each intranet address with an identifier. This allows two interconnected intranets to retain the same set of address values and keep them from being mistaken. One class A address, ranging from 10.0.0.0 to 10.255.255.255 is reserved for intranet usage. Again, since an intranet is a self-contained system, it only needs one class A network to designate the main network. Subnetworks use reserved class B and class C addresses. There are 16 class B addresses, from 172.16.0.0 to 172.31.255.255 and 256 class C addresses, which range from 192.168.0.0 to 192.168.255.255.

---

### Section 10.9  Review Questions

1. What is the benefit of using an intranet as opposed to the Internet?
2. What function does an extranet provide for Intranet systems?

---

## 10.10  TCP/IP, THE TECHNOLOGY BEHIND THE INTERNET

**user datagram protocol (UDP)**
application for sending unacknowledged messages.

The Transmission Control Protocol/Internet Protocol (TCP/IP) was originally designed as the transport medium for the **user datagram protocol (UDP).** Datagrams are characterized as unacknowledged messages. They were sent out on a network without any regard as to an acknowledgement of their reception. The advantage to this method of data transfer is speed. Datagrams can carry large amount of data quickly because there is no time used to wait for a packet of data to be acknowledged. The downside is the need for a highly reliable network to carry these datagrams so that concern of the failure to deliver packets is minimized. This was fine for use by the Internet since most of the data transferred between users and servers or other users does not require acknowledgement. When you request a web site, you get it or you don't. Your request itself is not acknowledged. There is no acknowledgement to e-mail transmissions either. As long as they are not bounced, there is no way to know if they arrived at their destination.

### Internet Protocol (IP)

The Internet Protocol (IP) portion of the TCP/IP suite is a network-layer routing protocol responsible for delivery of messages within the same network or between different, but interconnected networks. IP interprets Internet addresses and guides transport-layer packets through the system. The IP header that precedes the IP payload (TCP datagram frame)

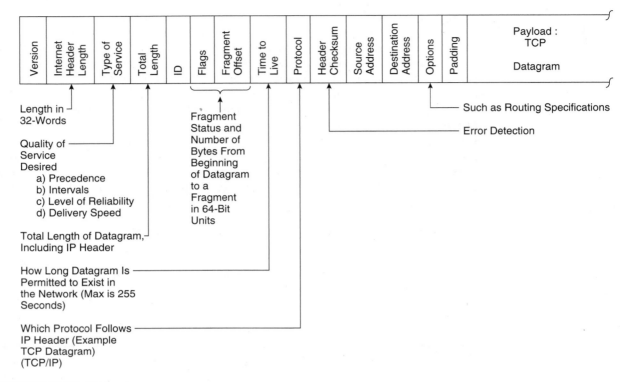

**FIGURE 10-12** IP Header

is shown in Figure 10-12. It begins with a version number, which is currently IPv4 but is being replaced by IPv6 as described under the Internet Addresses Section 10.5 of this chapter. The length of the IP header in 32-bit words is indicated next, followed by an indication of the quality or type of service desired. This indication includes precedence, level of reliability, and delivery speed information.

An indication of the total length of the datagram plus IP header is in the next field followed by an ID field and a fragment field. The flags and fragment fields combine to indicate fragment status and the number of bytes from the beginning of the datagram to a fragment in 64-bit units. The life time of a datagram message (how long it is allowed to exist in the network without finding a destination), is included next. The maximum possible time allotted is 255 seconds. The protocol of the payload area fills the next field with a checksum error detection for the header coming after it. Finally, IP source and destination addresses with options such as routing specifications complete the header with padding necessary to keep the header on 32-bit bounds.

## IPv6

One of the main difficulties with the Internet has been its own growth. The limit of available IP addresses fixed by the 32-bit address size is quickly approaching. The current IP

| N-Bits | 80 minus N-Bits | 48 Bits |
|---|---|---|
| Subnet or Subscriber Prefix | Subnet ID | IEEE 802 MAC Address |

**FIGURE 10-13** IPv6 Unicast Address

**dynamic host configuration protocol (DHCP)**
application for adding and deleting hosts on an IP network.

**neighbor discovery protocol (NDP)**
replaces ARP used in IPv4.

standard, IPv4, possesses this limit through this specified address. A later version, known as IPv6 was created, in part, to alleviate this problem by increasing the address size to 128-bits. This new version of the IP protocol does more than just provide for more address allocation. One functional addition included in IPv6 is *autoconfiguration,* which enables an IPv6 host to configure its IP addresses with little or no human intervention. This provides for an easier way for hosts to add and delete accounts through an application called **dynamic host configuration protocol (DHCP).**

The IPv6 address, as shown in Figure 10-13 is formed by concatenating a media access control (MAC) address with a subnetwork ID, a subnetwork prefix that is learnt by the host using neighbor discovery from routers on the same subnetwork. The **neighbor discovery protocol (NDP)** replaces the address resolution protocol (ARP) used under IPv4. IPv6 starts the process by issuing a multicast message to inquire what the local subnetwork router code is. The response to this message from the routers connected to the subnetwork is then used by the protocol to complete the address by adding the router address to the host address to form the prefix for the IPv6 address. This is concatenated with the user's MAC address to complete the Internet IPv6 address. After an address has been allocated or changed, the dynamic host configuration protocol updates domain name service (DNS) databases.

There are three types of IPv6 addresses, *unicast, multicast,* and *any cast.* Unicast addresses uniquely identify a single node or host interface. Multicast addresses are similar to group addresses in other protocols. They access selected sets of nodes. IPv6 replaces IPv4's broadcast address with an *all nodes* multicast address, which accesses all nodes on a subnetwork. The any cast address is a special form of unicast address that specifies traffic for the nearest node from the originating source node.

The format of an IPv6 address indicates eight fields of 16-bits each, separated by colons (:). Using X to designate a 16-bit field, the general format appears as:

$$X:X:X:X:X:X:X:X$$

Groups of fields with all 16-bits zero, use a shorthand notation of a double colon (::). Wherever these appear in an IPv6 address, you can replace it with enough zero fields to complete the 8-field address. For instance, an IPv6 address of:

$$0:0:0:0:0:0:0:1$$

can be written as:

$$::1$$

This address, ::1, is used by IPv6 for testing purposes as a *loopback* address to cause datagrams to be sent back to their source node.

IPv4 addresses are represented in an IPv6 format replacing the last two fields with decimal IPv4 address (which is 32-bits or two IPv6 fields):

X:X:X:X:X:X:d.d.d.d

The first six fields are in hexadecimal format, while the last two fields are converted to four decimal fields used by IPv4. IPv4 compatible addresses break down into two types with IPv6. The first, with all hexadecimal values zero, designates an IPv6 node that can accept IPv4 addresses as well. An IPv6 hexadecimal address of ::FFFF:d.d.d.d is an IPv4 node only, which cannot accept IPv6 addresses.

The any cast address format is shown in Figure 10-14. The subnetwork prefix identifies the subnetwork the message is destined for. A host address of zero specifies any cast rather than a unique address. Multicast or group addresses follow the format in Figure 10-15. The upper 8-bits are FF in hexadecimal. These are followed by four flag bits. The only flag used is the T flag. When it is 0, the multicast address is permanently assigned. When it is a 1, the address is transient and is only used at the time the message is sent. A 4-bit scope field follows, which identifies the scope of the multicast message. The scope can be node, site, organizational, or link local or global in nature.

Another change brought by the IPv6 protocol is the maximum size of the packets used. IPv4 set a maximum of 576 bytes and IPv6 increased the count to 1280 bytes maximum.

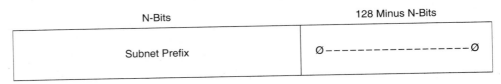

**FIGURE 10-14** IPVG Any Cast Address

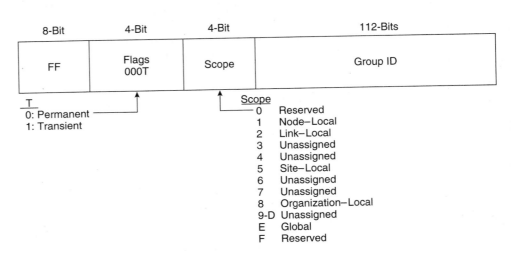

**FIGURE 10-15** IPv6 Multicast Address

## Transmission Control Protocol (TCP)

The transmission control protocol (TCP), on the other hand, is a transport-layer protocol that carries application-layer packets and services between two users. This protocol is responsible for setting up and terminating the session between the users, encapsulating information into a datagram structure, and transmitting and tracking those datagrams through the system.

TCP operations fall into five classes, depending on the depth of service required. The first class, Class 0 is the *simple class* in which one transport connection equals one network connection. There is no error detection or correction for this class of operation. Class 1, the *basic recovery class,* recovers from a network disconnect or reset by reassigning transport connections to another network. In the *multiplexing class 2,* several connections are multiplexed into a single connection allowing many channels of data to be transferred simultaneously. Class 3 combines classes 1 and 2, providing multiplexing and fault recovery capabilities. The last class, Class 4, includes the abilities and functions of all four classes and adds error detection and correction processes. Included in this class is the ability to recover from lost, duplicated, or bad packets or frames. The basic frame formats for the **transport protocol data units (TPDU)** of each class is shown in Figure 10-16.

The length field indicates the length of the TPDU frame in bytes. TPDU-type field indicates which class of service this frame applies to. Options are specific to TPDU-type and include options like CRC-4 error detection for class 4 frames, TPDU unit number, and an end of transmission indication. The remaining bytes of the TPDU frame contain the actual payload data. Example frames for all five classes are also shown in Figure 10-16.

The TCP header, illustrated in Figure 10-17, precedes TCP datagrams and includes valuable information. Source and destination addresses provide routing information for the packet. The order of the packets within the completed message framework are maintained by the use of sequence numbers. Acknowledge numbers also aid in verifying sequence in the case of the classes that include fault recovery functions. The number of 32-bit words in the header is known as the data offset number, because it indicates how much the beginning of the data is offset by the header. The sections of the header referring to urgent data indicate a priority message and how it is to be handled. An acknowledge flag indicates that this is an acknowledge frame for those classes who support fault recovery. The reset bit causes the transport connection to be reset at the end of a session. A synchronization (SYN) indication is used to establish a connection in combination with the acknowledge (ACK) bit:

$$SYN = 1; ACK = 0 \text{ is a connection request and}$$
$$ACK = 1 \text{ is a connection acknowledgement}$$

The final data segment is indicated by the FIN bit in the header. The window specifies the number of data bytes the sender is willing to accept, in return, from the host end. For those classes that perform error detection, a checksum character is provided in the header. The urgent pointer indicates the offset value of the current segment and informs the receiver to find the urgent data indicated by the urgent (URG) bit. There are three possible options that are specified in the options field. They are the following miscellaneous parameters:

$$0 = \text{end of option list}$$
$$1 = \text{no operation}$$
$$2 = \text{maximum segment size}$$

**transport protocol data units (TPDU)**
TCP data unit frame.

| Length | Transport Protocol Data Unit Type | Options | Data |
|--------|-----------------------------------|---------|------|

Frame Format

a

| Length | FOH | Unit Number | End of Transmission | Data |
|--------|-----|-------------|---------------------|------|

Class 0 and 1 Data Unit

b

| Length | FOH | Destination Reference | Unit Number | End of Transmission | CRC-16 | Data |
|--------|-----|-----------------------|-------------|---------------------|--------|------|

Class 2, 3, and 4 Data Unit

c

**FIGURE 10-16** Transport Protocol Data Unit (TPDU)

| Source | Destination | Sequence # | Acknowledge # | Data Offset (DO) | Unused | Urgent Data (URG) | Acknowledge (ACK) | Push (PSH) | Reset (RST) | SYN | Final (FIN) | Window (WIN) |
|--------|-------------|------------|---------------|------------------|--------|-------------------|-------------------|------------|-------------|-----|-------------|--------------|

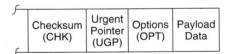

| Checksum (CHK) | Urgent Pointer (UGP) | Options (OPT) | Payload Data |
|----------------|----------------------|---------------|--------------|

**FIGURE 10-17** TCP Header

## Section 10.10 Review Questions

1. What is the chief characteristic of a datagram message?
2. Which open systems integration (OSI) model layers do TCP and IP belong to?
3. What is the maximum time that a datagram packet can exist on the Internet without finding a destination?
4. Is TCP/IP a connectionless or connection-oriented protocol suite?

# SUMMARY

No one has to expound on the popularity of the Internet. It is highly used and it is here to stay. The so-called information highway touted so loudly in the early eighties has arrived on the back of the TCP/IP protocol suite. e-mail, newsgroups, web-browsing, sharing information, and transferring files are the primary uses of the Internet. Rising at a rapid rate is the use of electronic commerce on the Internet. Concerns about limiting Internet growth because of address limitations and security on the Internet are two major concerns of the world wide web's continued use.

Various methods of authentication and cryptography are being applied to alleviate the security concerns. An updated version of the Internet protocol (IPv6) addresses the growth issue by extending Internet addresses from 32 to 128 bits.

Many protocols and application programs are present to facilitate the handling of the volumes of information transferred through the Internet. They include, TCP/IP, FTP, browsers, GOPHER, ARCHIE, VERONICA, WAIS, and others.

The Information Age is here and it is growing by leaps and bounds. We are only beginning to see the implications of where this will lead.

# QUESTIONS

### Section 10.1

1. Which network was the foundation for the Internet?
2. What type of switching is used by ARPANET?
3. What is the Internet?
4. What is the name of the group that oversees the Internet? What subgroups are under that group and what are their responsibilities?

### Section 10.2

5. Define datagram.
6. What is the main advantage and disadvantage of using datagrams?
7. Which suite of protocols are used to handle Internet traffic?
8. Which protocol was added to handle news and mail servers on the ARPANET network?
9. What is the name of the first Internet graphical browser?

### Section 10.3

10. Name four basic applications the Internet is used for.
11. List four search engines and note what each one searches for.
12. List the causes for undeliverable e-mail messages.
13. What happens to undeliverable e-mail messages?
14. Name two e-mail protocols and one e-mail specification.
15. Which portion of the X.400 specification addresses service authorization?
16. Which X.400 portion applies to routing messages to dissimilar mail systems?
17. What is the purpose of X.500?
18. What is the purpose of newsgroups?

19. What is the term for a newsgroup message?
20. Which protocol is used for downloading files from the Internet?
21. What is the first page of a web site called?
22. What function does a hyper-link perform?
23. What term applies to the most frequently asked questions about the Internet?

### Section 10.4

24. What is the name of the central locations that assign Internet addresses to clients?
25. What two protocols facilitate direct access to the Internet?
26. Which DLL is used by WINDOWS to facilitate access to the Internet?
27. What is the purpose of each channel in a basic rate ISDN interface used for accessing the Internet?
28. What are the three types of ISDN interface methods used to connect between the Internet and client node?
29. Which ISDN interface connects the Internet to a customer's LAN?
30. What is the most costly way of accessing the Internet? What are the advantages of using this method?
31. How are cable modems used to gain access to the Internet?

### Section 10.5

32. What is the formal term for an Internet address?
33. Which protocol translates Internet addresses associated with web sites?
34. What are the main sections of the user Internet address?
35. What is meant by a domain top in an Internet address?
36. What does a domain top identify? Give four examples of a domain top.
37. Which class of Internet address is used for a system with few networks that are each heavily populated?
38. Give a purpose for a subnetwork masked address.
39. The decimal fields in an ISP Internet address identify two items. What are they?
40. How do the three classes of Internet addresses differ?
41. How are Internet broadcast messages identified?
42. What is the purpose of an IP multicast?
43. How does an IP router reduce multicast traffic?
44. Which protocol defines multicast-type packets?
45. Why aren't multicast messages acknowledged?
46. Identify the sections of an e-mail address.

### Section 10.6

47. Name the two basic methods for securing Internet messages.
48. Explain the difference between a symmetrical and an asymmetrical key system.
49. What is the difference between a public and a private key?
50. What method is used to check the validity of a cryptographic key?
51. What is a brute force attack? What is it used for?
52. How does a plain text attack work to determine the value of a code system?
53. What is the weakness of the DES method? How does triple DES overcome this weakness?

54. What size key is used by PGP? Why does this size make a more secure system than the DES system?
55. What is the size of a triple DES key?
56. Explain the general process for encrypting and decrypting a message.
57. What is Skipjack?
58. What is a Clipper Chip?
59. Why does the government want everyone to use Clipper?

### Section 10.7

60. Define smart card.
61. What is SIM used for?
62. What kind of system is SIM used on?
63. What is a hash value as it applies to a digital signature?
64. What is the name of the signature standard used for companies that have transactions with government agencies?
65. What is a certificate of authority?
66. Who issues certificates?
67. Define the three classes of certificates.
68. What data bases are maintained by CAs?
69. What is the name of the digital signature system used by credit card companies?
70. What two CAs are used by credit card companies?
71. What is a web of trust?
72. What is the benefit of a web of trust?
73. What is a negative aspect of a web of trust?

### Section 10.8

74. What is the purpose of a firewall? Where are firewalls located?
75. List the different types of firewalls.
76. What method is used by firewalls to limit access to predefined web sites?
77. Which firewall type filters packets at the application level?
78. Which firewall limits access by comparing requests to an acceptable list of access types (browsing, e-mail, etc)?

### Section 10.9

79. What is an intranet? How does it differ from the Internet?
80. What functions does an extranet provide?

### Section 10.10

81. A few acronym definitions should become second nature to the student of data communications. TCP/IP is one of them. What does the acronym TCP/IP stand for?
82. What OSI model layers do IP and TCP operate at?
83. What is the chief responsibility of the IP protocol?
84. What type of packets was TCP designed to carry?
85. Briefly describe each class of the TCP protocol.
86. What is the purpose of the sequence and acknowledge number fields in the TCP header?

87. What parts of the TCP header is used to request and establish a connection?
88. Besides address size change, what else does IPv6 address?
89. How many bits make up an Internet address in IPv6? How does this compare to IPv4?

## RESEARCH ASSIGNMENTS

1. Use the Internet to write a short report on Internet security. Use at least three Internet references for your report.

2. Explore the different types of web browsers available. Select at least three commonly used browsers like Netscape, Internet Explorer, and MOSAIC. Compare their features, strengths and weaknesses.

3. Research other applications for TCP/IP.

4. Write a paper about the various ways to gain access to the Internet. Include how the access is accomplished, the services available, and the costs about each access method in your report.

5. Research the concepts of electronic commerce in relation to the Internet.

6. Write a report detailing support services on the Internet like distance learning and electronic libraries.

## ANSWERS TO REVIEW QUESTIONS

**Section 10.2**
   1. ARPANET.
   2. TCP/IP.
   3. MOSAIC.

**Section 10.3**
   1. E-mail, newsgroups, downloading files, chat, electronic commerce, web browsing.
   2. Simple mail transfer protocol.
      Post office protocol.
      Internet mail access protocol.
      X.400.
      Eudora.
   3. Hypertext markup language, JAVA.
   4. File transfer protocol (ftp).
   5. ARCHIE, VERONICA, GOPHER, WIDE AREA INFORMATION SERVER, WHOIS.

**Section 10.4**
   1. An ISP is a provider of Internet addresses and services to a user.
   2. ISDN modem, America OnLine (AOL) or other online service, TCP/IP interface, SLIP, PPP, WINSOCK, leased line direct access, cable modem.

3. Direct TCP/IP connection to an ISP. Fast digital rate and access to all Internet functions. Also quite expensive.

**Section 10.5**
1. 32 bits.
2. Host, subnetwork.
   sub-domains, domain names, domain top.
3. Universal resource locator.
4. .edu, .com, .mil, .gov, .org.
5. Multicasting.

**Section 10.6**
1. Firewall, authentication, encryption.
2. Brute force attack.
3. Pretty Good Privacy (PGP).

**Section 10.7**
1. Same key used to encrypt and decrypt messages.
2. Key itself has to be delivered to all users.
3. Encode using private key and decode using public key for verification of sender. Encode using public key and decode using private key for privacy of message sent.
4. Randomly try different key combinations until match is found.
5. Smart card contains secure information about user to verify authorized access.
6. Billing information.
7. Unique digital code that identifies a user.
8. Authenticates digital signatures.
9. Expands access to a system to authorized users by providing public keys to trusted individuals, who, in turn, supply their keys to other trusted people— access is indirect through a hierarchy of trusted users.

**Section 10.8**
1. Firewalls are used to restrict access to the Internet.
2. Bastion host.
3. Filtering checks if user has access to a Internet site and filters out those that are unauthorized.

**Section 10.9**
1. Intranets are secure self-contained networks.
2. Extranet is used to either connect intranets together or as a remote access into an intranet.

**Section 10.10**
1. Unacknowledged packet, quick, but no guarantee of delivery.
2. Transport and network.
3. 255 seconds.
4. Connectionless.

# CHAPTER **11**

# Fiber Optic Communications

## OBJECTIVES

After reading this chapter, a student will be able to:
- Have a firm understanding of the basics of fiber optic systems.
- Understand how light propagates in a fiber cable.
- Know the benefits and disadvantages of fiber optic cables.
- Determine the specifications and usage of light sources, including LEDs and LASERs, and light detectors.
- Understand the working of fiber optic networks, including FDDI, fibre channel, and SONET.

## OUTLINE

11.1 Introduction
11.2 Basic Concepts of Light Propagation
11.3 Fiber Cables
11.4 Light Sources
11.5 Optical Detectors
11.6 Fiber-Cable Losses
11.7 Wave Division Multiplexing
11.8 Fiber-Distributed Data Interface
11.9 FDDI-II: Isochronous Traffic
11.10 The Fibre Channel
11.11 SONET

## 11.1 INTRODUCTION

Early long distance data communications and networking owe their existence to the telephone network. Without it, any form of communications would have been considerably more difficult. Because the telephone network was in place when data communications came into being, it was not necessary to create an interconnecting network of cables and switch stations. The telephone company uses combinations of twisted-pair copper wire, known as Unshielded Twisted Pair (UTP), and coaxial cables for much of its land-based interconnections. It has been shown that these media restrict the rate of data flow due to limitations imposed by bandwidth specifications. Equally as important is the effect of various impairments and degradations inherent to UTP and coaxial cabling. Because of these factors, the distance between sites, without the use of repeaters, was severely limited.

The introduction of fiber optic cables, capable of transporting light rather than electrical signals, brought a new era to the way we communicate using networks. The advantages realized by the use of fiber optic cables include increased bandwidth, less sensitivity to electrical and magnetic interference, and increased security. Distances between sites is increased because of the reduced effects of signal degradation, not a severe problem with fiber optic cables. To understand why these advantages exist is to understand the nature of light as a transmitting medium. Light rays are high frequency waves with a short wavelength in the micron ($\mu$m) range. Wavelength is to light waves as time period is to electrical signals. The higher the frequency, the shorter the time period of an electrical signal. For fiber optic communications, light is expected to travel through a medium of glass formed into a cable. Light in the visible range exhibits wavelengths between 0.4 and 0.7 $\mu$m and is highly attenuated by glass. This attenuation does not make for efficient transfers of data through optic cables. Ultraviolet light waves, which are attenuated more than visible light, also are not used for fiber optic communications. On the other hand, light waves with wavelengths between 0.85 and 1.6 $\mu$m which is in the infrared range, are much less affected by glass and travel very efficiently through it.

Since glass is used as the conducting medium and light waves are such high frequencies, fiber optic transmission is immune to electrical interference. Glass, being an insulator, will not conduct electrical signals or radiations. When using coaxial or twisted pair cabling, currents can be induced from one metallic wire to another by the magnetic fields created by the current in the wire. Light rays are propagated free of these magnetic effects. This results in a lack of magnetic interference such as crosstalk between neighboring lines and effects of static electricity from causes like electrical storms or humidity.

The transmission of light is not affected by electromagnetic effects or reactive elements of capacitance and inductance that occur in coaxial or twisted-pair cabling. These reactive elements place bandwidth restrictions and present propagation delay considerations to signals using them. Instead, light travel through a medium is restricted by the light signal's wavelength and the medium's opaqueness. The important result of the difference is the increased bandwidth available to signals traveling via light versus electrical energy. Fiber optic cables can manage data rates exceeding 1,000 Mbps due to this increased bandwidth.

One additional advantage of fiber optic cables is their small size and light weight. A complete cable measures between 0.1 and 0.5 millimeters in diameter with core sizes ranging from 5 to 600 $\mu$m in diameter. The size does have a negative side though. Because of the smallness and the need to align cable ends precisely to sending and receiving devices and to each other (in the case of a splice), couplers and splices used for fiber optic systems have to very exact. This leads to difficulty in the assembling and installation of fiber optic cable networks. Additional problems exist when a cable has to be repaired. Many individual fiber cables are bundled together into a single cable jacket. This is done to give the cable bundle a rugged existence insuring a longer life for the individual cables. However, if something does go wrong causing a major break in the cable, it is a massive and expensive task to repair the break. Each individual cable has to be precisely aligned through a splice so that light traveling from one side is easily propagated to the spliced cable. One upside to the problem of physical operating with fiber cables, they are difficult for someone else to tap into. The result is a more secure cabling system, free from illegal wiretaps and other intrusions.

To summarize, despite the difficulty presented by installing and repairing fiber optic cables, the benefits of their use are:

1. Larger bandwidth than conventional cables.
2. Immunity to **electromagnetic interference (EMI).**
3. Immunity to **electromagnetic pulses (EMP)** due to atomic explosions or electrical storms.
4. Less susceptible to weather conditions (does not corrode or rust).
5. More difficult to tap into, making it more secure.

One testament to the benefits of fiber cable for communications purposes is illustrated obliquely by the "pin drop" ad campaign for one of the long distance carriers. You know the one—"So quiet you can hear a pin drop." This attested to the lack of noise and interference on the long distance lines used by that company, which are fiber optic. It also attests to the lack of degradation of the signal caused by the pin dropping, but I wouldn't blame that entirely on the fiber cable, although it did play a large part in reducing loss of signal due to distance.

**electromagnetic interference (EMI)**
interference caused by a steady transmission of power.

**electromagnetic pulses (EMP)**
large induced pulse on a cable.

---

### Section 11.1  Review Questions

1. Why was the telephone system used when long distance networks were first created?
2. List the prime advantages of using fiber cables.
3. What is the main drawback to using fiber cables?
4. Which type of light is used for long distance communications? Why is it used?

---

## 11.2  BASIC CONCEPTS OF LIGHT PROPAGATION

**rays**
unit of light traveling in a line.

**refraction**
bending of light ray so that it continues to travel in the same direction but at a slightly different angle.

**reflection**
return of light in a direction opposite of its original course.

One theory of light propagation is that light travels from a source to a receptor as a collection of ray elements, simply called **rays.** These rays have direction and varying strength characteristics depending on their energy content. As long as these rays travel through a single material (or medium), they move in a straight line. Once they encounter a second medium, they are bent, either continuing forward at a new angle or bounced backward going in a direction not quite exactly from where they came. The forward bending is known as **refraction** and the backward bending, **reflection.** The amount of refraction or reflection is dependent on the density and composition of the two media involved.

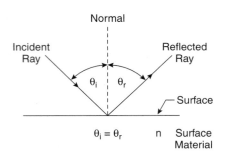

**FIGURE 11-1** Incident and Reflected Rays

## Reflection and Refraction

**refractive index**
value that determines the amount a light ray will be reflected or refracted when it passes from one material to another.

A **refractive index** ($n$) has been developed to aid in predicting the effect of light travel as rays attempt to pass from one medium to another. This index uses air as a standard medium with a refractive index of 1. Refractive indexes for other material are found by dividing the speed of light through air by the speed of light through the other material. Other common refractive indexes are quartz (1.46), water (1.39), fused silica (1.46), silicon (3.5), and glass (1.5).

Figure 11-1 illustrates a light ray being reflected as it meets a surface after traveling through air. The initial ray striking the surface is called the **incident ray.** An angle measure between the incident ray and a **normal** reference line that is perpendicular to the surface is called the **incident angle,** $\theta_i$. The reflected ray and its angle of reflection ($\theta_r$) are also shown. To achieve reflection, the refractive index of the surface material must be greater than the incident medium, in this case, air. If the reverse is true, as in the case of an underwater light source projecting a beam of light into the air, refraction occurs (Figure 11-2). Angles of incidence and refraction are measured compared with a reference that is the perpendicular normal to the meeting line of the two media as shown in Figure 11-2, where $n_1$ represents water and $n_2$ air.

**incident ray**
light ray that initially strikes a surface.

**normal**
line perpendicular to the surface by which all light ray angles are measured.

**incident angle**
the angle at which a light ray meets a surface.

**Snell's law**
defines the relationship between the ratio of critical and incident angles to the ratio of refractive indexes between two meeting surfaces.

A formula, known as **Snell's Law** is used as a measure of the relationship between the ratio of refractive indexes and the ratio of the incident and refractive or reflective angles. That equation is:

$$\frac{\sin \theta_r}{\sin \theta_i} = \frac{n_1}{n_2} \tag{11-1}$$

where $n_1$ is the refractive index of the incident medium and $n_2$ is that of the refractive or reflective medium. $\theta_i$ is the angle of incidence and $\theta_r$, the angle of refraction or reflection.

Two important angular values are the critical angle ($\theta_c$), which is the measure of the angle of maximum reflection from the normal (a line perpendicular to the reflecting surface), and *Brewster's angle* ($\theta_B$), which is a measure of the minimum angle of reflection. Light rays reflecting on a surface must meet that surface at an angle between those two values or else they will be refracted. The equations to determine their value are:

$$\sin\theta_c = n_2/n_1 \tag{11-2}$$

$$\tan\theta_B = n_2/n_1 \tag{11-3}$$

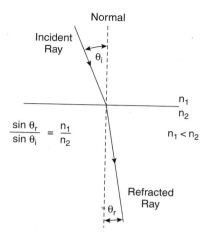

**FIGURE 11-2** Refracted Ray

---

### EXAMPLE 11-1

Given the angle of incidence as 12.5°, what is the reflective/refractive angle of the light ray if it enters from air to water? Is this a reflected or refracted ray?

**SOLUTION**

The refractive index for air is 1 and for water, 1.35. We can use Equation 11-1 to determine the refractive/reflective angle, $\theta_r$. First cross multiply and divide by $n_2$ to isolate $\sin\theta_r$ on one side of the equation. Then find the arc-sine of both sides:

$$\theta_r = \sin^{-1}[n, 1 \times (\sin\theta_i)/n_2] = 11.37°$$

Example 11-2 and 11-3 can be used to find the range of reflective angles for this example. The critical angle is 48.75° and the Brewster angle is 36.94°. That means that this ray is refracted into the water rather than reflected, since 11.37° is below the critical angle of reflection. Had the results lain between the critical and Brewster angles, then the light ray would have been reflected rather than refracted.

---

The importance of reflection and refraction cannot be overemphasized. In a fiber-optic cable, light is directed into the core of the cable. The most ideal situation is for the light to enter right at the center line of the core. When this happens, the light does not approach the walls of the core and travels fairly unimpeded through the cable. However, to align a light source so that all the light from it is directed down the core's center is extremely difficult and cost prohibitive. As such, quite a bit of the light from a source is going to enter the cable at angles away from the center line. Figure 11-3 shows a basic fiber cable entry point. Light enters the core and is slightly refracted due to the differences in refractive indexes of the core material and air. The rays of light travel in the core

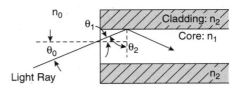

**FIGURE 11-3** Angle of Incidence ($\theta_1$), Refraction ($\theta_r$), and Reflection ($\theta_2$)

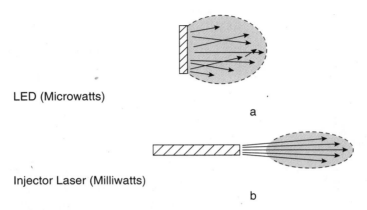

LED (Microwatts)

a

Injector Laser (Milliwatts)

b

**FIGURE 11-4** Radiating Patterns for LED and Injection Laser

**cladding**
material surrounding a fiber core, which has a refractive index that causes light rays to be reflected back into the core.

until they meet the core wall, which is encased in differently doped material called the **cladding.** By changing the doping, which is a process of inserting foreign material into a pure material, the refractive index ($n$) is changed. The purpose of the cladding material surrounding the core is to cause the light rays to be reflected and be retained in the core as they travel down the cable as shown in Figure 11-3. The reflected rays continue traveling down the cable's core, bouncing off the cladding walls as it moves along. Eventually, the light will exit the cable at the receiving end.

## Light into the Cable

Many methods have been developed to direct light into a fiber-optic cable. Each involves the use of a light source and some method to align the light source with the cable entry point. Two commonly used light sources are **light-emitting diodes (LEDs)** and **injection laser diodes.** The principal differences between these two sources are the amount of light emitted by each and the cone of emitted light projected from their surfaces. Light-emitting diodes emit a much lower level of light (-15 dbm power level) than do laser diodes (-6 dbm). Lasers also concentrate their light power into a tighter cone pattern as shown in Figure 11-4. The light-emitting diode pattern is much broader, since the light is emitted from the surface of the diode. Lasers project light from their edge, forming a more intense and narrow cone.

**light-emitting diodes (LEDs)**
semiconductor diode that emits light when it conducts.

**injection laser diodes**
semiconductor device that emits a concentrated cone of light.

To couple light from the source into the cable requires the source to be held firmly to the cable. If the coupling shifts at any time, the amount of coupled light is reduced. Devices used to connect light sources to cables vary in shape and size according to the type

Cross-Section of Connector

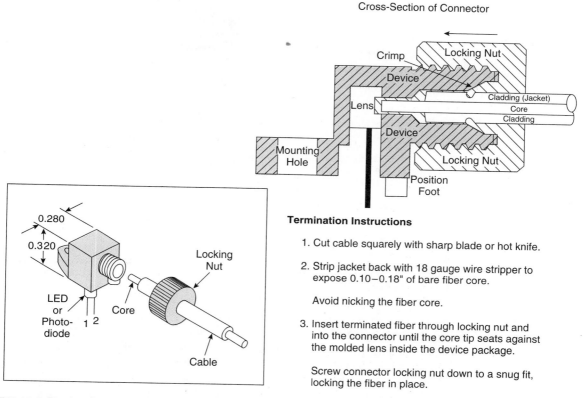

**Termination Instructions**

1. Cut cable squarely with sharp blade or hot knife.

2. Strip jacket back with 18 gauge wire stripper to expose 0.10–0.18" of bare fiber core.

   Avoid nicking the fiber core.

3. Insert terminated fiber through locking nut and into the connector until the core tip seats against the molded lens inside the device package.

   Screw connector locking nut down to a snug fit, locking the fiber in place.

**FIGURE 11-5** Device Connector for Fiber-Optic Cable. *(Courtesy of Motorola, Inc.)*

of light source being employed and the actual diameter of the fiber cable used. An example of a connecting device is shown in Figure 11-5. The light source is mounted inside the connector. The screw end is slipped onto the cable, which is then inserted into the connector. The screw cap is then brought to the connector and screwed on. As the cap is hand-tightened onto the connector, it crimps down onto the cable and holds it in place. The end of the fiber cable must be cut flat and clean to assure best coupling.

**Section 11.2  Review Questions**

1. What happens to a light ray when it passes between two pieces of material with different refractive indexes?
2. What information is conveyed by the critical angle and Brewster angle?
3. What is the purpose of a fiber cable's cladding?
4. What are the two primary differences between LEDs and laser diodes?
5. How does the connector illustrated in Figure 11-5 attempt to assure dependable interface between a light source and the fiber cable?

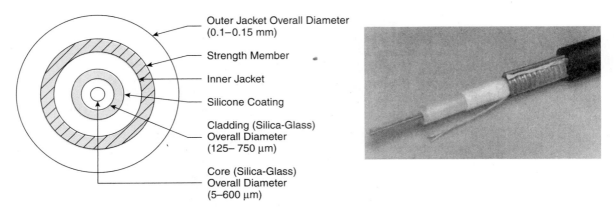

Outer Jacket Overall Diameter
(0.1–0.15 mm)

Strength Member

Inner Jacket

Silicone Coating

Cladding (Silica-Glass)
Overall Diameter
(125– 750 µm)

Core (Silica-Glass)
Overall Diameter
(5–600 µm)

**FIGURE 11-6**  Fiber Cable Components *(Photo Courtesy of Anixter Brothers, Inc.)*

## 11.3  FIBER CABLES

Figure 11-6 shows a cross section of fiber cable. The core, constructed of silica (silicon dioxide or $SiO_2$) occupies the center with a diameter between 5 and 600 microns (µm). A silica cladding with a different refractive index acts as a reflective wall surrounding the core. It brings the cable diameter to between 125 and 750 µm. The cladding is covered with a silicone coating and an inner jacket to provide strength against the bending the inner cable will experience as a result of installation and environmental shifts during its lifetime. An additional strength member and an outer jacket surround the inner jacket to bring the overall cable size to between 0.1 and 0.15 millimeters in diameter.

A fiber-optic cable is very small in diameter, which accounts for its light weight. Thousands of these fiber cables can be bundled together and packed into a tough sheathing. This provides for the ruggedness required when laying miles and miles of fiber cables. This is done to protect against breaks in the cable under normal usage. Attempting to repair a break in a fiber optic bundle by "splicing" cables together is much too expensive. Each cable would have to be accurately aligned in the splice so that the light from one section is coupled efficiently to the other. In most cases, a break in a fiber bundle requires that the section of bundled cables be replaced rather than spliced.

Application of imaging for fiber optics involves determining where to place a light source and a photo receptor at either end of the fiber cable. Figure 11-7 illustrates the concept of using several light rays and a photodetector as a receptor. The receptor element is limited in size, so some of the light sent through the lens (placed at the output of a cable) does not strike the detector surface. Note that ray a is bent at such an angle that it misses the photodetector. Ray b is shown having the maximum angle of incidence that still allows the ray to be detected. This angle is measured in relationship to a normal line, which is parallel to ray c. Ray c is the ideal situation—a light ray projected along the center line of the lens or cable.

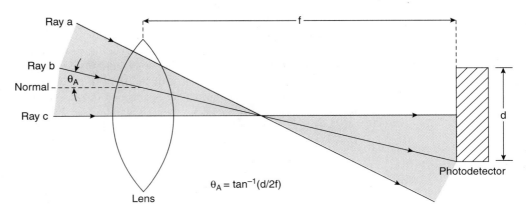

**FIGURE 11-7** Acceptance Angle ($\theta_A$)

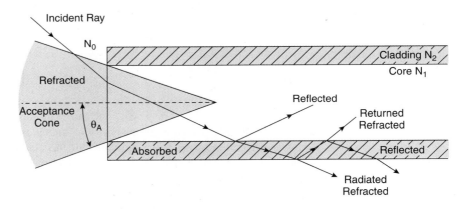

**FIGURE 11-8** Incident Ray within Acceptance Cone

**acceptance angle**
angle at which light rays can enter a fiber core.

**acceptance cone**
group of all light rays that can enter a fiber cone.

**focal point**
The point at which light converges on one side of a lens after passing through that lens.

The angle formed by ray b and the normal is called the **acceptance angle.** All incident light rays approaching the lens at this angle or less, when grouped together form an **acceptance cone.** Acceptance cones define the range of incident angles that rays can approach and successfully enter the core. The maximum angle of acceptance ($\theta_A$) is found by using the following formula:

$$\theta_A = \tan^{-1}(d/2f) \tag{11-4}$$

where $d$ is the diameter of the photodetector or cable core and $f$ is the **focal point** of the lens.

The effect of the acceptance cone, refraction, and reflection as considerations for a fiber-optic cable is illustrated in Figure 11-8. A single incident ray is shown approaching the cable within the acceptance cone. The differences between the refractive indexes, $n_0$ and $n_1$ cause the incident ray to be refractive slightly. This could cause the ray to miss the focal point as it enters the core. This is not crucial. What is important is that the ray entered within the acceptance cone. Any rays trying to enter outside the cone will be refracted into and completely absorbed by the cladding. The illustrated ray continues on a straight line within the cable core until it strikes the cladding (refractive index $n_2$). The difference in the refractive indexes causes most of the light to be reflected back into the core. However, the cladding is not completely opaque, so some of the light energy is absorbed. This absorbed amount will travel through the cladding until it meets the silicon coating. Here some of the absorbed energy is radiated into the silicon and some is reflected back into the cladding.

The original ray continues through the fiber core, with most of its energy being reflected against the cladding until it emerges at the distant end. The radiated energy that results from the energy absorbed into the cladding reenters the core at a refracted angle. Most of it is reflected back into the cladding. The level of absorbed energy is sufficiently low, so that for most applications, it can be ignored. The bulk of the light ray's energy travels through the cable. Notice that the closer the ray enters the cable to the normal angle (parallel to the normal or center line), the fewer times it meets the cladding and the less energy will be lost due to absorption. As can be seen, the placement of the light source to the fiber is of considerable importance.

## Propagation Effects

**pulse spreading**
signal distortion caused by different propagation times for each light ray traveling through a cable.

Another aspect that effects the results of data transmission using fiber optics is the propagation time through the core. Light that enters the core at the center line travels in an unimpeded straight line through the core (as long as the core itself is straight). Light entering at any other angle will eventually hit the cladding and be "bounced" down the cable. These rays travel at greater distances than the one traveling down the center. The larger the angle of incidence, the further the rays have to travel before exiting the core. As a result, the rays emerge at different times producing a phenomenon known as **pulse spreading,** which causes the replicated electrical information to be distorted by the varying arrival times of the light rays. Fortunately, the distortion is not great, but it does present a limiting factor in the length of the fiber cable and the data rates that can propagate through it. If the length is too long, then the spreading can cause the loss of digital bits and create data rates.

---

**EXAMPLE 11-2**

Illustrate the effects of pulse spreading on this data pattern: 110010101.

**SOLUTION**

The effects of pulse spreading cause the time periods for a data bit to be longer and the amplitude to become lower. A normal pattern for the sequence is shown in Figure 11-9.

Pulse spreading effects on the final are shown in Figure 11-10. Notice how some of the data bits can be "lost" due to pulse spreading.

---

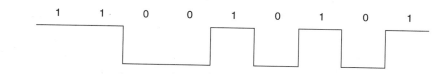

**FIGURE 11-9** Example Bit Pattern

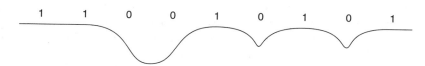

**FIGURE 11-10** Severe Pulse Spreading Effects

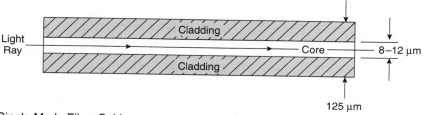

Single Mode Fiber Cable

a

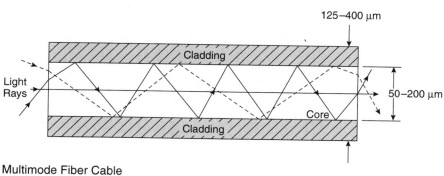

Multimode Fiber Cable

b

**single mode fiber**
cable with a narrow acceptance cone that concentrates light travel through the core's center.

**multimode**
cable with a wide acceptance cone that permits light rays to enter at many different incident angles.

**FIGURE 11-11** Comparison of Single- and Multimode Fiber Cables

## Fiber-Optic Cable Modes

Fiber-optic cables are constructed to operate in one of two modes, **single mode** or **multi-mode** (Figure 11-11). The use of a particular mode depends on the requirements for the communications network. Tight coupled systems using laser diodes for sources employ single mode cables, which are used with highly directional light sources. The single mode cable concentrates its light energy passage to the center of the core, which is very narrow (6 to 12 μm in diameter). Recall from the previous discussions that a light ray concentrated at the core's center moves the quickest through the core with the least distortion and attenuation.

Multimode cores on the other hand, are designed to accept the entrance of light at a fairly wide acceptance cone. The core diameter is between 50 and 200 μm. Each ray entering at a specific angle experiences a given amount of attenuation loss and propagation time, which characterize its mode within the cable. While the advantage of the multimode cable is a wider acceptance angle, allowing more light energy to enter the cable, the disadvantage is the distortion of the demodulated signal. The distortion occurs because of the pulse spreading caused by the arrival of different light rays at the end of a fiber cable, at different times due to the actual distances they must cover through the core.

## Refractive Indexes in Fiber Cores

Besides having the option of two different modes, fiber-cable cores also come with different indexing types, *step index* and *graded index*. In a step index core used up to this point in this chapter, the incident ray enters the core, is refracted slightly, and travels through the core as it is reflected from one side of the cladding to the other (Figure 11-12). It has been shown that the cladding absorbs some of the energy each time a ray strikes it. A graded index core, shown in Figure 11-13, bends the ray back toward the center, reducing the physical distance it travels. Many of the rays entering the core never reach the cladding, reducing the absorption of light energy by the cladding. Graded indexes are constructed by slightly altering the refractive index of the core as you move away from the center. Since each layer has two different indexes, light rays are refracted as they try to transverse through the core. Recall, that in the step index, after the initial refraction due to the light entering the core from air (a change in refractive index), that the ray traveled pretty much in a straight line until it hit the cladding. By grading the cores indexes, light rays are gently refracted back toward the core's center and now travel more in a arc path rather than a straight line as shown in Figure 11-11. The amount of variation in the refractive index between each layer of a graded core is computed using the following formula:

(11-5)

$$n_r = n_o \sqrt{1 - [(n_0^2 - n_c^2)/(2n_0^2)]}$$

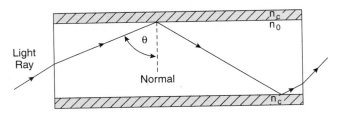

**FIGURE 11-12** Step Index Fiber Cable

where $n_r$ is the variance of refractive index, $n_0$ is the refractive index at the center of the core, and $n_c$ is the refractive index of the cladding.

---

### EXAMPLE 11-3

What is the refractive index of the first layer above the center in a graded index core with a refractive index at the center of 1.2 and a cladding index of 1.1?

### SOLUTION

First find the refractive variance using Equation

$$n_r = 1.2\sqrt{1-(1.44-1.21)(2.88)} = 1.15 \tag{11-6}$$

Next divide the current layer index (1.2 for center) by the rate change to get the index of the next layer: $1.2/1.15 = 1.043$

---

### Section 11.3  Review Questions

1. What is the importance of the acceptance angle?
2. Which ray arrives at the end of the fiber cable first? Why do all other rays experience propagation delays?
3. Which fiber-optic cable mode is more efficient in transferring data?
4. Which mode is less expensive?
5. What are the advantages of graded indexes compared to step indexes?
6. What is the rate of index change for a graded index core with a center index of 1.4 and a cladding index of 1.08?

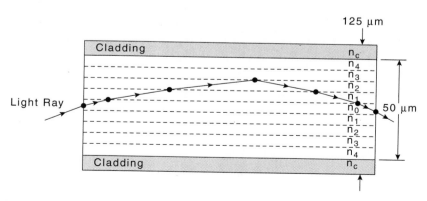

FIGURE 11-13 Graded Index Fiber Cable

## 11.4 LIGHT SOURCES

The two primary sources of light used in fiber-optic systems today are light-emitting diodes (LEDs) and laser diodes. Within these broad areas lie different devices for each type of light source.

### Light-Emitting Diodes

Light-emitting diodes work on the principle that an electrical current flowing through a semiconductor PN junction causes light energy, in the form of photons, to be released from the semiconductor material. The amount of photon energy that is released is set by the form and amount of impurity doping of the semiconductor material and the amount of current passing through the diode. The more current, the more photon energy is released. Light-emitting diodes come in two basic forms, surface-emitting and edge-emitting. For surface-emitting diodes, the light pattern is almost hemispherical, allowing light to be dispersed in a wide area. The light from this type of pattern, illustrated in Figure 11-14, can be easily coupled into an optic fiber cable. Given the large area of dispersion, a certain amount of light will enter the core within its acceptance angle. On the down side, much of the light power emitted from the diode is lost outside of the acceptance angle.

**aluminum gallium arsenide (AlGaAs)**
surface emitting diode material.

An example of a high-speed surface-emitting LED is Motorola's MFOE1201/2 diode. This LED is made of the material **aluminum gallium arsenide (AlGaAs).** The material selected is based on the wavelengths the diode is expected to transmit light at. This light-emitting diode is specified to operate at digital rates up to 200 Mbps or analog rates up to 100 MHz. Physically, it is mountable into a fiber-optic connector as shown in Figure 11-15.

Edge-emitting diodes are constructed in a way that forces light to be emitted in a much smaller conical pattern as shown in Figure 11-16. This allows more power to be directed to, and coupled into, the fiber core. A tighter pattern also requires tighter alignment of the coupling between the light-emitting diode and the cable. Light-emitting diodes, because of their lower power and wider dispersion are suitable for short distance connections, which minimizes attenuation losses due to the length of the cable. For longer distances, injection laser diodes are used.

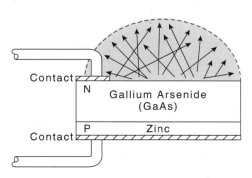

**FIGURE 11-14** Surface-Emitting LED

## Lasers

Lasers generate highly concentrated light beams by forcing much of the light energy produced by them into a tight, highly directional dispersion pattern. The light produced by various forms of lasers has found use in eye surgery and other applications requiring

**MFOE1201**
**MFOE1202**

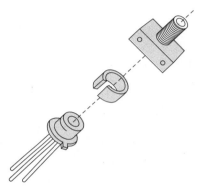

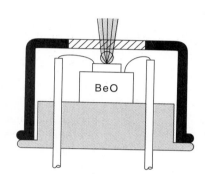

Compatible with amp #228756-1, Amphenol #905-138-5001 and OFTI #PCR001 Receptacles using Motorola alignment bushing MFOA06 (included)

**Package Cross-Section**

**FIGURE 11-15** Photodiode *(Courtesy of Motorola, Inc.)*

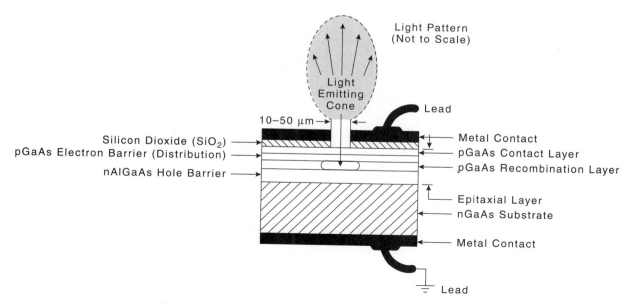

**FIGURE 11-16** Edge-Emitting LED

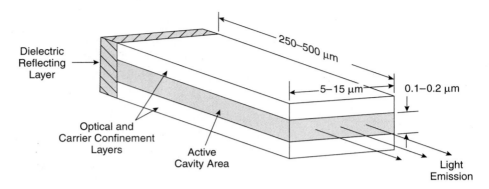

**FIGURE 11-17** Laser Diode Construction

intense and finite use of energy. Injection lasers used for fiber-optic communication do not have the power required for eye surgery.

The construction of an injection laser diode is shown in Figure 11-17. They are produced on a small-size scale and require much less power than those lasers used for eye surgery. Injection lasers produce a tight directional dispersion pattern that can be directed into the fiber, entering at an angle and position close to the core's center. Losses due to attenuation are reduced, since the bulk of light energy travels down the center of the cable and is not partially absorbed by reflection/refraction caused when rays hit the cladding wall. This reduced loss increases the efficiency of energy transfer through the cable. Fiber cable and splicing losses (those incurred because of the connectors used to splice cables together) for lasers are typically 0.5 dB per kilometer (km) of cable, compared with 2 dB per km using light-emitting diodes.

---

**Section 11.4 Review Questions**

1. What is the advantage of a surface-emitting diode compared to an edge-emitting one?
2. What advantage does an edge-emitting LED have over a surface-emitting LED?
3. What is the chief benefit of using lasers compared to LEDs?

---

## 11.5 OPTICAL DETECTORS

On the receiving side of the fiber-optic cable is a light-sensitive detector encased in a special adapter as shown in Figure 11-18. The connector must hold the alignment to the end of the cable tight to the lens of the detector to allow the maximum amount of light to be sensed by the detecting device. A photosensitive detector like a photodiode or phototran-

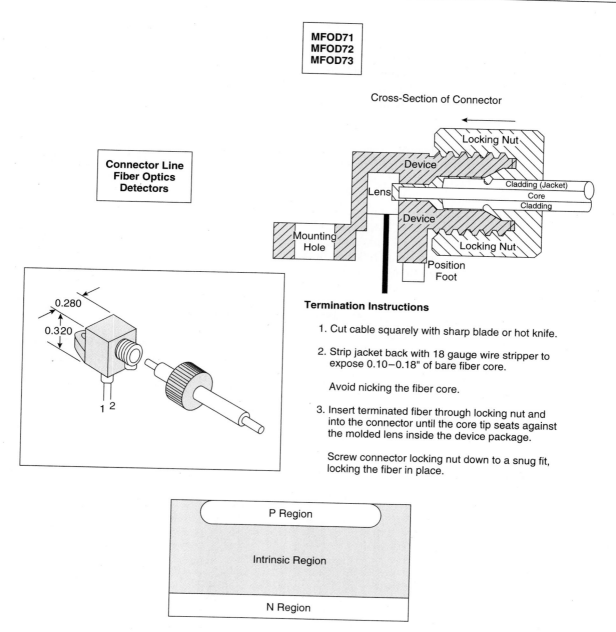

**Termination Instructions**

1. Cut cable squarely with sharp blade or hot knife.

2. Strip jacket back with 18 gauge wire stripper to expose 0.10–0.18" of bare fiber core.

   Avoid nicking the fiber core.

3. Insert terminated fiber through locking nut and into the connector until the core tip seats against the molded lens inside the device package.

   Screw connector locking nut down to a snug fit, locking the fiber in place.

**FIGURE 11-18** PIN Photodetector *(Courtesy of Motorola, Inc.)*

sistor operates on the principle that light striking an exposed section of a PN junction of semiconductor material will cause the release of free electrons and cause electron and hole flow activity to begin. The more light applied, the higher this activity, which translates to greater electrical current or lower device resistance. In essence, as the light increases, the effective resistance of the semiconductor is reduced by the increased activity.

**pin**
a type of diode that has a large
amount of intrinsic material
between the P and N channels.

A typical example of a photo detector is Motorola's MFOD71/2/3 PIN photodiode, also shown in Figure 11-18, used for short distance cables. The **PIN** designation refers to the inclusion of a large amount of intrinsic material between the P and N material that form the detector. This intrinsic material assures a low leakage current when there is an absence of light and the diode is off. The lack of leakage current allows for fast responses of the diode to the presence of light, since the diode does not have to first overcome this leakage current.

Detector response to light changes is one of the most significant specifications for these devices. The MFOD7X series response time is less than five nanoseconds. A 1.0 MHz PIN TTL receiver using the MFOD71 detector is shown in Figure 11-19. Potentiometer R5 is used to set the sensitivity to light of the photodiode circuit by establishing a switching reference for comparator U1. The amount of current through R3 will vary in response to the light it detects. A reference voltage for comparator U1 is set by the voltage divider of R5 and R4. Current changes through the photodiode increase or decrease the bias into the bases of Q1 and Q2. Changes in the conduction of Q2, resulting from the bias change, are reflected in the emitter current through R3. U1, acting as a comparator, switches from ground (0V) to Vcc (+5V) as the voltage across R3 decreases below the reference voltage across R5. Thus, as the light intensity changes between one level representing a logic 1 to another representing a logic 0, the TTL output of the comparator changes between +5V and 0V. In this way, an alternating light source between on and off, can be translated to digital TTL voltages.

For longer fiber cable applications, a photodarlington transistor, such as Motorola's MFOD2302 (Figure 11-20) is used. This particular device detects infrared light sources operating at medium frequency ranges over fiber cable distances up to

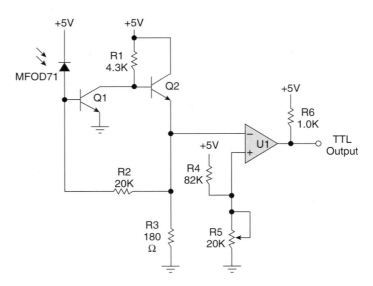

**FIGURE 11-19** 1.0-MHz PIN Receiver *(Courtesy of Motorola, Inc.)*

1,000 meters. In comparison, the MFOD7X series is designed for cable systems with maximum lengths from 25 to 32 meters. Darlington pair transistors have the advantage of operating with a low input. This input is then amplified twice through the beta gain of each transistor. As expected with any system, light energy is lost over long distances, making the Darlington detector an excellent device for detecting these reduced light levels.

### Section 11.5  Review Questions

1. What happens when a light source is applied to a semiconductor photodetector?
2. What is used in a MFOD2302 to allow the detector to work on fiber systems with cable distances up to 1,000 meters?

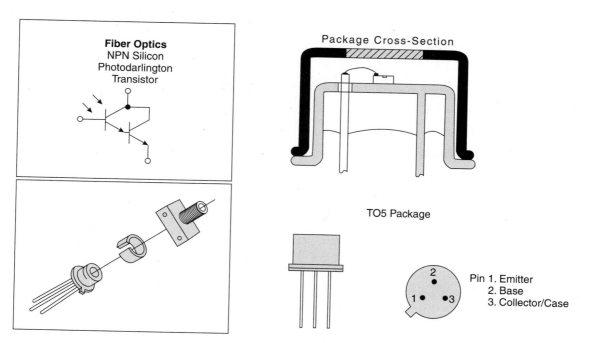

MFOD2302

**Fiber Optics**
NPN Silicon
Photodarlington
Transistor

Package Cross-Section

TO5 Package

Pin 1. Emitter
2. Base
3. Collector/Case

**FIGURE 11-20**  Long-Range Infrared Detector *(Courtesy of Motorola, Inc.)*

## 11.6 FIBER-CABLE LOSSES

While fiber-optic cables have many advantages over copper or coax cables, they also exhibit losses and problems that copper or coax do not have. One such type of loss has already been discussed—the absorption attenuation of the cable itself. Larger losses result from the physical connections that bring light sources and detectors into alignment with the cable. Additional losses are experienced with splices, which are connectors that marry two cables together. Even the best alignment of light sources to a cable can still result in some of the light power being lost by occurring outside of the cable's acceptance cone. This type of loss increases as the misalignment of light source to cable worsens. The same type of problem occurs at the detector end. Misalignment here causes some of the light power to miss the detector's window. Figure 11-21 illustrates splice-type misalignment, which is the same as alignment problems at the source or detector end.

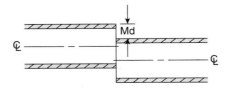

Lateral or axial misalignment

a

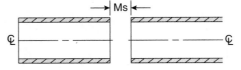

Longitudinal or Space Gap

b

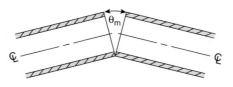

Angular Misalignment

c

**FIGURE 11-21** Mechanical Misalignments of Fiber Connections

## Connector and Cable Misalignment

The lateral or axial (Figure 11-21$a$) misalignment is the most common problem when connecting cables together or when connecting sources and detectors to cable ends. The fiber ends remain straight, but the center lines do not match with the other cable or device. Figure 11-21$a$ is exaggerated for clarity. Md is the amount of distance of the misalignment. Light rays exiting from the left side of the connection enter at a different place in the right hand cable. Some of the rays fail to make the transition, while others are refracted at a new angle.

A second type of misalignment is the longitudinal or space gap shown in Figure 11-21$b$. Light is forced to exit the cable on the left and reenter the next cable section on the right after traveling over the gap distance (Ms). Once again, some of the exiting light reaches the other side outside of the acceptance cone and is lost. Additionally, the air in the gap presents a different refractive index and will cause the light rays trying to bridge the gap to alter direction by being refracted or reflected.

The third type of misalignment (Figure 11-21$c$) is an angular misalignment. Here the cables are not matched in a straight line, with the edges separated at an angle of $\theta_m$. Light power is lost by the combination of rays falling outside of the acceptance cone again and those entering the righthand cable at a new incidence angle.

These misalignments are reduced by precision-fitting couplings and splices and by carefully following the connecting processes outlined by manufacturers of these devices. Care must be taken when cutting the end of the fiber to assure the edge is created as straight as possible. Once the cable is inserted in the coupler, an outer knurled knob is screwed over the coupling, crimping the cable end in place (Figure 11-22). This aids in alignment and prevents the coupling from shifting as the cable is moved about.

## Effects of Bends in the Cable

Fiber-optic cables are not always laid out in a straight line. More realistically, a cable is expected to undergo many bends as it is physically positioned. A bend in the cable is illustrated in Figure 11-23. Breaks in the cable do not occur if the bending radius remains greater than 150 times the fiber-cable diameter. Tighter bends place too great a stress on the glass core and can create microbends or result in a breakdown of the material. A light ray traveling through the core is reflected, as described earlier, as long as the angle of incidence ($\theta_i$) remains greater than the critical angle ($\theta_c$). However, upon approaching a bend, the angle of incidence ($\theta_2$) as the ray strikes the cladding in the bend changes. If this angle becomes less than the critical angle, the ray is refracted and energy is lost into the cladding. Less light is reflected back into the core. As the bend becomes tighter, $\theta_i$ becomes smaller, and more light is refracted and less reflected. Care must be taken to minimize bends and, when forced to use them, to keep the bend as wide as possible.

## Absorption Losses and Scattering

An additional, although comparatively small, absorption loss is experienced due to the intrinsic nature of any material. Impurities in the core cause some light energy to be

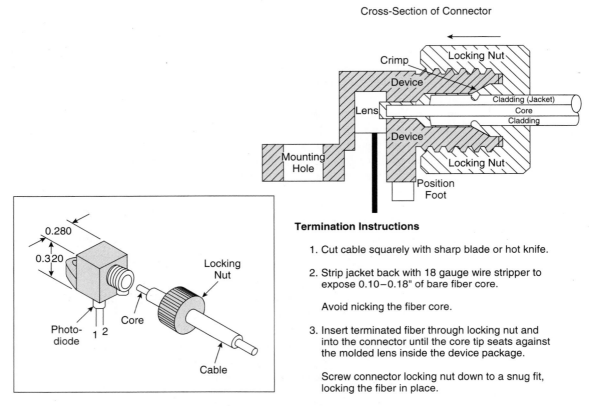

**FIGURE 11-22** Device Connector for Fiber Detector *(Courtesy of Motorola, Inc.)*

### Termination Instructions

1. Cut cable squarely with sharp blade or hot knife.

2. Strip jacket back with 18 gauge wire stripper to expose 0.10–0.18" of bare fiber core.

   Avoid nicking the fiber core.

3. Insert terminated fiber through locking nut and into the connector until the core tip seats against the molded lens inside the device package.

   Screw connector locking nut down to a snug fit, locking the fiber in place.

absorbed as the ray transverses the cable. A larger consideration is given to the effect of scattering in a core. During the manufacturing process, doping molecules move freely through the hot liquid glass. As the glass is cooled and formed into the shape of a core, these molecules stop moving and settle randomly throughout the core material. This creates uneven molecular density within the core, resulting in localized changes in the refractive index of the core. Additional pockets are created by small air bubbles trapped in the fiber during the cooling process. A ray, traveling through the cable, strikes these pockets and part of the ray's energy is scattered away from the bulk of the ray, resulting in a loss of light power at the exit end of the cable.

### Section 11.6 Review Questions

1. What are the two losses experienced by misaligning a splice between two cables?
2. List the three types of fiber-optic cable splice misalignment.

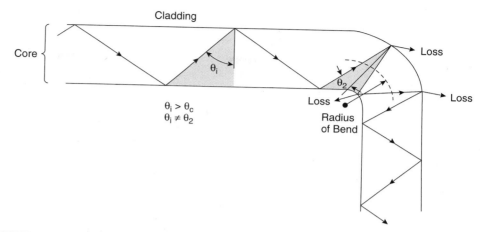

**FIGURE 11-23** Effects on Reflection Angle Due to Bend in Fiber Cable

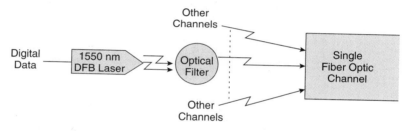

**FIGURE 11-24** WDM Multiplexer

3. What is the general rule regarding the tightest bend that should be made in a fiber cable?
4. In practice, what two guidelines govern the use of bends in a fiber cable?
5. Why do losses occur in bends in a fiber cable?
6. How does scattering cause a loss of light energy?
7. What causes impurities to be scattered throughout the core?

**wave division multiplexing (WDM)**
multiplex method to place several fiber optic channels on to a single line.

**distributed feedback Bragg (DFB)**
a semiconductor laser used in multiplexed systems.

## 11.7  WAVE DIVISION MULTIPLEXING

A method for multiplexing several fiber optic channels into a single transmission channel is known as **wave division multiplexing (WDM).** The process of multiplexing, as shown in Figure 11-24, starts with a number of different laser inputs being focused into a single fiber-optic cable. Each digital channel is used to modulate a 1550 nm (nanometers) **distributed feedback Bragg (DFB)** semiconductor laser. The unique characteristic of this

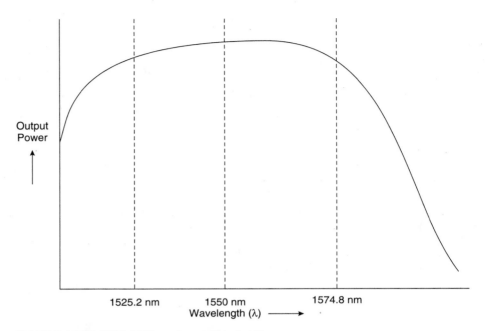

**FIGURE 11-25** DFB 1550-nm Laser Bandwidth

type of laser is that its output spans a wavelength distance from 1525.2 nm to 1574.8 nm as shown in the wavelength chart in Figure 11-25.

Each laser's output is sent through a 1.6-nm narrowband optical filter set at different wavelengths throughout the 1550-nm laser range, which could produce 31 different light signals. The wavelength multiplexer provides for a 0.8 nm peak-to-peak channel separation between channels selected for multiplexing illustrated in Figure 11-26. This reduces the total possible number of channels from 31 to 16, but also prevents channels from overlapping each other. The 16 filtered optical signals are entered into a single optical fiber.

A wavelength demultiplexer reverses the process, as shown in Figure 11-27, accepting the numerous light signals and filtering each back to an individual channel. These individual signals are routed to their destinations after filtering.

**Section 11.7  Review Questions**

1. How many lasers are required to multiplex 16 channels into a single fiber cable?
2. What is used between each laser and the fiber cable?
3. What is special about the DFB laser that makes WDM possible?

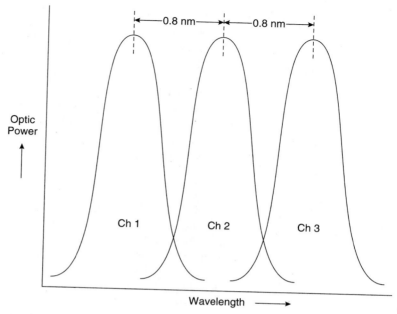

**FIGURE 11-26**  Channel Spacing

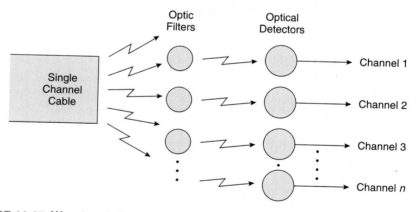

**FIGURE 11-27**  Wavelength Demultiplexer

## 11.8  FIBER DISTRIBUTED DATA INTERFACE

**American national
standards institute (ANSI)**

**fiber distributed data
interface (FDDI)**
a fiber-optic network protocol
based on dual ring topology.

The **American national standards institute (ANSI)** has developed ANSIX3T95, a 100-Mbps network standard for fiber optic ring networks more commonly known as **fiber distributed data interface (FDDI).** The FDDI token ring is composed of two counter-rotating rings—a primary ring and a secondary one that passes tokens and messages in the opposite direction to the primary ring. This secondary ring may be used as a separate data channel or as a backup ring in case of failure of the primary ring. In contrast to standard ring configurations in which the token is held by the sending station until it detects the return of its messages, the FDDI ring provides for release of the token immediately after the data transmission by the sending station instead of waiting for a forwarding acknowledgement from the next station in the ring. Furthermore, the ring is equipped with optical bypasses to allow stations in the ring to be bypassed in the case of failure.

## FDDI Specifications

**physical media dependent (PMD)**
specifies physical properties of FDDI network.

**physical (PHY)**
physical protocol of the FDDI suite.

**media access control (MAC)**
one of the four protocols of the FDDI suite.

**station management (SMT)**

The maximum frame size specified for FDDI is 4.5 Kbytes, and the network is managed by four protocols, **physical media dependent (PMD), physical (PHY), media access control (MAC),** and **station management (SMT).** The PMD protocol specifies the physical properties associated with fiber optics, including bandwidth, transmitter and receiver specifications, optical waveform characteristics, and power requirements. Specific parameters require a 500 MHz per kilometer minimum bandwidth when using multimode cables along with a 1,300-nanometer wavelength LED transmitter. Fiber losses are specified as 3db to 4db per kilometer depending on the multimode cable in use. Single mode cable losses are to be held to 0.5db per kilometer.

The FDDI specification has been expanded to allow use of dual ring transmissions using category 5 unshielded twisted-pairs (UTP) of copper wire. Denoted as TP-PMD, this specification permits 100-Mbps transmissions up to 100 meters using UTP wire.

## FDDI Physical Layer

**dual attached stations (DAS)**
station that is connected to both rings of a network.

**single attached stations (SAS)**
station that is normally connected only to the primary ring of a dual ring network or interfaced through a concentrator.

**wiring concentrator**
hub for centralizing a number of single attached stations.

**open systems interconnection (OSI)**

**non-return-to-zero (NRZ)** and **NRZ-AMI** or **NRZI non-return-to-zero alternate mark inversion** digital encoding formats.

Clock synchronization, symbol alignment, and data encoding are detailed in the PHY protocol. Ring diameter sizes are limited to a maximum 200 KM for multimode fiber, 30 KM for single mode, and 100 M for UTP copper wire. The PHY specification limits the number of stations that can be attached to the two rings. **Dual attached stations (DAS)** are those that connect to both rings as shown in Figure 11-28. Five hundred of these stations may be connected into a single FDDI network. **Single attached stations (SAS),** connected to the primary ring, using a DAS **wiring concentrator,** are restricted to a 1,000 units per network.

The encoding scheme used by FDDI is referred to as NRZ 4B/5B, with NRZ referring to standard non-return-to-zero encoding shown in Figure 11-29. 4B/5B is a scheme that converts 4 bits per symbol into 5-bit digital codes. The PHY and PMD protocols address the physical layer of the **open systems integration (OSI)** standard network model.

## FDDI MAC Framing

Media access control (MAC) is responsible for frame formation and error detection. Error detection is accomplished using CRC-32 as a frame check sequence. MAC configures the data stream into a timed token protocol, converting the **non-return to zero (NRZ)** format shown in Figure 11-30 to a 125-Mbps **NRZ-AMI** or **NRZI (non-return-to-zero, alter-**

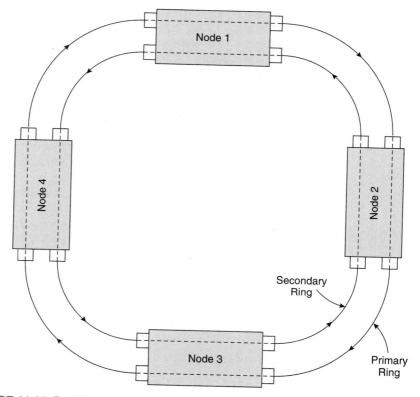

**FIGURE 11-28** Dual Attached Stations

Non-Return-to-Zero (NRZ) Form

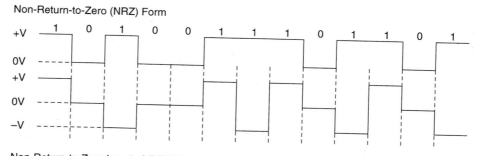

Non-Return-to-Zero Inverted (NRZI) Form

**FIGURE 11-29** NRZ and NRZI Binary Formats

**nate mark inversion)** data stream shown in the lower part of Figure 11-30. Signals vary from low to high as in any digital format, but with this method, 0 or space logic, is at 0V, while logic 1s or mark levels alternate in positive and negative polarities. This scheme incorporates error detection capabilities by a process known as *polarity violation.* Any occurrence of two simultaneous positive or negative polarities violates the constraints of the encoding scheme denoting an error has occurred. MAC is also responsible for maintaining an error recovery scheme to assure data integrity. MAC applies to the data-link layer of the OSI model.

A major job of the MAC protocol is to format FDDI frames and tokens. The basic token format is shown in Figure 11-31 and begins with a preamble that is used like all protocols that deal with synchronous digital data to assure clock recovery, bit, and character synchronization. A starting delimiter follows and serves the same purpose as a starting flag in the HDLC data-link protocol it established the beginning of the actual token frame. Delimiters in the FDDI protocol suite are formed in the same manner as those for token bus or ring, by using polarity violation or nondata bits. Frame control is used to indicate the frame type (free token, information frame, management frame, etc.). Finally, the token frame ends with an ending delimiter.

FDDI information frames begin in the same form as the token frame, with a preamble, start delimiter, and frame control field. These are followed by destination and source addresses, a payload field for data, a **frame check sequence (FCS)** and ending delimiter. The frame status field contains reservation and acknowledgement bits, which are used in the same manner as they are with token ring, to determine if another station needs to add a message or if a sent message has been received and copied. The similarity in format between the FDDI information frame and an HDLC information frame is no accident. Both serve the same basic function of establishing the data link and passing information from one station to another.

**frame check sequence (FCS)**
error detection characters.

## Station Management Protocol

Station Management (SMT) protocol, which lies outside of the OSI model, oversees the network and the three other protocols. This involves the start and recovery of network ring operation, including station insertion and removal from the network. Initialization,

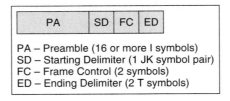

PA – Preamble (16 or more I symbols)
SD – Starting Delimiter (1 JK symbol pair)
FC – Frame Control (2 symbols)
ED – Ending Delimiter (2 T symbols)

**FIGURE 11-30** FDDI Token Format

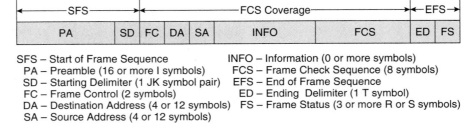

SFS – Start of Frame Sequence
  PA – Preamble (16 or more I symbols)
  SD – Starting Delimiter (1 JK symbol pair)
  FC – Frame Control (2 symbols)
  DA – Destination Address (4 or 12 symbols)
  SA – Source Address (4 or 12 symbols)

INFO – Information (0 or more symbols)
  FCS – Frame Check Sequence (8 symbols)
EFS – End of Frame Sequence
  ED – Ending Delimiter (1 T symbol)
  FS – Frame Status (3 or more R or S symbols)

**FIGURE 11-31** FDDI Frame Format

configuration, and statistics involving the network are included in the management protocol. SMT defines error detection and fault isolation algorithms, including usage of optical bypasses to circumvent bad stations without losing network operation.

Additional elements of the FDDI specification describe requirements for FDDI backbones and switches, which allow larger networks to be subdivided into smaller ones. FDDI hubs, adding central control and management to the FDDI ring network extend the distances for these networks by allowing connections for up to twelve 100-Mbps workstations, file servers, or backbones. The hub or star topology manages several FDDI ring networks.

## Supernet—An FDDI Application

Advanced Micro Devices (AMD) has developed a five-chip application set for FDDI. Uses for this network include, but are not limited to, interconnecting mainframes, minicomputers, and high-speed storage devices; backbone networks that tie together different types of networks, such as token ring, Ethernet, star or hub, and so on; and front-end networks, such as workstation environments. Supernet, which includes the hardware and software for this system, is designed to maintain a maximum of 500 stations over a network with a maximum fiber cable length of 200 km. It uses two counterrotating rings passing data at 100 Mbps. The secondary ring application for Supernet is used to back up the primary ring.

There are two classes of stations connected to an FDDI network, class A and class B. Class A stations, called dual attached stations (DAS), are connected to both the primary and secondary rings (Figure 11-32). Class B stations, known as single attached stations

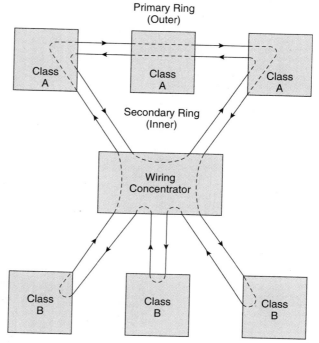

**FIGURE 11-32** FDDI Ring Network

(SAS), on the other hand, are interfaced only to the primary ring and require a specialized DAS, called a wiring concentrator, to make the interface. If a station fails, the secondary loop is used to loop back data along the primary ring as illustrated in Figure 11-33. Data in the primary link reaches toward a break between node 1 and node 2. The software at node 1 detects the break and causes the data to be switched to the secondary ring and looped back through nodes 4 and 3 to node 2. The ring integrity is completed by having the data looped onto the primary ring from the secondary ring at node 2 where the break was also sensed.

In the event of a station temporary ring failure, the station management applications programs respond by making the ring whole again. This may include bypassing the faulty station or dropping it from the ring entirely. In the case where the secondary ring is used as a backup ring and the failure on the primary ring is catastrophic, SMT causes the switch between rings to be accomplished. Once the necessary corrections are made, a station is selected to transmit a **beacon token** around the ring. When the originating station detects the return of the beacon, it is assumed that the ring is whole again and normal transmission can be resumed.

**beacon token**
used to check integrity of the primary FDDI ring following the repair of a station on that ring.

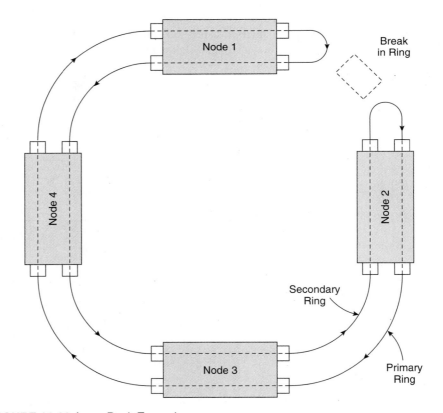

**FIGURE 11-33** Loop-Back Example

**unrestricted token**
FDDI free token that can be captured by any station on the ring.

**restricted token**
reserved FDDI tokens for higher priority stations.

**rotation time**
access algorithm for the FDDI network.

**transmission valid time**
restricts the length of time a FDDI station can send a packet.

**fiber-optic distributed data network (FDDN)**
military fiber optic network.

**militarized fiber-optic transmission network (MFOTS)**
high-speed military fiber optic network.

Two forms of token passing are performed on an FDDI network. The first, called **unrestricted token,** may be captured by any station on the network. The second, **restricted token,** is reserved for stations with a specific priority code. Other stations on the network must wait until these stations are finished before capturing a token and adding messages to the network. Access to the network is through a function called **rotation time.** All the stations on the ring negotiate for rotation time based on their required network bandwidth. The larger the bandwidth requirement by a station, the more required rotation time gets to access, receive, and send messages. Stations with the shortest rotation time capture tokens first. It is presumed that they will hold the token for the shortest time. Besides rotation and hold time, there is also a **transmission valid time,** whose sole purpose is to assure that packets do not exceed the maximum size specified for a particular network.

## An Avionics FDDI Network

Northrop Corporation has developed a fiber network utilizing FDDI for the military. It is called **fiber optic data distributed network (FDDN)** and is responsible for fast data transfers within an aircraft. FDDN is fault-tolerant, providing high-performance processing of advanced aircraft functions, including guidance and missile control. Fault tolerance refers to a system that is capable of detecting a fault and maintaining system integrity by bypassing the fault so that communications within the system are not lost.

The network, illustrated in Figure 11-34, consists of several operational levels. These include the host interface responsible for interconnecting the host system into the network and the transfer level, which assures the dependable transfer of data from the host to the FDDI interface. Some system error detection and recovery is included in the transfer level. The FDDI interface delivers the data from the host transfer level to the central network through a fiber-optic interface based on the FDDI standard. The resulting data stream is further encoded and formatted as it enters and leaves the **militarized fiber-optic transmission (MFOTS)** network. Within this network, processed data are routed to their proper destination at speeds up to 10 Mbps. The high volume and speed required by advanced aircraft systems make for a suitable and desirable environment for a fiber-optic network.

**Section 11.8 Review Questions**

1. How is an FDDI network protected against the failure of a station on the ring?
2. What is the purpose of a concentrator on an FDDI network?
3. What is meant by a polarity violation?
4. What is the protocol that oversees the network?
5. Describe how a dual ring FDDI network recovers from a station failure on the primary ring.
6. What is the difference between a restricted and unrestricted token?
7. What is the purpose of rotation time?
8. What is meant by a system that is fault tolerant?

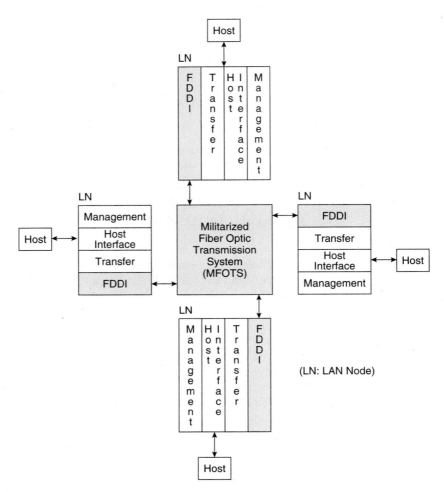

**FIGURE 11-34** FDDN Network

## 11.9 FDDI-II: ISOCHRONOUS TRAFFIC

**isochronous service**
network service that can handle data, voice, video, and other types of services.

An additional specification incorporated into the FDDI suite of protocols adds circuit-switched isochronous data service on top of FDDI's packet-switched data service. **isochronous service** includes integrated data, voice, and video traffic. Some uses for this service include image transmission, real time interactive video, as well as basic data transfers. The I-MAC (isochronous media access control) sublayer of this protocol addition addresses how the isochronous service is to be interfaced to standard FDDI processes, including adapting between standard FDDI and isochronous FDDI data rates. Also detailed in the I-MAC protocol are methods for multiplexing circuit-switched channels.

FDDI-II has two modes of operation, the *basic mode,* which handles packet-switched data only and allow networks to operate like regular FDDI networks, and the *hybrids mode.* In the hybrid mode, both packet- and circuit-switched data services are used. Most of the FDDI ring's bandwidth is allocated to isochronous services with whatever bandwidth leftover set aside for packet-switched data.

### Section 11.9  Review Questions

1. What type of service is added to FDDI in FDDI-II?
2. What is the difference between the basic and hybrid modes?

## 11.10  THE FIBRE CHANNEL

Another use of fiber optics for networking is a high-speed network used for the transport of data, voice, and video called the *Fibre Channel.*

Typical data rates used by the fibre channel include 133, 266, and 1062 MBps, using point-to-point, looping, or switched connection topologies. Arbitrated loops are fibre channel ring connections that provide shared access to bandwidth via a form of arbitration between stations on the loop. One hundred and twenty-seven *loop* (L) ports may be attached to a single network. However, only two ports may communicate at any given time, forcing the remaining ports to be idle until the loop becomes available. When a L port wants to gain access to the loop, it sends a request onto the loop when it is free of traffic. The requesting port establishes a bidirectional connection with the destination port and communication can then commence. If more than one L port attempts to gain access simultaneously, the access is resolved using arbitration based on the sending stations' addresses. The L port with the lowest IEEE 48-bit address has the highest priority for purposes of loop access. These are reserved for host stations, followed by switched-fabric connections, and finally *node (N) ports* with the highest address and lowest priority.

**fabrics**
switched-fibre channel network.

Switched-fibre channel networks are called **fabrics** and are used to interconnect devices like workstations, personal computers, servers, routers, mainframes, and storage devices. An originating port calls the fabric switch by entering the destination port address into a fibre-channel frame header. The fabric switch sets up the desired connection based on the source and destination addresses. There are no permanent virtual circuits hogging any of the bandwidth, allowing for fabric switches to handle 16 million addresses within the network. Because the switch sets up connections between stations, there is no possibility of contention for, or blocking of access to, the system.

The fibre channel protocol is subdivided into several functional layers, which are similar to the layers of the OSI model. Layer FC-0 addresses the physical layer of the network. Three types of cabling are specified under FC-0, including 2 KM of multimode fiber carrying data at 200 Mbps. Other cables specified include 10 KM of single mode fiber and 100 meters maximum distance using shielded twisted-pair copper wire.

FC-1, akin to the data-link layer of the OSI model, specifies type of data encoding used for clock recovery and byte synchronization purposes. The scheme used is called 8B/10B, which codes a 8-bit symbol into 10 bits of data. A comma character is used to ensure proper word and byte alignment. FC-1 also addresses methods used for error detection and correction.

The transport layer of the OSI model is incorporated into the FC-2 layer, which deals with framing protocols and flow control. FC-2 sets up addresses and signaling functions, such as those that define the connection between two ports, one an originator (source) and one a responder (destination). Sequencing of packets is an additional responsibility of this layer. Formatting headers and setting frame sizes and formats are other functions of the FC-2 layer.

Traffic management, including flow control, link management, buffer memory management, CRC-32 error detection resolution, and managing full-duplex operation of the network rounds off the FC-2 layer.

The FC-3 layer specifies areas associated with the network layer of the OSI model. This layer, called the common services layer, uses multiple links to transmit data boosting the use of available bandwidth. The last layer, FC-4, is known as the multiple-services interconnect layer and addresses concerns of the remaining upper levels of the OSI model. This layer can handle any incoming protocol payload and convert it into fibre channel use and format. This layer deals with both individual (user) channels and network traffic. Examples of upper level protocols that can be mapped into the fibre channel are:

Internet Protocol (IP)
ATM Adaption Layer 5 (AAL-5)
IEEE 802.3 (Ethernet)
Small Computer Systems Interface (SCSI)

Fibre-channel networks have found uses as backbones for larger networks, for handling imaging and multimedia data, and for video and teleconferencing.

The fibre channel is further subdivided into networking classes. Class 1 provides acknowledged connection services, which guarantee delivery of packets. Class 2 handles switched-frame service, which is connectionless (bandwidth on demand) but still guarantees delivery of packets. This class guarantees the delivery by using confirmations of the receipt of data as acknowledgement of packets sent. Class 3 is a one-to-many connectionless datagram service that does not guarantee delivery of data. Finally, class 4 service, which is connection-based, guarantees bandwidth allocation and specific latency factors, is used to support isochronous services.

An interesting application has arisen for fibre-channel technology. It is being used to allow fast access to large storage devices connected together in a *Storage Area Network (SAN)*. These storage devices include disks, disk arrays, tape drives, etc. The SAN's purpose is to link these devices to their own network, thereby freeing much needed bandwidth on the communications network. SAN provides faster, unencumbered access to these storage devices compared with accessing them as a node on a general purpose network. Fibre-channel technology can be used for a SAN to interconnect devices through hubs or switches and transfer data at a rate of 100 Mbps over copper cable up to 30 meters or at 1.06 Gbps using fiber-optic cable up to 10 kilometers in length.

## 11.11  SONET

**synchronous optical network (SONET)**
fiber optic transport network.

**synchronous digital hierarchy (SDH)**
European version of SONET.

**synchronous transport signal (STS)**
SONET multiplexing function.

An fine example of an optical network that accepts multiple electronic channels, converts them into optical signals, and multiplexes them into a single channel to transport them over long distances using fiber-optic cables is the **synchronous optical network (SONET).** The standards that define SONET in the United States are mirrored in a similar European body known as the **synchronous digital hierarchy (SDH).**

SONET is a multiplexed transport mechanism that uses a systemwide clocking signal to synchronize all activity on the network. The **synchronous transport signal (STS)** multiplexing function, shown in Figure 11-35, accepts multiple electronic signals, converts them into optical signals, and multiplexes them into a single STS signal. The lowest defined STS layer is STS-1 that operates at 51.840 Mbps. Other STS levels are multiplexed STS-1 lines, so that, for instance, an STS-3 would contain three STS-1 signals, making its data rate $3 \times 51.480$ Mbps = 155.520 Mbps. STS-3 is the lowest level of the European SDH hierarchy.

**optical carriers (OC)**
SONET physical system for STS-1 signals.

The physical link at each level describes the physical specifications required to carry the STS signals. These are classified as **optical carriers (OC).** For example, an OC-1 defines the specifications to carry the STS-1 signal. The remaining units in Figure 11-35 show the functional units of the OC-1/STS-1 system.

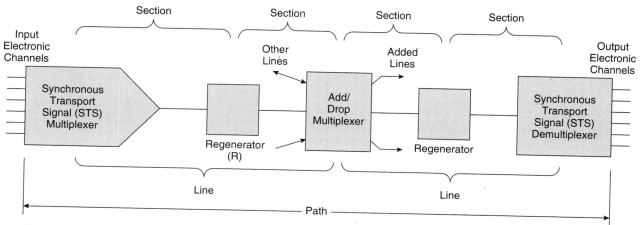

**FIGURE 11-35** SONET Topology

Besides doing the conversion and multiplexing process, the *STS multiplexer* is defined as the entry and exit points of a SONET link. The inputs and outputs to the multiplexer can be either end stations or network node access points. The multiplexed STS signal is sent along a section connection to a *regenerator,* which extends the length of link by regenerating optical signals that could have experienced some losses.

From the regenerator, the STS signal is fed along the next section to an add/drop multiplexer, which allows other STS lines to be added or dropped from the link. Thus, additional channels can be multiplexed or demultiplexed from the signal path without having to demultiplex the entire signal first. Add/drop multiplexers do this by examining header information to identify individual links and determine if they are headed toward the same destination as the current STS signal or if they are to be routed elsewhere. In the latter case, the individual lines can be added to another STS line through another add/drop multiplexer linked to the one in the current path.

The process is reversed on the received end, with regenerators performing the same function as they did earlier. A STS demultiplexer separates the individual lines and converts them back to electronic form to be sent on to their connecting nodes.

## SONET Connections

**section**
SONET connection between devices.

**lines**
SONET connection between multiplexers.

**path**
SONET end-to-end connection.

Also shown on Figure 11-35 are the classifications of SONET connections. A **section** connects any two optical neighbor devices such as STS multiplexers, regenerators, and add/drop multiplexers. The connections from one multiplexer to another are referred to as **lines.** Lines can contain from one to many sections. Finally, the end-to-end connection is a SONET **path,** which is formed from one or many lines. In its simplest form, a connection directly from a STS multiplexer to a STS demultiplexer would have one section, one line, and one path, and they would all define the same thing.

## SONET Frame

A SONET frame is defined as a matrix comprised of 9 rows, each with 90-byte size columns, as shown in Figure 11-36. The frame is transmitted, starting with the byte in the first column of the first row and then followed sequentially by the remaining bytes in the row. The transmission continues with each row following in order. The first three columns contain section and line overhead information. Section overhead occupies the first three rows and line overhead, the next six. Path overhead information is found in column four bytes and the remaining columns carry user data and other information.

Overhead defines address and multiplexing indications at each connection level, section, line, and path. User data also contains possible billing and payment information connected with supplying the service to a user.

## Virtual Tributaries

**virtual tributary (VT)**
manage lower data-rate services on SONET.

While SONET is designed to transport STS-1 and above signals, it can also be used to handle lower data-rates services through a mechanism called a **virtual tributary (VT).**

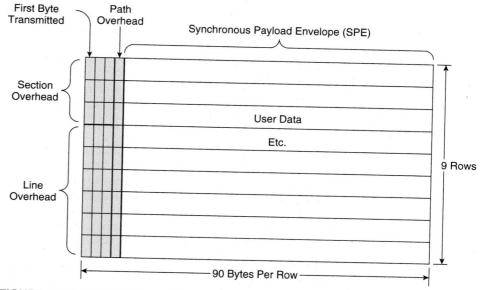

**FIGURE 11-36** SONET Frame (STS-1)

The payload area is subdivided into partial areas, each carrying these VTs. The full STS line then transports these multiple VTs as a single transmission. At the received end, they are removed and returned to their original protocol state. The types of VTs currently defined and serviced by SONET are:

VT1.5:  DS-1 or T1 line at 1.544 Mbps
VT2:  E1 line at 2.048 Mbps
VT3:  DS-1C (fractional T1) at 3.152 Mbps
VT6:  DS-2 at 6.312 Mbps

**Section 11.11 Review Questions**

1. What is the basic STS-1 data rate?
2. What is used in the SONET system to transfer ISDN basic rate frames?
3. What is the main difference between SONET and SDH?
4. What is OC-1?

# SUMMARY

Fiber-optic applications in the digital and data communications field are a continually expanding area. Older technologies are becoming updated by the replacement of coaxial and twisted-pair cables with fiber-optic cables. For example, the transatlantic phone cables, laid at the turn of the twentieth century, are being replaced by fiber-optic cables. This will increase the number of transatlantic connections by a large factor. Localized networks have been expanded beyond the confines of their original specifications. Characteristics of fiber cables, such as their large bandwidth, allow higher data rates to be transmitted greater distances.

Use of fiber-optic technology is finding numerous applications in distributed networks, such as FDDI, and integrated services networks, such as ISDN and the fibre channel. We have seen in earlier chapters, how fiber-optic technology has been incorporated into older network protocols and standards. One specific example of that adaption is fast Ethernet (100 Mbps), isochronous (mixed services) Ethernet, and gigabit Ethernet, all using fiber optics as the media to handle Ethernet traffic.

FDDI and other fiber-optic systems are finding extensive use as network backbones because of their high bandwidth. Separate systems, such as the fibre channel are developed as high-speed networks to handle integrated data, voice, and video information. Fiber-optic cables and systems are rapidly replacing existing copper wire installations for telephone communications. The use of fiber optics shall continue to increase as needs for faster and larger networking requirements increase. This is the technology of today and tomorrow.

SONET is a fiber-optic transport network that is used to transfer multiplexed electronic channels after they have been converted to optical signals by the SONET system. Even though SONET is designed to transport data at high rates, it also can handle lower T1 and E1 digital lines using virtual tributaries in the SONET payload.

# QUESTIONS

### Section 11.1

1. Give some advantages of using light as a means of conveying digital data.
2. What are the advantages of using fiber cable compared to copper cable?

### Section 11.2

3. Define the meaning of reflection.
4. Define the meaning of refraction. How does it compare to reflection?
5. What is used as a reference for refractive indexes?
6. What is the refractive index of air?
7. Define incident ray.
8. Define the term *normal* with respect to light rays.

### Section 11.3

9. How do fiber-optic cores and claddings differ?
10. What is the purpose of cladding in a fiber cable?
11. How does pulse spreading occur in a stepped index core?
12. What are the effects of pulse spreading on received information?
13. What is the difference between single and multimode fiber cores?
14. Compare the advantages and disadvantages of single and multimode fiber cores.
15. Calculate the critical and Brewster angles for surfaces that have refractive indexes of 1.25 for $n_2$ and 1.65 for $n_1$.

16. Calculate the critical and Brewster angles for surfaces that have refractive indexes of 1.10 for $n_2$ and 1.35 for $n_1$.
17. A ray with an incident angle of 15° approaches a surface with a refractive index of 1.63 from one with a refractive index of 1.15. Is the ray reflected or refracted?
18. A ray with an incident angle of 20° approaches a surface with a refractive index of 1.35 from one with a refractive index of 1.05. Is the ray reflected or refracted?
19. Compute the acceptance angle of a photodiode with a window diameter of 0.75 mm placed in front of a lens with a focal point of 1.05 mm.
20. Compute the acceptance angle of a photodiode with a window diameter of 0.82 mm placed in front of a lens with a focal point of 1.10 mm.
21. Compute the refractive variance of a multimode fiber core with a center index of 1.42 and a cladding index of 1.15.
22. Compute the refractive variance of a multimode fiber core with a center index of 1.36 and a cladding index of 1.10.

### Section 11.4

23. What are the advantages of light emitting diodes (LEDs) compared to laser diodes?
24. What are the advantages of laser diodes compared to LEDs?
25. Compare edge- and surface-emitting diodes used as light sources for fiber-optic communications.
26. Why are edge-emitting diodes preferred over surface-emitting diodes?

### Section 11.5

27. How is an optical detector used to detect the difference between a logic 1 and logic 0?
28. What effect does light energy have on the exposed window of a photodetector?

### Section 11.6

29. What is the greatest loss in fiber cables besides absorption attenuation?
30. What steps are taken to minimize losses due to connector misalignment?
31. What are the three types of alignment problems in fiber-optic cable connections?
32. What types of losses does misalignment cause?
33. What is the rule of thumb regarding the amount of bend that can be made in a fiber cable?
34. What types of problems do excessive bends in a cable cause?
35. What is the best advice regarding the use of bends in fiber cables?
36. What is the radius of the tightest bend in a cable whose diameter is 250 µm?
37. What is the radius of the tightest bend in a cable whose diameter is 325 µm?
38. What causes scattering to occur in fiber cable?
39. What two types of scattering occur in a fiber-cable core?
40. What is the resulting effect on light rays due to scattering?

### Section 11.8

41. What are the tasks of the FDDI media access control (MAC) protocol?
42. What is the primary responsibility of the station management protocol (SMT)?
43. What are the uses of the physical (PHY) and physical dependent (PMD) FDDI protocols?

44. What is the difference between dual and single attached stations used in the FDDI topology?
45. What is the purpose of a wiring concentrator?
46. What methods are used to interconnect smaller FDDI networks into one large network?
47. How are dual-ring systems used to maintain a link when a station fails?
48. What is meant by a system being fault-tolerant?
49. What is a polarity violation?
50. What does polarity violation normally indicate?
51. When are polarity violations purposely used?
52. Define rotation time. What is it used for?

**Section 11.9**

53. Define isochronous service.
54. Give three uses for isochronous service.
55. What type of traffic is handled by the FDDI-II network?
56. How does the FDDI-II network differ from standard FDDI network?
57. What types of switching are used by FDDI-II?
58. List the responsibilities of the I-MAC sublayer.
59. What are the operating modes of FDDI-II? What type of data does each support?

**Section 11.10**

60. What type of topologies are usable for fibre channel networks?
61. How is access achieved in the fibre channel system?
62. What is a fabric?
63. Which connections to the fibre-channel have the highest priority?
64. What is the chief responsibilities of the layers of the fibre channel network (FC-0 to FC-4)?
65. What are the networking classes of the fibre channel?
66. Which classes of fibre-channel service guarantee data delivery?
67. Which fibre-channel classes are connection-based?
68. Which fibre-channel classes are connectionless service?
69. Which class of fibre-channel service supports isochronous traffic?
70. List some uses for the fibre channel network.

**Section 11.11**

71. What is SONET?
72. How does SONET differ from SDH?
73. What is STS-1 service?
74. Define: line, segment, and path for SONET networks.
75. What does a regenerator do?
76. What are the functions of a STS multiplexer?
77. What is the purpose of an add/drop multiplexer?
78. Define virtual tributary. What is its purpose?
79. What are the elements of a STS-1 frame?

# RESEARCH ASSIGNMENTS

1. Design a fiber-optic system using the specifications for light sources, detectors, and the cable itself found in various manufacturer data manuals. Support your design with the technical information from the data specifications. Your system is to successfully transmit and receive a serial data stream of information at a rate specified by your instructor.

2. Research current fiber-optic applications. Suggested areas include ISDN and BISDN applications, fast and gigabit Ethernet, and long distance telephone usage.

# ANSWERS TO REVIEW QUESTIONS

### Section 11.1

1. Telephone system was already in existence.
2. Large bandwidth.
   Not susceptible to electromagnetic interference.
   Low attenuation.
   Small and rugged.
3. Expensive to make, install, and repair.
4. Infrared—efficiently travels through glass.

### Section 11.2

1. It is bent—refracted or reflected.
2. They determine the maximum and minimum angles of reflection.
3. Causes light rays to be reflected back into the core.
4. The differences between LEDs and laser diodes is the amount of light each delivers and the type of dispersion cone pattern emitted.
5. Holds light source to cable end, restricting movement that might cause misalignment and losses.

### Section 11.3

1. The acceptance angle determines the angle of incidence a ray can use to enter the fiber core.
2. Rays that travel along or parallel to the center line of the core arrive with the least propagation delay. Other rays must travel longer distances and rely on reflection when they hit the cladding wall.
3. Single mode.
4. Multimode.
5. Graded index cores bend rays in stages—away from the cladding—reducing the need to reflect them off the cladding wall. Step indexes rely on reflection to keep the ray in the core. Some of its energy is lost due to absorption by the cladding.
6. 1.1372.

**Section 11.4**

1. Surface: wide area dispersion which is easily coupled into fiber cable.
2. Edge: tight direction dispersion—no loss due to light power outside of acceptance cone.
3. Lasers generate highly concentrated beams of light for an efficient transfer of light energy.

**Section 11.5**

1. Causes electron and hole flow.
2. Photodarlington.

**Section 11.6**

1. Light falling outside acceptance cone.
   Absorption in cladding due to refraction.
2. Lateral or axial.
   Longitudinal or space gap.
   Angular.
3. Bend radius > 150 times cable diameter.
4. Minimize number of bends and use looser bend.
5. Absorption of energy in the cladding.
6. Uneven molecular density, causing local changes in the refractive index.
7. Trapped impurities or air bubbles as core cools down.

**Section 11.7**

1. Sixteen.
2. Narrowband optic filters.
3. Wide wavelength bandwidth.

**Section 11.8**

1. Use of the secondary ring to maintain ring integrity when a station fails.
2. Connect single attached stations to a dual ring network.
3. Two consecutive logic ones with the same polarity ( + or − ) when using a bipolar alternate mark inversion (BPAMI) code scheme.
4. Station Management Protocol (SMT).
5. Data approaching the failed station is routed onto the secondary ring and travels in the reverse direction until it encounters the failed station again. It is then routed back onto the primary ring and continues on to the next station.
6. Any station can grab an unrestricted token and add a message to it. A priority scheme is used to determine who can use a restricted token first.
7. Rotation time is used to allocate access time by assigning a rotation time period based on a station's requirements for bandwidth usage. Stations with low rotation times (low bandwidth needs) receive higher priorities.
8. It can recover from a station failure and maintain communications.

**Section 11.9**

1. Circuit-switched isochronous service.
2. Basic-packet-switched service only.
   Hybrid—mix of packet- and circuit-switched services.

Section 11.10
1. FDDI—dual ring topology.
   Fibre channel—looping or switch connection topology.
2. 133, 266, and 1062 Mbps.
3. Class 1: acknowledge connection-oriented service.
   Class 2: switched frame service.
   Class 3: one-to-many connectionless datagram service.
   Class 4: connection-oriented isochronous service.

Section 11.11
1. 51.48 Mbps.
2. Virtual tributary.
3. SDH starts at STS-3 level.
4. Optical carrier—physical specification for STS-1 service.

# CHAPTER **12**

# Wireless Communication Systems

## OBJECTIVES

After reading this chapter, the student will be able to:
* tell what is required for microwave wireless communications.
* describe the cellular phone topology and operation.
* trace the history of satellite communications.
* define wireless networks, their specifications, and operation.

## OUTLINE

12.1  Introduction
12.2  Microwave Communications
12.3  Cellular Mobile Telephone Service
12.4  Personal Communications Systems
12.5  IEEE 802.11: Wireless LANs Using CSMA/CA
12.6  Cellular Digital Packet Network
12.7  Satellite Communications
12.8  Methods of Satellite Communications
12.9  Satellite Networking

## 12.1  INTRODUCTION

Wireless communications dates back to the late 19th century when Marconi demonstrated the transmission of intelligence over radio waves. Since then, radio and television transmissions have become second nature functions in all of our lives. In addition, mobile telephones have joined the evergrowing use of wireless communications. Another aspect of communicating without wires has been the telephone companies' long use of satellites to complete long distance calls throughout the world. Besides long distance communications, applications for wireless connections have found their way into networking and the transfer of volumes of data. Infrared and laser technology have aided in the local network arena, while microwave transmissions and satellites have made wide area and global area networking possible.

---

**Section 12.1  Review Question**

1. List the general categories of wireless communication applications in use today.

---

## 12.2  MICROWAVE COMMUNICATIONS

**wave guide**
transmission medium for microwave signals.

Microwave transmissions use ultra-high frequency carrier signals to complete a link from one point to another. These signals are propagated from the transmitter to the antenna using **wave guides,** which are hollow rectangular tubes as shown in Figure 12-1. These tubes form a resonant cavity, easily passing a selected frequency signal. These cavities are tuned by inserting or extracting a "slug" into the cavity, causing its resonance to change. Frequencies other than the resonant frequency are highly attenuated.

**standing wave ratio (SWR)**
measure of quality of a microwave transmission line.

One problem with wave guides is the concept of standing waves, which are reflected or echoed signals as seen in Figure 12-2. They occur when wave guide and terminating equipment impedances do not match or when a wave guide is slightly out of tune. The reflected wave is in opposition to the transmitted wave, causing the main wave to be attenuated. A figure of merit for wave guides is the **standing wave ratio (SWR),** which is a measure of the ratio of the standing wave power to the main signal power. The lower the ratio, the less attenuated is the transmitted main signal.

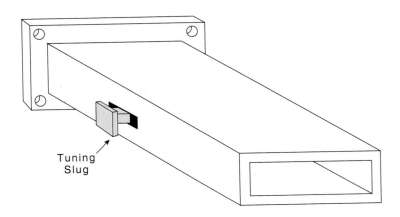

Tuning
Slug

**FIGURE 12-1** Wave Guide

Figure 12-3 shows a microwave transmission, which is generated between antenna towers set no more than 26 miles apart in a line-of-sight configuration. Each tower receives the signal, amplifies and otherwise regenerates it, and sends it to the next tower. Transmitted signals are often degenerated by weather and other factors. For example, temperature layers cause some of the signal to be refracted and add in-phase opposition to the transmitted signal. Other portions of the transmitted signal are reflected from the earth's surface and also effect the transmitted signal. The term for these effects is called **multipath fading.** In addition to attenuating the signal, these unwanted signals also cause frequency-dependent phase and amplitude distortions. Multipath fading is reduced by including a space diversity antenna 10–15 meters below the main receiving antenna. The signals picked up from the space diversity antenna are used to cancel some of the unwanted multipath signals.

Microwave systems are usually used to send and receive multiplexed channel signals on their high-frequency carriers, providing one way to handle large volumes of data transfers without using wires. Another way is an extension of the cellular mobile communications telephone system.

**multipath fading**
attenuation and distortion created by the refraction and reflection of transmitted signals.

### Section 12.2  Review Questions

1. Describe what a standing wave ratio provides for a user.
2. What type of transmission is used by microwave systems?
3. What is the purpose of space diversity antennas?

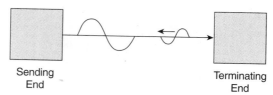

Sending
End

Terminating
End

**FIGURE 12-2**  Standing Wave

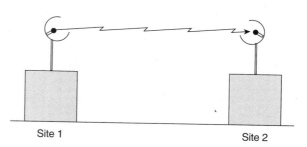

Site 1

Site 2

**FIGURE 12-3**  Line-of-Sight Microwave Transmission

# 12.3 CELLULAR MOBILE TELEPHONE SERVICE

The Cellular Mobile Telephone System was developed to allow dial-up telephone service using mobile telephone handsets and radio transmissions. The cellular system is characterized by intermittent connections between users and flexible communication times. Some users, particular in the business world, make many frequent calls on their cellular phone. Other users do not have a need to use the phone often, but when they do, they communicate for a long time, keeping one channel busy for an extended time period.

## Cellular Topology

**cells**
geographical division for mobile communications.

**mobile telephone switching office (MTSO)**
central office for a mobile phone cell.

The cellular system uses many different types of media to complete calls, including microwave and satellite transmissions. This is all transparent to the user who is happy to have the ability to call anywhere, anytime. In order to accommodate the wide area of cellular use, the system is designed so that the serviceable area is divided into **cells,** illustrated in Figure 12-4. Cells have varying sizes, but, in general, range from a one-to-twelve mile radius around a **mobile telephone switching office (MTSO).** The MTSO is responsible for linking calls together and linking neighboring cells together. In this way, when a caller is cruising along a freeway with that mobile phone stuck in their ear, the call is uninterrupted as they go from one cell to another. The MTSO of one cell "hands off" the call to the next MTSO. As long as the fool does not have or cause an accident by paying more attention to his call than the traffic, the call itself should not be disrupted. Cells are of varying flexible sizes. As the number of subscribers grows in a particular cell, it can be divided into a set of smaller cells, each with its own MTSO.

The notion of cells allows transmission power of local mobile services to be low, in the neighborhood of 25 watts for the MTSO and less than 3 watts for the handset. Despite contrary fears given credibility by TV and newspaper accounts, there is no evidence that this low signal power will cause brain damage, give you cancer, or make you sterile. In actuality, there is more danger from the loss of concentration to other activities while

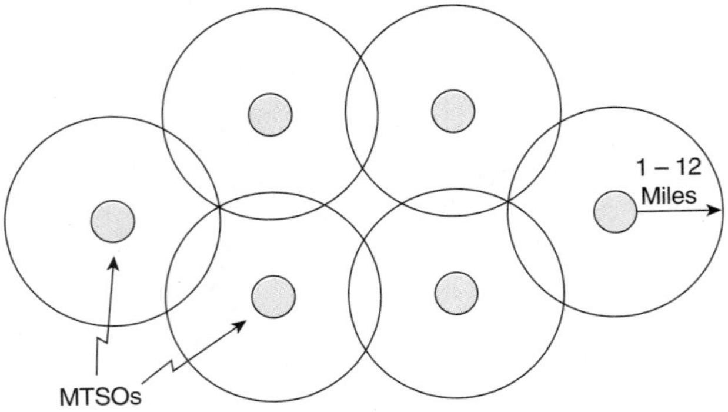

**FIGURE 12-4** Cellular System Cells

calling than from any effect of handset transmission. The indiscriminate use of mobile phones by automobile drivers has forced local governments to look into the problems caused by the use of cellular phones in a moving vehicle. This use forces the driver to:

1. Use one hand for the phone.
2. Divide their attention between driving and their phone conversation.
3. Add additional hazard while dialing a call.

The results of the loss of attentiveness and reduced control of their vehicle has caused a number of drivers to place themselves and others on the road in a hazardous situation. These situations have resulted in accidents, and occasionally, a fatality.

## Analog Cellular Service

Analog cellular phones use frequency modulation (FM) techniques to modulate and transmit voice signals on radio frequency (RF) carriers. Each analog cellular service within a cell uses pairs of channels for full-duplex operation. Besides handing off calls to neighboring cells and managing local connections, MTSOs were also responsible for interconnecting the mobile units to long distance carriers.

## Digital Cellular Service

The advent of digital cellular service expanded the capabilities of mobile telephone networks. Digital cellular service uses time division multiple access (TDMA) to divide analog voice channels into three separate time-based channels, allowing three calls to be made using one channel. Digital cellular provided a measure of security by using **code division multiple access (CDMA)** processes to encode calls before they are transmitted. Only the intended receiver would have the key to unlock the transmission sent. Unrecognized CDMA signals are treated as background noise by the receiver.

**code division multiple access** or **cellular digital multiple access (CDMA)** uses digital encoding to multiplex several channels on to one line.

## Secure Cellular Transmissions

Security of transmission has always been a problem with mobile phones. When you call someone, you are sending a signal into the air. Criminals with scanners and other sophisticated equipment pick up your signal and either determine your cellular phone number or listen to your conversation or both! To respond to this growing problem, a number of methods were put into use to foil the would be criminals who misuse the cellular phone system. One method, eluded to toward the end of the last section, is to use a method of coding your calls. A number of encryption methods are available and the ones that work the best require a good-sized code key that is changed on a regular basis.

Another method to prevent an invasion of your phone privacy is called **frequency hopping.** As you are talking, the carrier frequency carrying your conversation is continually changed in an apparently random manner. Actually, the changes are not random, only the pattern used is randomly selected. Both the sender and receiver have to be using the same pattern for the signal to get through. Even though an intruder might

**frequency hopping** constantly changing carrier frequency during a mobile phone transmission.

lock in on one of your signals, he or she cannot keep up with the continually changing carrier frequency.

Still, one more approach is to transmit portions of your message on several different carrier signals at the same time. This is called **direct sequencing** and requires several transmitters and receivers to send and receive the different signals. Again, an intruder may scan in and lock on one of the signals being transmitted, but would not get much of a message since the other portions would go undetected.

**direct sequencing**
mobile transmission that divides and transmits message portions at different frequencies.

## A Global Cellular Network

A cellular system for worldwide telephone service is known as the **global system for mobile communications (GSMC).** This system allows subscribers to use the same mobile telephone anywhere in the world where GSMC is in operation. This cell-based system lets a user "roam" from one location to another applying a hand-off process from one cell to another. The system interfaces with existing telephone systems and is composed of several functional units shown in Figure 12-5.

The **mobile station (MS)** is the telephone handset itself. It is usable by inserting a **subscriber identity module (SIM)** card, which allows the subscriber access to their account. The SIM card is a form of cellular **smart card** that contains information authenti-

**global system for mobile communications (GSMC)**
worldwide mobile phone network.

**mobile station (MS)**

**subscriber identity module (SIM)**

**smart card**
used to authorize access to a mobile phone network.

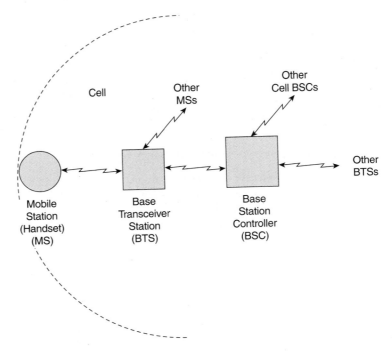

**FIGURE 12-5** GSMC Hierarchy

**home locator register (HLR)**
database of subscriber home addresses.

**visitor location registers (VLR)**
identifies where a user is or has been.

**base transceiver station (BTS)**
central office of a GSMC cell.

**base station controllers (BSC)**
controls handoffs to BTS stations.

cating the user and where to send the bills. Identification information on the card is checked against a **home locator register (HLR),** which is a record of subscriber's home or base identifier (that's where the bills are sent). **Visitor location registers (VLR)** identify where a user is or has been. This facilitates the hand-off process and also sets the amount of the billing.

The **base transceiver station (BTS)** supplies services similar to the MTSO, covering one cell. **Base station controllers (BSC)** manage multiple BTSs by controlling hand-off procedures between cells using the visitor location register described above.

---

**Section 12.3  Review Questions**

1. What is a cellular service area called?
2. What two types of cellular systems are in use today?
3. What type of multiplexing is used to allow multiple calls on a single channel in a digital cellular system?
4. What types of measures are used to secure mobile telephone calls?
5. For the GSMC system, what is the term that applies to the handset?
6. What is the purpose of a smart card?

---

# 12.4  PERSONAL COMMUNICATIONS SYSTEM

**personal communication system (PCS)**
wireless mobile telephone network.

**air interface**
set of parameters that define the behavior of wireless connections.

A standard for mobile telephone and networking has been formalized as the **personal communication system (PCS),** which uses an **air interface** as a set of parameters to define the behavior of wireless connections. Those parameters include frequencies used, accessing methods, and coding schemes for encryption.

## PCS Frequency Spectrum

The PCS radio spectrum is configured as a set of seven frequency blocks within the 2-GHz spectrum. The specifications for each block are listed in Table 12-1.

PCS also includes a narrowband channel working at 900 MHz for carrying advanced paging and data message services. Three bands are defined in ranges of 901–902, 930–931, and 940–941 MHz bands. Within these bands, there may be a subdivision into channels that can be 50-KHz wide paired with another having a bandwidth from 12.5 to 50 Khz, an unpaired standalone 50-KHz channel, or several wide bandwidth channels.

## Personal Communication Network

A third type of PCS system is an unlicensed PCS service operating within a 40-MHz bandwidth in the 2-GHz band. This service is a low-power, limited-range service used for

**TABLE 12-1**

**PCS Frequency Blocks**

| Block | Bandwidth (MHz) | Central Frequency (MHz) | Maximum Power mW/cm² | Minimum Transmit Distance (Meters) |
|-------|-----------------|-------------------------|----------------------|-------------------------------------|
| A | 15 | 1857.5 | 1.238 | 0.802 |
| B | 15 | 1872.5 | 1.248 | 0.798 |
| C | 10 | 1885.0 | 1.257 | 0.795 |
| D | 5 | 2132.5 | 1.422 | 0.748 |
| E | 5 | 2137.5 | 1.425 | 0.747 |
| F | 5 | 2142.5 | 1.428 | 0.746 |
| G | 5 | 2147.5 | 1.432 | 0.746 |

**personal communication network (PCN)**
wireless LAN.

wireless LANs and is specified as a data network standard called **personal communication network (PCN)** for the following applications:

* Person-to-person calling.
* Messaging.
* Closed user group communication.
* Mobile communications.
* Fax and laptop interfaces.
* Cordless telephone.
* Wireless PBX.
* Backbone interconnection to existing services.

PCN has all the characteristics of a PCS system, including hand-off processes, portable handsets, rural area coverage in cells, and use in or outside buildings and structures for networking purposes. Lightweight portable handsets work anywhere that the service is in place—be it the home, office, or on the road.

## Cellular Digital Multiple Access

**cellular digital multiple access (CDMA)**
same as code division multiple access.

A more common term for PCN, based on the underlying technology known as **cellular digital multiple access (CDMA)** (also known as code division multiple access, which has the same acronym), uses a code to multiplex 64 channels into a wideband (1.25 MHz) channel as seen in Figure 12-6. A correlator at the receiver picks out the correct code and treats all remaining signals as background noise. In this way, a single channel is selected from the 64-channel group. CDMA has a higher capacity than standard TDMA and FDMA channels. Additional advantages of CDMA systems is lower fade margins, adaptability to other environments, and multiple levels of diversity. CDMA uses antenna diversity on uplink transmissions to satellites, frequency diversity for wideband signals, time diversity for interleaving and coding channels, and path diversity for multipath and hand offs.

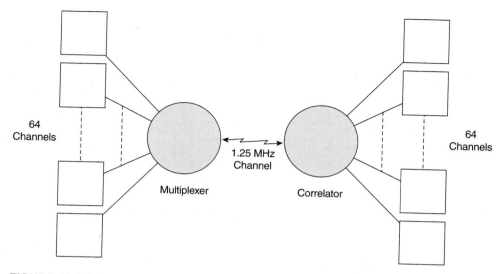

FIGURE 12-6 Multichannel Transfer

The reason that CDMA has evolved into an acronym for two different items is based on advertising for the digital cellular phone system. The purpose was to hype CDMA as a replacement for the current analog system used by mobile phone subscribers. The gimmick is to stress the improved quality of the calls and reduced noise resulting from using a digital cellular phone. Somehow, code division multiple access doesn't quite handle the message advertisers wanted to convey. However, CDMA was a catchy acronym, so the solution was to give it a different meaning and thus was born, cellular digital multiple access.

**Section 12.4 Review Questions**

1. List some applications for the PCN system besides mobile telephone service.
2. Which type of cellular system, analog or digital, is CDMA?

## 12.5 IEEE 802.11: WIRELESS LANS USING CSMA/CA

The IEEE 802 committee for networking includes the 802.11 specification for wireless LANs. The basis for local area wireless networks is the Ethernet specifications of 802.3, using *Carrier Sense Multiple Access with Collision Avoidance (CSMA/CA)* instead of CSMA/CD. The reason behind collision avoidance rather than collision detection is that it is difficult to detect collisions on a medium such as air. Access uses a form of arbitration to settle who gets immediate access. The controlling station monitors the airwaves for a signal from any of the network nodes. Upon detecting a request for access, the central node makes sure that no one else is also trying to gain access. This being the case, the central node grants access to the requesting station. If additional requests are detected at the same time, the central node resolves which station gains immediate access and which one will follow.

## Infrared Systems

**line-of-sight**
infrared connection that requires the receiver to be exactly aligned with the transmitter.

**diffuse**
infrared connections that allow signal to bounce off wall or object.

Wireless LAN connections are made using infrared or radio frequency transmissions. Infrared connections are limited to two types, **line-of-sight** and **diffuse.** Line-of-sight infrared systems (Figure 12-7a) require that the transmitter and receiver of the infrared light be straight and in line with each other. Diffuse systems allow an infrared signal to bounce off a wall or other object (Figure 12-7b). Alignment is still critical using a diffuse system, but it can be used to direct a signal around a corner or obstruction. Line-of-sight systems handle traffic between 4 and 16 Mbps, while diffused systems are limited to 1- to 10-Mbps transmission rates.

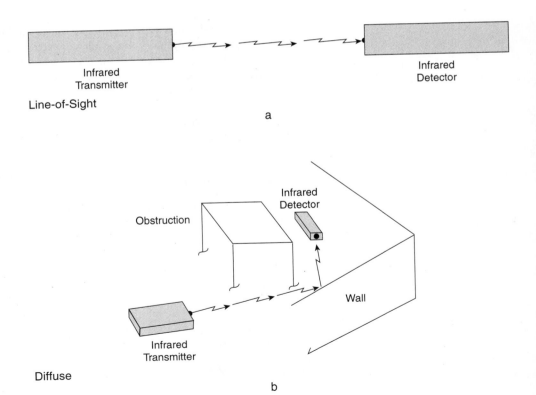

**FIGURE 12-7** Types of Infrared Connections

## RF Systems

Radio frequency systems remove the need for a direct line-of-sight path between transmitter and receiver. They use spread spectrum techniques to modulate data and allow signals to occupy more of the radio band than necessary. This helps to assure that connections are made within a given frequency band. Frequency hopping and direct sequencing methods are employed along with spread spectrum to make the network transmissions more secure.

## Media Access Control

**distributed foundation wireless media access control (DFWMAC)**
wireless MAC specification.

IEEE 802.11 defines media access control functions for wireless networks under a protocol named **distributed foundation wireless media access control (DFWMAC).** The applications defined under this protocol are file transfer, program loading, transaction processing, multimedia data, and manufacturing access control. DFWMAC identifies two basic types of network configurations, *infrastructure-based networks* and *ad hoc networks.* Infrastructure-based networks allow users to roam through a building while maintaining connection with a computer network. Ad hoc networks allow any number of users to set up a link quickly, facilitating conference applications.

**basic service area (BSA)**
DFWMAC cell topology.

The structure of DFWMAC, illustrated in Figure 12-8, is built around the concept of a **basic service area (BSA),** which is a single cell within the infrastructure based network

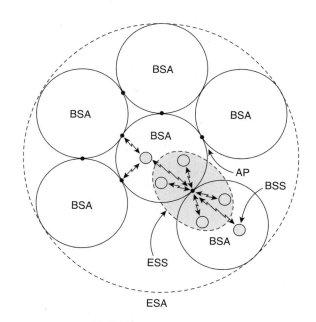

BSS – Basic Service Set
AP   – Access Point
BSA – Basic Service Area
ESS – Extended Service Set
ESA – Extended Service Area

**FIGURE 12-8** DFWMAC

**extended service area (ESA)**
multiple BSAs.

that contains a number of discrete groups of wireless stations. An **extended service area (ESA)** is formed from multiple BSAs to cover a larger area. The BSAs are interconnected by **access points (AP)** and a wired distribution system. A **basic service set (BSS)** is defined as a group of stations associated with the same access point. The **extended service set (ESS)** is a set of stations within multiple BSS's connected via a distribution system.

**access points (AP)**
interconnect point between BSAs.

Access control for DFWMAC is performed at the lowest level using a *distributed coordination function (DCF)*, which operates as an asynchronous communications between multiple stations. CSMA/CA with acknowledgement as described above forms the access method. Media access control (MAC) acknowledge frames assure the integrity of individual packets as they are sent through the network. Failure to receive an acknowledgement causes the sending station to retransmit the lost packet after a random backoff time.

**basic service set (BSS)**
group of stations associated with same access point.

**extended service set (ESS)**
multiple BSSs.

### Section 12.7  Review Questions

1. Why is CSMA/CA used in place of CSMA/CD in wireless LANs?
2. What are the two types of infrared connections used for wireless LANs?
3. What are the two types of DFWMAC network configurations?

## 12.6  CELLULAR DIGITAL PACKET NETWORK

**cellular digital packet network (CDPD)**
large wireless network.

Building large wireless digital networks requires combining cellular systems with existing wired digital networks. In North America, one such system is known as the **cellular digital packet network (CDPD).** This system uses 30-KHz radio frequency bandwidth channels to transmit packet data over a connectionless network. Connectionless circuits have the advantage of using bandwidth allocations only as they are needed. CDPD operates full duplex, using two channels as a communication pair.

Data is sent over cellular channels not currently in use to carry voice communications. In the event that a currently used channel for data transmissions is needed for a voice communication, the data is frequency hopped to another available channel and the original channel is turned over to the voice use. A **mobile data base station (MDBS)** retrieves data packets from the wireless network and channels them to destination networks within the system. The MDBS uses channel sniffing to continual monitor cellular traffic. Part of its duties is to tap off a small amount of transmitted signal (sniff the channel) and check it for voice activity. Upon detecting voice use, the MDBS signals the need for a frequency hop and sees that it is carried out.

**mobile data base station (MDBS)**
interface between mobile networks and other systems.

Each cell in a cellular system contains up to 60 voice grade channels that can be used for voice or data transmissions. Once the MDBS has the data packets from the wireless network, it transfers the packets to a wired system through a **mobile data intermediate station (MDIS).** This unit is also responsible for authenticating users, billing, and data encryption. It supplies the routing between wireline networks and wireless base stations.

**mobile data intermediate station (MDIS)**
user authentication and billing.

Besides encryption, authentication, and frequency hopping, the CDPD system also uses a method called **channel sealing** to aid in security. When the voice system detects interference on a channel, it seals the channel to voice traffic, thus blocking off the inter-

**channel sealing**
voice system seals a channel when it detects interference with the transmission.

ference by an unwanted intruder. A CDPD system can be set up as a dedicated system by forcing the voice system to recognize CDPD transmissions as an interference. This causes the voice system to seal channels to voice traffic, but allows them to be used for CDPD data packet transmissions. This is a method to get around the forced frequency hopping caused by higher priority voice use. CDPD can be used as an online wireless service providing access to the Internet and E-mail transfers as well as basic packet data transfers.

Another large network use for wireless systems is to carry ATM cells over a wireless network. This can be handled using the personal communications system (PCS), RF radio links, or infrared connections (for short hops). A base station can be set up to interconnect between LANs or WANs and wireless subnets using ATM cells as the routing medium across these networks. The base station converts packets from mobile units to wired ATM networks and maintains routing information for the distribution of the packets.

**advanced radio data information service (ARDIS)**
Motorola's private carrier service.

Motorola, as a private carrier, has developed **advanced radio data information service (ARDIS)** to carry data information. The system has 1,000 base stations tied into 31 network controllers. Each base station controls its cell area, which is overlapped with neighboring cell areas to avoid loss of coverage. The network controllers decide which transmit site best can handle the current transmission. Others nearby are turned off for the duration of the transmission (0.5 to 1 second) to reduce interference. The ARDIS RF link runs at a data rate of 4800 bps with half of the overall traffic used for overhead. This results in a user throughput of 2048 bps.

### Section 12.6 Review Questions

1. What two types of networks are combined in a CDPD system?
2. Which has priority on a CDPD channel, voice or data?
3. What methods are used to secure a CDPD channel?

## 12.7 SATELLITE COMMUNICATIONS

Much of today's communications between distant localities involves the use of communications satellites. When the Union of Soviet Socialist Republics (USSR) launched the first successful satellite, SPUTNIK I, into orbit in 1957, the Soviets, as well as everyone else, had no idea to what extent satellites would effect our daily lives. Worldwide sports events and news coverage are now presented as soon as they occur. Business deals and up-to-the-minute information concerning the world brings forth quicker and stronger changes in the world's political, financial, and social communities. Worldwide area networks (WAN) such as the Internet are being established and maintained through an interface of satellites and ground communication systems.

**transponder**
satellite repeater that receives at one frequency and transmits at another.

Communication satellites are essentially electronic repeaters located many miles above the Earth's surface. They are wideband **transponders,** receiving transmission from one Earth station and sending data to another Earth station. The satellite receives transmissions at one carrier frequency, amplifies the weakened signal, and retransmits the information at a different lower carrier frequency. Satellites receive their operating power from solar cells attached to either fins or wings connected to the satellite's body or from solar cells connected directly to the satellite's body. Figure 12-9 illustrates the *Mariner 10*

**FIGURE 12-9** Satellite Communications

satellite, which uses wing-type extensions to contain solar cells needed to power the satellite. Also illustrated in Figure 2-1 are the transmission paths between the satellite and Earth stations. Transmissions from ground to satellite are known as **uplink** transmissions and in the reverse direction, **downlink** transmissions.

## GEO Satellites

Many communication satellites orbit the Earth about the equator in a **geosynchronous** or **geostationary orbit (GEO)** 22,282 miles above the Earth. They travel at a rate of approximately 6,879 miles per hour, which synchronizes their orbital motion with the rotation of the Earth. In this way, the satellite appears to be stationary at one spot above the Earth. Because of the apparently motionless aspect of the satellite in respect to Earth stations, tracking of the satellite is easy. Lower, elliptical orbit satellites require Earth station antennas to track them as they came into view until they disappeared over the horizon. To achieve constant coverage using lower orbit satellites requires the ability to hand off communication from one satellite to another as each came into and left the range of the current tracking station.

GEO satellites are capable of receiving and transmitting to Earth stations that are in a line of sight with the satellite. In other words, if a person were to take a panoramic picture of the Earth from the satellite, Earth stations that are located in the area shown in the picture, known as an area of coverage, could use that satellite. The *Jupiter* satellite in Figure 12-10 illustrates an earth coverage beam. The *Pioneer* satellite, also shown in the figure as an intermediary satellite, is an exploratory type of satellite and is not parked in a geostationary orbit.

A single satellite using an earth coverage beam can blanket about 42.4% of the Earth's surface using a geosynchronous orbit. Three GEO satellites circling the Earth, equally spaced above the equator, can cover 90% of the world. The chief areas not covered by these satellites are the Northern and Southern polar regions, which are sparsely populated. It should be noted that signal strengths received by Earth stations within one area of coverage vary. Additional concern is directly associated with the distance of the satellite from the Earth's surface. Signals experience about a 270-ms round-trip propagation delay because of this distance and the various layers of atmosphere they must travel through.

Another type of beam sent by satellites is the **spot beam,** which concentrates signal strength into a narrow transmission beam. Spot beams emanating from the Pioneer satellite in Figure 12-10 have the advantage of increased signal strength when received by the Earth station due to the concentration of beamed power into a narrow corridor. Spot beams are commonly employed for satellite transmissions to areas such as Hawaii and Puerto Rico, where signal strength is concentrated to the islands and not lost in the surrounding oceans. Some satellites use steerable spot beams, which allow the beam to be directed to different areas on the Earth's surface as needed.

## LEO and MEO Satellites

Geostationary satellites orbit the Earth above the equator at a distance of 36,000 kilometers (22,238 miles) and appear to be in a stationary position with reference to the Earth.

While they are easy to track and maintain a constant line-of-sight connection, they experience serious propagation delays. A typical one-way delay of 0.25 seconds effects both voice and data communications. **Medium Earth orbit (MEO)** and **low Earth orbit (LEO)** satellites were designed primarily to reduce this propagation delay. Instead of being stationary above the Earth, they have an orbital cycle that causes them to appear

**medium Earth orbit (MEO) and low Earth orbit (LEO)**
lower orbit satellites.

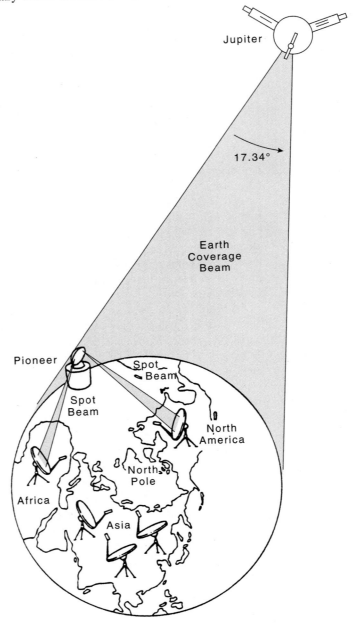

**FIGURE 12-10** Spot and Earth Coverage Beams

over a given area of the earth for a specific time period. To maintain communications, transmissions are handed off from one satellite as it disappears over the horizon to another satellite that has replaced it over the same earth coverage area.

MEOs orbit the Earth at approximately 1,800 miles above the Earth's surface, as shown in Figure 12-11, and experience a propagation delay of 0.1 second, two and one-half times less than GEO satellites. They are chiefly used for mobile voice low-speed (9.6 K to 38.4 K bps) data communications. One example of a system that uses MEOs is TRW's *Odyssey* satellite network that placed 12 satellites in three orbital planes circling the Earth. Each satellite contains 2,300 full duplex circuits. As satellite orbit heights from the Earth are reduced, the number of satellites needed to maintain constant communications increases.

Low Earth orbit satellites (LEOs) are used today to create satellite networks based on satellite constellations known as clusters (Figure 12-12). LEOs are closer to the Earth, from a low of 1,850 KM to a medium distance of about 18,000 KM. They do not appear stationary like GEO satellites, but appear over the horizon twice a day for a duration of 1.5 to 10 hours, depending on the system. Transmission and reception coverage operates

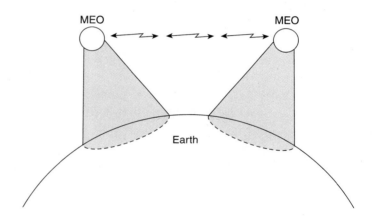

FIGURE 12-11  MEO Satellite Coverage

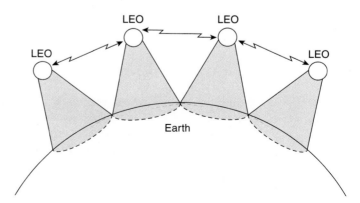

FIGURE 12-12  LEO Satellite Coverage

by blanketing an orbit with numerous satellites so that as one disappears over the horizon, another takes its place. Again, depending on the application, these satellites transmit and receive in either the C, Ku, or Ka bands. They are affected by interference as well as attenuation caused by the absorption of signals by rain.

**cluster**
group of satellites within an orbital plane.

A typical **cluster** of LEO satellites has 21 different orbital planes, each containing a group of 40 satellites per plane. This arrangement is needed to maintain a 40° position above the horizon by at least one satellite at all times. Each satellite must be able to transmit and receive from those satellites in front, back, and to each side, as well as to and from the Earth stations in order to route data to a desired location.

Being closer to Earth, the main advantage of LEO systems is less latency (the time between the sending and the receiving of messages), since it takes less time for data to reach the satellites from Earth stations. The obvious drawback to LEOs is the number of satellites required in orbit to maintain a data path. Networks using LEOs are connectionless, packet-based systems using bursty traffic for teleconferencing and ISDN-integrated data transfers.

One example of a LEO system in use is Loral Aerospace and Qualcom's Globestar satellite cluster. This system uses 48 satellites in 8 orbital planes 1,401 kilometers above the earth. Each satellite contains 2,800 full-duplex circuits for communications between Earth and satellite and between satellites. A second example of a LEO system is the IRIDIUM communications network developed by Motorola.

## Satellite Frequency Allocation

Regulation and allocation of transmissions between the Earth and satellites is accomplished by various licensing and regulatory agencies throughout the world. In the United States, the chief regulatory agency is the Federal Communications Commission (FCC), which assigns specific frequency ranges for commercial and military satellites. International satellite frequency assignments are made to all countries with satellite services. Table 12-2 charts the FCC and international frequency assignments for United States commercial communications satellites. The chart also defines the frequency range for standardized band allocation to allow reference between satellite types and frequency bands.

The uplink and downlink transmission frequency references by band can be illustrated using the Satellite Business System (SBS) used for business communications as an example. Uplink transmissions for SBS satellites range from 14 to 14.5 GHz in the K-band. Downlink frequencies depend on the version of SBS in orbit. There are three sets, all in the K-band: 10.95–12.2 GHz, 12.45–12.7 GHz, and 12.7–13.2 GHz. SBS satellites that use the first or second set of downlink frequencies are referred to as operating at 11/14 GHz. The third set of downlink frequencies satellites are referred to as 12/14 GHz satellites. As you may surmise, the 11 or 12 number refers to the downlink frequency and the 14 to the uplink frequency. At times these satellites are referred to as 14/12 or 14/11 GHz sets, where the uplink frequency precedes the downlink frequency. The order of reference does not matter since the uplink frequency will always be higher than the downlink allocation.

**TABLE 12-2**

**Frequency Allocation by Band; FCC Frequency Allocation for Communications Satellites; International Allocation for U.S. Communications Satellites**

| Band | Frequency Range | FCC Terrestrial Common Carrier Band (GHz) | International Satellite Frequency Bands | | Bandwidth (MHz) |
| --- | --- | --- | --- | --- | --- |
| | | | Downlink (GHz) | Uplink (GHz) | |
| P | 225–390 MHz | | | | |
| J | 350–530 MHz | | | | |
| L | 390–1,550 MHz | | | | |
| S | 1.55–5.2 GHz | 2.11–2.13 | | | 20 |
| | | 2.16–2.18 | | | 20 |
| | | | 2.50–2.535 | 2.655–2.69 | 35 |
| | | | 3.4–3.7 | | 300 |
| | | 3.7–4.2 | 3.7–4.2 | 4.4–4.7 | 500 |
| C | 3.9–6.2 GHz | | | | |
| X | 5.2–10.9 GHz | 5.925–6.425 | | 5.925–6.425 | 500 |
| | | | 7.25–7.75 | 7.9–8.4 | 500 |
| K | 10.9–36.0 GHz | 10.7–11.7 | 10.95–11.2 | | 500 |
| | | | 11.45–11.7 | | 500 |
| | | | 11.7–12.2 | 14–14.5 | 500 |
| | | | 17.7–21.2 | 27.5–31 | 3500 |
| Ku | 15.35–17.25 GHz | | | | |
| Q | 36–46 GHz | | | | |
| V | 46–56 GHz | | | | |
| W | 56–100 GHz | | | | |

The choice of 12/14 GHz for business satellites evolved from the need to locate Earth stations close to the businesses they serve. This could mean that an Earth station antenna might be located on top of a building in the middle of a densely populated city. Concern about interference with microwave and other transmissions in the city environment led to the 12/14 GHz allocation for business satellite communication. Another consideration was the volume of the traffic expected for business satellites versus telephone communications and entertainment satellites. The 12/14 GHz systems can transmit signals occupying one-half the beamwidth required for satellites using 4/6 GHz frequencies. On the down side, the higher transmitting frequencies for business satellites are more susceptible to interference from weather conditions. particularly those resulting in heavy moisture in the air (heavy rains, clouds, or fog). These conditions tend to attenuate 12/14 GHz signals more that those at lower frequencies. The most common use of satellite communications today is the long distance telephone connections, television transmissions, teleconferencing, world wide fax, Telex, E-mail, and Internet connections.

## History of Satellite Communications

Satellite communications were preceded by the launching, in late 1958, of a satellite that broadcast a Christmas message recorded by President Dwight D. Eisenhower. The first actual two-way communications came into being in 1960 with the passive Echo satellites. Unlike today's satellites, these earlier models merely reflected the signal sent to them. This was achieved because the satellites were parked into a low orbit due to limitations of the rocket technology of the day. These low orbits decayed comparatively rapidly, giving these early satellites a short lifespan.

Two breakthroughs in technology led to the present day communications satellite technology. The first involved improved rocket booster engines that could launch satellites into higher, nondecaying orbits. The second was the development of efficient solar panels to supply power for the satellites. This changed them from passive reflectors to active transponders. The first transponder type satellite, Courier, was sent into orbit late in 1960 by the Department of Defense (DOD). The first commercial communications satellite, Telstar (which inspired a successful pop music instrumental recording), came online in July 1962. From that time, many additional communications satellites have been placed into orbit. The **international telecommunications satellite organization (INTELSAT)** series began with the launch of the Early Bird satellite in 1965. Intelsat arose from a United Nations resolution to develop worldwide satellite communications on a nondiscriminatory basis. Intelsat I, better known as Early Bird, has two transponders capable of handling 240 voice channels and one television signal within a 25-MHz bandwidth. It was placed in a geostationary orbit and radiated using a 360° omnidirectional antenna. Unfortunately, much of the power transmitted by Early Bird went into space instead of toward the earth. Later versions of the Intelsat satellites incorporated all that upgrade technology could provide, including earth coverage beams and an increase to 1,200 voice channels using two transponders. The Intelsat V satellite is a multichannel (12,500 channels), high bandwidth (2,300 MHz) system. It consists of twenty-seven transponders and weighs a mere 950 kilograms.

The United States Communication Satellite Corporation (COMSAT), although owned through stockholder shares, regulates the use and operation and sets tariffs for U.S. satellites. It operates as a monopoly and sells user time on satellites to many diverse users. Presently, the United States uses the largest share of satellite time (24%), followed by Great Britain (13%) and France (6%). Other users include Germany, Italy, Canada, Iran, and South American countries. COMSAT has a counterpart in Russia called INTERSPUTNIK, which controls all aspects of satellite use for Russia and Eastern European countries.

Western Union's Westar, begun in 1974, incorporates twelve transponders retransmitting 1,200 voice channels and one color television channel at a digital data rate of 50 Mbps. Two of the transponders are reserved for backup systems in case of failure of one of the other ten units. Radio Corporation of America (RCA) launched SATCOM satellites in 1975, and in 1976 AT&T joined the satellite communications family with its COMSTAR. The first commercial satellites launched by Canada in 1972 were designated by the name Anik, which is Eskimo for "Little Brother." Some of these satellites are illustrated in Figure 12-13.

**international telecommunications satellite organization (INTELSAT)**
an international commercial satellite organization.

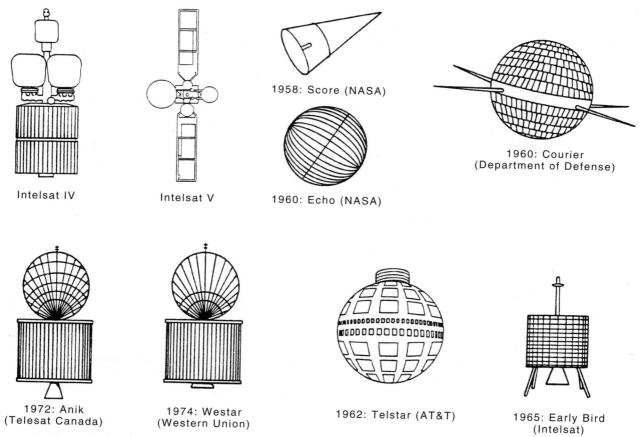

Intelsat IV

Intelsat V

1958: Score (NASA)

1960: Echo (NASA)

1960: Courier
(Department of Defense)

1972: Anik
(Telesat Canada)

1974: Westar
(Western Union)

1962: Telstar (AT&T)

1965: Early Bird
(Intelsat)

**FIGURE 12-13** Early Satellites

Russia, then still the Union of Soviet Socialist Republics, launched their first set of Domestic Satellites (DOMSAT), the Molniya (lightning) in 1966. These satellites beamed television and radio coverage to remote areas within the vast Soviet Republics territory. Four of these satellites were originally spaced equidistant in a nonstationary orbit. As one satellite lost contact with Earth stations, the next in line took over.

Since the mid-seventies, the proliferation of communications satellites have littered the space waves above the Earth as a blanket. In addition to communications applications, many satellites have been launched for military, security, and weather observation purposes. Some satellites have been placed into orbit for the sole purpose of giving man a closer and clearer look at the bodies that occupy the universe above our heads. Being placed beyond the Earth's restricting atmosphere, they have returned startling and interesting pictures from beyond our world.

Concentration on GEO satellites has been replaced by work on medium- and low-orbit (MEO and LEO) satellites. The purpose behind this concentration is to reduce the round-trip delay caused by the distance between Earth and the satellites. One leader in this push toward LEO satellites is Motorola, which has designed and begun to implement the IRIDIUM satellite system. Initial plans are to blanket the low orbit space with 66 satellites plus six spares. These satellites will orbit the earth in 6 planes at a height of 785 kilometers above the Earth. They will have a 100-minute orbit and use the process of hand off from satellite to satellite to complete and maintain communications.

Not to be outdone by Motorola, a consortium headed by McGraw Cellular and Microsoft have designed a system of their own and formed a company, Teledesic to implement the plan. Thrusting toward providing a global Internet as well as mobile telephone service, the Teledesic system will use 840 satellites orbiting 700 kilometers above the Earth. Each satellite will have 8 cross links to maintain connection between neighboring satellites as well as Earth stations.

Other systems in present use include a 48 LEO (1,401-kilometer altitude) satellite network known as Globestar, developed and maintained by Loral Aerospace and Qual-Comm and TRW's Odyssey, based on 12 MEO satellites at a height of around 10,300 kilometers. With all of these satellites circling the Earth along with those designed for other uses, it's a wonder that collisions do not occur at a regular rate!

### Section 12.7 Review Questions

1. What capability must a LEO have to maintain a transmission path?
2. What is the prime advantage of LEOs?
3. What is meant by a geostationary orbit?
4. Define uplink and downlink transmissions.
5. Supply an advantage and a disadvantage of a GEO satellite.
6. Why is the 12/14 GHz range preferred for business communications?
7. Which satellite was the first
   a) One to beam a message to Earth?
   b) To manage two-way communication?
   c) With active transponders?
   d) Commercial communications satellite?
   e) GEO system of satellites?
   f) Canadian communications satellite?
8. How will the Iridium system differ from early GEO systems?

## 12.8 METHODS OF SATELLITE COMMUNICATIONS

**frequency division multiple access (FDMA)**

**time division multiple access (TDMA)**

Two technologies first used to send multiple-channel data using satellites are **frequency division multiple access (FDMA)** and **time division multiple access (TDMA)**. FDMA methods based on frequency division multiplexing (FDM) techniques illustrate how multiple channels are developed and used to modulate increasingly higher-frequency carriers

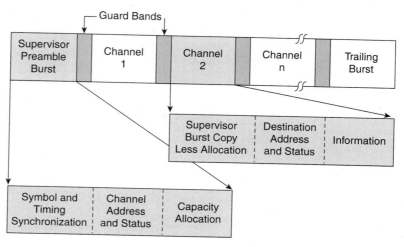

**FIGURE 12-14** TDMA Frame

until the transmission frequencies reach the gigahertz (GHz) range required for satellite transmissions.

## Time Division Multiple Access

A form of time division multiplexing (TDM), called time division multiple access (TDMA), is used as an alternate method to FDMA. As with standard TDM principles, channels are assigned time slot allocations. These time slots are separated by headers and guard bands. A typical TDMA frame is shown in Figure 12-14. A supervisor burst of information precedes the transmission of channel information. This burst is used to establish symbol and burst timing. The first guard band follows the supervisory burst, then the first channel's transmission.

At the beginning of the channel block is a header (or preamble), which repeats the supervisor burst and adds source and destination addresses. Channel information completes the channel's block. Channel blocks are separated by guard bands to prevent overlapping between channels. A closing burst indicates the end of the transmission and reports status information about the transmission capacity and possible error occurrences.

Figure 12-15 shows a functional block diagram of a TDMA communications system. On the transmit side, the channels are multiplexed into their time slots and the control supervisor burst is appended. The data stream modulates a 140-MHz intermediate carrier that is up-converted into a 14-GHz carrier. The 14-GHz signal is sent to the satellite uplink, arriving with a power of approximately 1 picowatt. The satellite transponder demodulates the 14-GHz carrier back to the 140-MHz intermediate carrier. It then up-converts the 140-MHz signal to a 12-GHz carrier to be beamed down to Earth on the downlink. The 12-GHz is received by an Earth station and is down-converted to 140-MHz and

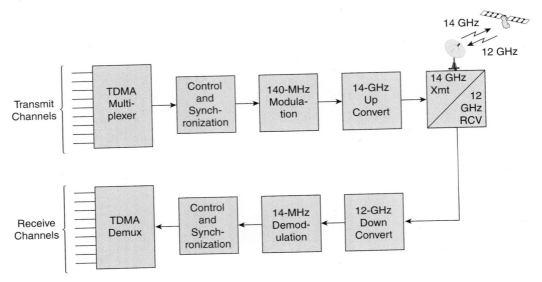

**FIGURE 12-15** TDMA Communications System

demodulated. Finally, the demodulated information is demultiplexed into individual channels and sent onto their destinations.

### Earth Station Antennas

One additional aspect of satellite links is the Earth station's capacity to receive high-frequency, low-power signals from the satellite and to transmit high-power, high-frequency signals to the satellite. These are accomplished using large aperture or dish antennas, which are generally parabolic to allow maximum gathering of the incoming low-power signals sent by the satellite. Some form of feed horn is used to direct transmitted signals from the Earth station to the parabolic reflector, so that for the transmit function, signals can be directed in a more directional and concentrated form as they are sent to the satellite. Figure 12-16 illustrates the physical geometry of a typical parabolic reflector antenna.

The feed horn, for transmit purposes, is located at a radius ($r$) from the reflector portion of the antenna. Using the central axis ($c$) as a reference line, signals leave the horn at angle $\theta$. The geometry of the antenna is designed so that the reflected signal leaves in a direction parallel to the central axis. To assure this, the following relationship between the geometric elements of the antenna has to be met:

$$\tan(\theta/2) = d/(4r) \tag{12-1}$$

where $d$ is the distance between the reflector's edges as shown in Figure 12-16.

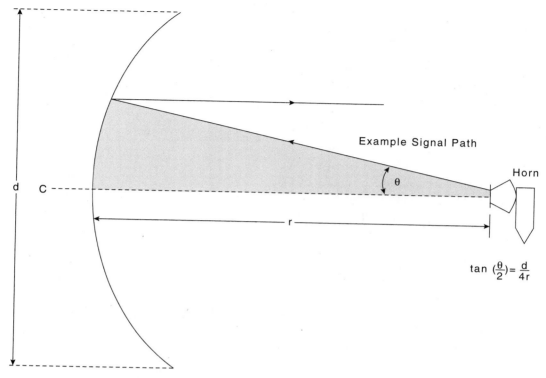

**FIGURE 12-16** Parabolic Reflector Antenna

---

**EXAMPLE 12-1**

What is the maximum angle a signal can leave the horn for a reflector measuring 3 meters from edge to edge, if the horn is placed 1.5 meters from the reflector?

**SOLUTION**

From the data, $d$ is 3 meters and $r$ is 1.5 meters. Substituting these values into Equation 12-1 and solving the arc tangent will yield the angle:

$$\tan(\theta/2) = (3 \text{ m})/[(4)(1.5 \text{ m})] = .5$$

Taking the tangent of .5 yields an angle of 26.5°. Since $\theta/2 = 26.5°$, than $\theta$, the angle the signal can leave the horn, is 53.12°.

---

Other antenna configurations with varying geometries are in use today. Despite their different configurations, the methods for designing them are identical. A relationship

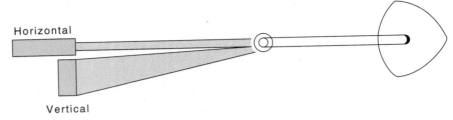

**FIGURE 12-17** Horizontal and Vertical Polarization

between the existing signal, the reflector geometry, and any intermediate devices (mirror reflectors, corner horns, etc.) exists, which determines the antenna's actual dimensions based on its application.

## Satellite Polarized Transmissions

Besides the type of beam (spot, steerable, or earth coverage) discussed earlier, different manners of transmitting signals are used with satellites. To expand the number of channels transmitted at one time, polarized transmissions was developed. Electromagnetic radiations, such as those emanating from an antenna, have a polarizing quality. Antennas can be made to radiate signals that are oriented 90° apart. Referencing one signal as horizontal, the other would be vertical, as illustrated in Figure 12-17. This means that two signals of the same frequency can be transmitted simultaneously without interfering with each other if one is polarized horizontally and one vertically. One example of the use of polarized transmissions is RCA's SATCOM satellite, which manages twenty-four transponders, each with a bandwidth of 36 MHz. Half of the transponders transmit signals polarized horizontally, while the other half use vertical polarization.

**circular polarization**
antenna that radiates signals in a spiral form.

Another, less common form of polarization is called **circular polarization.** Signals are radiated in spiral form (Figure 12-18) from the radiating antenna. Referencing the direction of a signal as clockwise, a second signal, at the same frequency, can be radiated in a counterclockwise spiral. An important point about polarization is that while it allows additional transmission to occur simultaneously, it also divides the radiated power between each pair of polarized radiations.

### Section 12.8  Review Questions

1. What is the purpose of the TDMA supervisor burst?
2. What is the purpose of the trailing TDMA frame burst?
3. Why are the downlink and uplink transmission frequencies different?
4. Repeat the calculation of Example 12.1 for an antenna 5 meters from edge to edge and a distance between horn and reflector of 1.75 meters.
5. What is the benefit of polarized transmissions?
6. What is the drawback of polarized radiations?

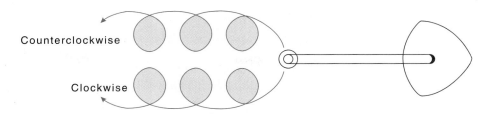

**FIGURE 12-18** Circular Polarization

# 12.9 SATELLITE NETWORKING

**consultative committee for space data communications (CCSPC)**
satellite communications standards organization.

**space communications protocol standards (SCPS)**
set of satellite communications standards.

Like Earth networks, satellite communications have standard organizations responsible for setting specifications for satellite networking. One such organization is the **consultative committee for space data communications (CCSPC),** which has generated a set of standards known as the **space communications protocol standards (SCPS).** In this set of documents are specifications for:

1. End-to-end applications.
2. Monitor and command space vehicles and payloads.
3. Space communications using satellites as repeaters.
4. Intersatellite links.
5. OSI model level 3 to 7 communication functions.
6. Coding, framing, and multiplexing of packets.

Monitoring and controlling the space vehicle and satellite payloads include telemetry information, which deals with status information about the vehicle, command information about controlling the space vehicle or payload, and mission data (sending data to Earth).

## Space Standards

Space standards also include specifications about the networking environment and data-handling protocols used for the communications link. Important concerns about connectivity between satellites and between satellites and Earth, include propagation delay limits, methods for error detection, and the data-handling capacity of the communications links. Space protocols detail how the data itself is formatted, multiplexed, and transferred. Programs should be uncomplicated and have low overhead to maintain a high-data throughput. Routing through satellites and Earth stations have to handle delays and bursty type traffic. The links between satellites may change dynamically as packets are routed through a network. This is particularly true for MEO and LEO satellite networks. Other concerns detailed in these protocols deal with security issues and data protection. Access control, authentication, integrity, and confidentiality methods fall under these categories.

## Transport Level

Transport level protocol areas specify how satellite networks interface with land-based networks. Using satellites to expand networks like the Internet requires these systems to be able to interconnect with TCP/IP, ATM, and other land-based networks. Delivery through this interconnection has to be reliable and deal with packet segmentation, multicasting, congestion, and protocol translation processes.

## Network Level

Network level issues address the routing through the satellite network including various multiple routing options that deal with hand offs between satellites as well as Earth station linkages. Prioritization of traffic and survival of packets are items of concern at this level, which, like the transport level, must deal with congestion, data corruption detection and reporting, and possible multicasting processes.

## ACTS System

advanced communications technology satellite (ACTS)

An example of a satellite network is the **advanced communications technology satellite (ACTS)** system. These satellites operate in the Ka band (30/20 GHz) with a 2.5-GHz available spectrum. They orbit the Earth in a geosynchronous orbit above the equator and use multibeam spot beams to connect with Earth stations. These spot beams have a beamwidth of 0.5° which allows each beam to cover approximately a 150-mile diameter of the Earth's surface.

The circuitry on board the satellites include microwave circuit switches as well as transponder demodulator and modulator units. The circuit switches can switch microwave data that carries data rates from 9.6 Kbps to 1.544 Mbps. The chief modulation technique is QPSK (quad phase shift keying) of TDMA bursts of data. Onboard demodulators transform the incoming TDMA bursts into a digital stream, which is then demultiplexed into 1,728 64-Kbps channels. These channels are fed into two baseband processors for subsequent circuit-switch routing to other satellites within the system or to Earth stations. Satellites that receive a number of channels from neighboring satellites, multiplex these channels into TDMA bursts for downlink transmissions. Services provided by the ACTS system include ISDN and ATM traffic as well as portable and mobile communications links.

## IRIDIUM, Motorola's LEO Network

The IRIDIUM system developed by Motorola uses 66 satellites to cover the world. They orbit in six equally divided planes or rings at a height of 785 KM. To satisfy the interconnection requirement, each satellite has four crosslinks, one forward, one back, and one to each side of the satellite. Each satellite carries 3,840 full-duplex circuits for transferring data.

System control facilities based on broad-processing capabilities include message routing, tracking, telemetry, altitude control, communications network management, and performance evaluation. Gateways are used to interface between IRIDIUM ground stations and external communications networks. To access the IRIDIUM system, a subscriber requires an identification number. There are several numbers used to assist in securing the access. The first is the **mobile subscriber integrated services digital network (MSISDN)** number, which is the permanent number assigned to a subscriber. It is the telephone identification number of the unit dialed to reach an IRIDIUM subscriber, much like your home telephone number. The actual number put on the airwaves is called a **temporary mobile subscriber identification (TMSI)** number. This number is always different each time a subscriber sends or receives messages. The permanent MSISDN number is never sent on the airwaves.

When using a remote access smart card, there is a **mobile subscriber identity (MSI)** number associated with its use. When the card is inserted into an IRIDIUM phone, the MSI is read from the card to verify that the user is authorized to use the system. All the IRIDIUM 12-digit numbers are composed of three fields, a 4-digit country code, a 3-digit geopolitical code, and a 5-digit IRIDIUM subscriber identification number.

**mobile subscriber integrated services digital network (MSISDN)** subscriber access number.

**temporary mobile subscriber identification (TMSI)** actual number on the airwave.

**mobile subscriber identity (MSI)** used with remote access smart card.

## Teledesic Broadband LEO Network

A consortium headed by McGraw Cellular and Microsoft formed a company, called Teledesic, to develop a broadband LEO system for real time wireless bandwidth-on-demand service. 840 satellites are planned to be launched into a 700-KM orbit. Each satellite will use 8 crosslinks to assure rapid and dependable hand off from one to the other to complete a network link. The crosslink frequency band is 59–64 GHz and data will be transferred between 155 Mbps and 1.244 Gbps. The network will support packet switching and ATM communication service.

## MSAT Satellite Network

**mobile satellite (MSAT)** integrated services satellite network.

**Mobile satellite (MSAT)** satellite network is an integrated service between the United States and Canada. This network provides for mobile communications throughout both countries with the following applications:

* Law enforcement.
* Public safety.
* Fleet management.
* Airline reservations.
* Federal aviation administration (FAA) control.
* Fishing fleets, barges, and tugs.
* Trucking and rail freight.
* AMTRACK—American passenger rail system.
* Point-to-point telephone links.

**mobile telephone service (MTS)**
handles voice, fax, and data services.

**mobile radio service (MRS)**
extends MTS coverage.

**mobile data service (MDS)**
data switch and private data service.

**very small aperture terminal (VSAT)**
limited satellite network.

MSAT is composed of four functional units. **Mobile telephone service (MTS)** handles voice, fax, and data services over the phone system. It extends its coverage to include land, marine, and aviation communications. **Mobile radio service (MRS)** is a private version of MTS. **Mobile data service (MDS)** encompasses data-switched and private data services. Packet-switched services include data broadcasts, data acquisition and control, and data routing services. MDS facilitates connection to public networks like the phone system and ISDN services. Circuit-switched services handle voice, asynchronous data streams, and fax data.

## VSAT Systems

**Very small aperture terminal (VSAT)** systems are used for batch and transaction processing including:

- Automated teller machines.
- Credit card purchases.
- Hotel chain check-in service.
- Airline reservations.
- Linking LANs to hosts at data rates to 256 Kbps.
- Two-way communications such as teleconferencing.
- Interactive data communications.

## IEEE 802.16 Broadband Satellite Network

The IEEE committee on networking has submitted a standard for broadband wireless metropolitan area network (MAN) using microwave, radio, and satellite links. Base stations will communicate via satellites in the 24 to 38 GHz range for United States systems, 28 GHz for Canadian-based systems, and 40 GHz European networks. The network specified is for public network data, voice, and video services over point-to-point or point-top multipoint architecture.

### Section 12.9   Review Questions

1. What functions are specified by the Space Communications Protocol Standards?
2. Which satellite types are used by the ACTS network?
3. What services are provided by the ACTS network?
4. What type of satellites are used in the IRIDIUM system?
5. How many links does one IRIDIUM satellite use?
6. Which IRIDIUM number is used by a subscriber to access the system?
7. How many crosslinks does each Teledesic satellite contain?
8. What types of data services are supported by the Teledesic system?

# SUMMARY

Wireless communications systems are expected to dominate the data communications industry for long distance links. Mobile and cellular telephone systems are already well established and an integral part of our lives. With Internet access made accessible through home TV sets, it won't be long before wireless communications dominates our daily living as well.

Wireless LANs are slow in their acceptance because of the limiting nature of infrared connections. The future for wireless LANs will no doubt be integrated with radio frequency links. Larger wireless networks, on the other hand, using microwave and satellite systems are booming. Older GEO systems with satellites parked in fixed places and LEO clusters bringing real time communications closer are blanketing the heavens above us. Worldwide information and communications are a reality that will continue to grow and improve as technology meets demand.

Satellite networks are now in place to expand worldwide communications applications. Besides handling ISDN and other routing traffic, these systems are being adopted for Internet access. Lower orbit satellite clusters are reducing the inherent propagation delay, which will eventually make Internet access practical.

# QUESTIONS

**Section 12.2**

1. How is a wave guide used to propagate a particular microwave signal?
2. How are waveguides "tuned" to the transmit frequency?
3. What is a standing wave?
4. What does a standing wave ratio convey?
5. What type of transmissions are microwave transmissions?
6. How does temperature layers within the atmosphere affect microwave transmissions?
7. How does multipath fading degrade a microwave signal?
8. How are multipath fading effects reduced?

**Section 12.3**

9. Give three characteristics of a mobile cellular telephone system.
10. What term defines the area covered by a mobile system's base station?
11. What are cellular base stations called?
12. What advantage is gained by using cells in mobile systems?
13. What are the proven health risks involved with using mobile telephones?
14. What is probably the greatest risk to using mobile telephones?
15. What types of cellular service, in broad terms, is currently available?
16. What happens to a cell once the number of subscribers within that cell becomes excessive?
17. What type of modulation is used with analog cellular phones?

18. Two types of multiplexing are used in the digital cellular phone system. What are they?
19. What methods are used to secure a mobile phone?
20. Explain the differences between direct sequencing and frequency hopping. How do each help in securing a mobile phone connection?
21. What is the term used to denote the mobile telephone used in the GSMC system?
22. Name the two databases that keep track of a user and his/her whereabouts for the GSMC system.
23. What term is used to describe the ability of a mobile user to move about through many cells and countries using GSMC?
24. What services do a smart card provide?
25. What is a GSMC base station called?
26. Which GSMC unit is responsible for maintaining hand offs among GSMC base stations?

**Section 12.4**

27. What does the air interface of the personal communications system (PCS) define?
28. How is the RF spectrum configured for PCS?
29. What is used by PCS for carrying advanced paging and data services?
30. List five applications for the personal communications network (PCN).
31. CDMA is an acronym with two different meanings. What are they?
32. How many channels are multiplexed using CDMA?

**Section 12.5**

33. Which IEEE 802 specification covers wireless LANs?
34. What method is used to access a mobile network according the IEEE standard? How does it differ from the 802.3 specification?
35. Explain the two methods used for infrared connections.
36. What is the benefit of using RF for a wireless LAN compared to infrared connections?
37. What type of techniques are used with RF wireless LANs to modulate carriers?
38. Two types of networks are defined under IEEE DFWMAC protocol. What are they and what are their main feature?
39. Define the following DFWMAC terms
    a) Basic service area.
    b) Extended service area.
    c) Basic service set.
    d) Extended service set.
40. What DFWMAC term refers to where BSAs are interconnected?

**Section 12.6**

41. Give three characteristics of the cellular digital packet network (CDPD).
42. List several duties of the CDPD mobile data base station (MDBS).
43. What is used by MDBS to continually monitor cellular traffic?
44. How many voice channels are assigned to each cell in the CDPD system?

45. Which CDPD unit is responsible for authenticating and billing users?
46. Define channel sealing. Why is it used?
47. How do CDPD cellular data networks take advantage of channel sealing?
48. List a number of basic uses for the CDPD network.

**Section 12.7**

49. What basic function does a communications satellite perform?
50. What is meant by uplink and downlink transmissions?
51. Describe a geosynchronous orbit. Include the prime advantages when compared to an elliptical orbit.
52. What is a LEO satellite? How does it differ from a GEO satellite?
53. What is the main advantage and disadvantage of using LEO satellites?
54. Define cluster in terms of satellites.
55. Minimally, how many crosslinks are used in each LEO satellite? What are they?
56. What is required of satellite transmissions to maintain links to Earth stations?
57. Where do satellite transponders get their operating power?
58. List the main functions of a satellite transponder.
59. Give a good reason why uplink and downlink frequencies are not the same.
60. What is the difference between a spot beam and an Earth coverage beam? Which would be used to transmit data to Easter Island?
61. Which United States agency assigns transmitting frequencies to satellites?
62. Give one advantage of using 12/14 GHz for business applications in a large city.
63. Which satellite system used for two-way communications was nothing more than a passive reflector of transmitted signals?
64. How did Courier satellites differ from Echo satellites?
65. Which United Nations organization is responsible for overseeing the Early Bird satellites for worldwide communications?
66. Which two forms of multiplexing are used in multichannel satellite communications?
67. What is the purpose of the supervisory and closing bursts in a TDMA frame?
68. Determine the angle at which transmitted signals leave the horn of a dish antenna that is 4.6 meters across with the horn 1.8 meters from the reflector.
69. What is gained by using horizontal and vertical polarized stations? What is the main drawback to polarized radiations?
70. How does circular polarization differ from horizontal and vertical polarization?
71. Why are Earth dish antennas generally parabolic in shape?
72. Which organization sets the standards for satellite networking?
73. What is the name for the set of satellite networks?
74. What functions are detailed in SCPS protocols?
75. What areas are specific to the transport layer of SCPS protocols?
76. What areas are specific to the network layer of SCPS protocols?
77. What type of satellites are employed in the ACTS system?
78. What type of beams are transmitted by ACTS satellites?
79. What type of modulation and multiplexing techniques are used by ACTS satellites?
80. What are the two competing LEO systems being currently placed into orbit?

81. How many satellites are used in the IRIDIUM system? What type of satellites are they?
82. How many duplex circuits are carried by a single IRIDIUM satellite?
83. What is the name of an IRIDIUM telephone number?
84. What must a user have to gain access to the IRIDIUM system?
85. Which companies are behind the Teledesic network?
86. How many satellites are planned for the Teledesic network?
87. List eight applications applied to the MSAT system.
88. What are the general services for each of the functional units of the MSAT system?

## RESEARCH ASSIGNMENTS

1. Research the latest developments in satellite technology. Include how each advancement in technology benefits this form of communication.

2. Write a report on the number of different satellites in use today and what their functions are. This would include communications (telephone, fax, data, etc.), television, ISDN, defense "spy," business applications, and so forth.

3. Write a report about security for wireless communications. Include areas covering authentication, cryptography, and transmission methods.

4. Determine what types of new wireless communication systems are in the works and write a detailed report on one of them.

## ANSWERS TO REVIEW QUESTIONS

**Section 12.1**
1. Radio, television, mobile phones, networking, etc.

**Section 12.2**
1. Standing wave ratio is a measure of transmitted versus reflected power.
2. Line-of-sight.
3. Reduce multipath fading.

**Section 12.3**
1. Cell.
2. Analog and digital.
3. TDMA—time division multiple access.
4. Frequency hopping, direct sequencing, encryption.
5. Mobile station.
6. User identification and authorization.

**Section 12.4**
1. Messaging, fax, networking laptops, user group communications, PBX, interconnecting other services.
2. Digital.

**Section 12.5**

1. Difficult to detect a collision in a wireless transmission.
2. Line-of-sight and diffuse.
3. Infrastructure and ad hoc.

**Section 12.6**

1. Cellular and wired digital networks.
2. Voice.
3. Encryption, authentication, frequency hopping, channel sealing.

**Section 12.7**

1. Orbit where satellites appear to be stationary in reference to a particular Earth coverage area.
2. Transmit to neighboring satellites as well as Earth stations.
3. Low latency.
4. Uplink is transmission from Earth station to satellite, downlink is the reverse direction.
5. Advantage of GEO—easy to maintain line of sight with specific Earth stations. Disadvantage—propagation delay between Earth and satellites.
6. Doesn't interfere with microwave or other transmissions.
7. a) Satellite that broadcast Eisenhower Christmas message.
   b) Echo
   c) Courier
   d) Telstar
   e) Early Bird
   f) Anik
8. Use LEOs and operate with satellite clusters intersatellite hand off.

**Section 12.8**

1. Establish symbol and burst timing.
2. Indicates end of transmission and reports status information about capacity and possible detected errors.
3. To avoid cross interference between uplink and downlink signals.
4. $\tan \theta/2 = 5/(4)(1.75) = .7413 \qquad \theta/2 = 35.5° \; \theta = 71°$
5. Allows more than one signal to be transmitted using the same frequency at the same time.
6. Divides transmitted power among the transmitted signals.

**Section 12.9**

1. End-to-end applications; monitor and command space vehicles and payloads; space communications using satellites as repeaters; intersatellite links; OSI model level 3 to 7 communication functions; coding, framing, and multiplexing of packets.
2. GEOs.
3. ISDN, ATM, portable, and mobile links.
4. LEOs.
5. 4
6. Mobile Subscriber Integrated Services Digital Network (MSISDN).

7. 8
8. Packet switching and ATM communication services.
9. Law enforcement, public safety, fleet management, airline reservations, federal aviation administration (FAA) control, fishing fleets, barges, and tugs, trucking and rail freight, AMTRACK, point-to-point telephone links.
10. Mobile Telephone Service, Mobile Radio Service, Mobile Data Service, Circuit Switched Service.
11. Automated teller machines, credit card purchases, hotel chain check-in service, airline reservations, linking LANs to hosts at data rates to 256 Kbps, two-way communications such as teleconferencing, interactive data communications.

# APPENDIX A

# Acronyms and Key Terms

**A**

**AAL**—ATM Adaption Layer. ATM network layer.

**ABR**—Available Bit Rate. Defines amount and type of traffic a channel will handle.

**Acceptance Angle**—Angle at which light rays can enter a fiber core.

**Acceptance Cone**—Group of all light rays that can enter a fiber cone.

**Access Unit**—Entry point into an ISDN network.

**ACIA**—Asynchronous Communications Interface Adapter. Motorola's version of an UART.

**ACK**—ACKnowledge.

**ACTS**—Advanced Communications Technology Satellite.

**ADC**—Analog-to-Digital Converter. Circuits or devices that convert between analog and digital signals.

**ADM**—Adaptive Delta Modulation or Asynchronous Disconnect Mode.

**ADPCM**—Adaptive Differential Pulse Code Modulation.

**ADSL**—Asymmetric Digital Subscriber Line.

**Air Interface**—Set of parameters that define the behavior of wireless connections.

**Alphanumeric Characters**—Printable characters in a character code, comprised of alphabet, numerical, and punctuation characters.

**AlGaAs**—Aluminum Gallium Arsenide. Surface-emitting diode material.

**AM**—Amplitude Modulation.

**AMI**—Alternate Mark Inversion.

**ANI**—Automatic Number Identification.

**ANSI**—American National Standards Institute.

**Answer Station**—Secondary station in a data link.

**AP**—Access Point. Interconnect point between BSAs.

**APPC**—Advanced Program-to-Program Communication. Allows programs from different computers to communicate with each other.

**APPN**—Advanced Peer-to-Peer Network. Application that monitors computers on an ATM network.

**ARCHIE**—Data-base of FTP sites.

**ARDIS**—Advanced Radio Data Information Service. Motorola's private carrier service.

**ARM**—Asynchronous Response Mode.

**ARP**—Address Resolution Protocol. Translates between non-ATM and ATM addresses.

**ARPANET**—Advanced Research Projects Agency NETwork.

**ARQ**—Automatic ReQuest for retransmission or Automatic Repeat reQuest.

**ASCII**—American Standard Code for Information Interchange.

**ASK**—Amplitude Shift Keying.

**ASN**—Abstract Syntax Notation. Encryption method used at the presentation level.

**Asynchronous Data**—Serial data that does not require a synchronizing clock or signal between sender and receiver.

**Asyncronous protocol**—Character-oriented asynchronous data-link protocol.

**ATDM**—Asynchronous Time Division Multiplexing

**Attenuation or Gain Distortion**—Measure of the change in received signal amplitude across the voice bandwidth.

**ATM**—Asynchronous Transfer Mode. Routing protocol that uses fixed size packets called cells.

**Attributes**—Those accents of a message that bring attention to a part of the message. They include boldface, underline, blinking, etc.

**ATU-C**—ADSL Terminal Units—Central office.

**ATU-R units**—ADSL Terminal Unit-Remote.

**AU**—Access Unit.

**AUI**—Attachment Unit Interface. Connect thicknet to Ethernet NIC.

**Authentication**—Type of security that acknowledges users, through the use of passwords, as having authorized access.

**B**

**B6ZS**—Binary 6 Zero Substitution. Another method used to prevent a long-string of transmitted zeros.

**B8ZS**—Binary 8 Zero Suppression. Scheme that prevents the transmission of a long stream of zeros.

**Back Channel**—Channel which is used to convey control data between two stations in a data link.

**Baseband**—Single-channel communication system in which that channel is the only one occupying the system's bandwidth.

**Basic Rate Interface (BRI)**—Defines an ISDN network service consisting of two B channels and one D channel.

**Basic Conditioning**—Standard telephone company specifications for attenuation distortion and propagation delay variances.

**Baud Rate**—Digital information transfer rate.

**BCC**—Block Check Character.

**Bearer (B) channel**—64-Kbps ISDN data channel.

**Beacon Token**—Used to check integrity of the primary FDDI ring following the repair of a station on that ring.

**BECN**—Backward Explicit Congestion Notification.

**BER**—*See* Bit Error Rate.

**BiPolar Violations (BPV)**—Occur when the voltage level of two successive 1s is the same when using alternate mark inversion data format.

**BISDN**—Broadband Integrated Services Digital Network.

**BISYNC or BSC (BInary SYNcronous Control)**—Character-oriented synchronous data protocol.

**Bit-Error Rate (BER)**—Measure of the number of errors in a stream of data.

**Bit-Error Rate Tester (BERT)**—Instrument for testing errors in a bit stream.

**Bits Per Second (bps)**—Rate at which raw serial binary data is sent and received.

**Block check character**—Error detection character.

**Blocking**—Keep a node from receiving messages not addressed to it.

**BNT**—Broadband Network Termination.

**BRI**—*See* Basic Rate Interface. Defines an ISDN network service consisting of two B channels and one D channel.

**BNZ-AMI**—Bipolar Non-Return-to-Zero with Alternate Mark Inversion.

**BPS**—*See* Bits per Second.

**BPV**—*See* BiPolar Violations.

**Broadband**—Multichannel communication system in which the system bandwidth is shared by the channels using it.

**Broadcast Address**—In a multipoint system, it is the address used by the primary to send a message to all of the secondaries in the system.

**Broadcast Storm**—Overload of traffic on a network due to the issuing of too many broadcast messages.

**BRZ-AMI**—Bipolar Return-to-Zero with Alternate Mark Inversion.

**Burst**—Heavy data traffic for a short duration.

**BSA**—Basic Service Area. DFWMAC cell topology.

**BSC**—Basic Station Controller. Controls handoffs to BTS stations.

**BSS**—Basic Service Set. Group of stations associated with the same access point.

**BTS**—Base Transceiver Station. Central office of a GSMC cell.

**Bus**—Network topology where all nodes are connected to a common data path.

**BW**—BandWidth.

**C**

**CA**—Certificate of authority. Verifies validity of digital signature and public key owner.

**CAN**—Campus Area Network

**CASE**—Common Application Service Element. Interconnects several buildings in a restricted area.

**Carrier Pilot**—Reference signal that sets the amplitude of a FDM group.

**CBR**—Constant Bit Rate. Defines amount and type of traffic a channel will handle.

**CCITT**—Consultative Committee for International Telephony and Telegraphy.

**CCSPC**—Consultative Committee for Space Data Communications. Satellite communications standards organization.

**CD**—Carrier Detect. Detect received signal after link is established.

**CDMA**—Code Division Multiple Access or Cellular Digital Multiple Access. Uses digital encoding to multiplex several channels on to one line.

**CDPD**—Cellular Digital Packet Data. Large wireless network.

**CDSL**—Customer Digital Subscriber Line.

**Cell**—ATM data packet or geographical division for mobile communications.

**CERN**—Center for European Nuclear Research.

**Channel**—Single line of communication.

**Channel Sealing**—Voice system seals a channel when it detects interference with the transmission.

**Character Code**—Binary code representing alphanumeric, formatting, and data-link characters.

**Charged Coupled Devices**—Converts light to electrical energy.

**Checksum**—Error-detection process that uses the sum of the data stream in bytes.

**CIR**—Committed Information Rate.

**Circuit Switch**—A network switching method that physically connects two nodes to facilitate communications.

**CLP**—Cell Loss Priority. Determines which cells can be discarded to avoid congestion.

**Cladding**—Material surrounding a fiber core, which has a refractive index that causes light rays to be reflected back into the core.

**Circular Polarization**—Antenna that radiates signals in a spiral form.

**CLSN**—Collision Detection SigNal.

**Cluster**—Group of satellites within an orbital plane.

**Cluster Controller**—The hardware and software that manages several stations or nodes.

**CMC**—Common Messaging Calls. Defines cells for sending and receiving e-mail messages.

**C-Message**—Voice bandwidth between 300 and 3 kHz.

**CMOS**—Complementary Metal Oxide Semiconductor.

**CORBA**—Common Object Request Broker Architecture.

**CODEC (COder/DECoder)**—Used to convert voice signals to digital codes.

**Collision**—Result of two or more messages sent on the same line at the same time.

**Companding**—Compression/Expanding of signals to reduce signal to noise effects of smaller signals.

**Concentrator**—Device used to connect several single attached stations to a dual ring network.

**Contention**—Two or more stations vying for use of a line.

**Conditioned Lines**—Lines which have been electrically altered to meet more stringent requirements.

**Continuous Wave Keying (CWK)**—Form of data transmission that uses the presence of a sine wave to represent a logic one and its absence, a logic zero.

**Convergence Layer**—ATM layer responsible for placing data into a common format.

**CPU**—Central Processing Unit. Controlling unit of a computer.

**CRC**—*See* Cyclic Redundancy Check.

**Crosstalk**—Coupling of one signal to a neighboring line.

**CSMA/CA**—Carrier Sense Multiple Access with Collision Avoidance—same as CSMA/CD with collision avoidance.

**CSMA/CD**—Carrier Sense Multiple Access with Collision Avoidance—same as CSMA/CA with Collision Detection—Ethernet Bus access.

**CSU**—Channel Service Units. Equipment to facilitate direct ISP service.

**CTS**—Clear To Send. Modem response to a request to send.

**CWK** —*See* Continuous Wave Keying.

**Cyclic Redundancy Check (CRC)**—Error detection method that uses a pseudo division process.

**D**

**DAA**—*See* Data Access Arrangement.

**DAC**—Digital-to-Analog Conversion.

**DARPA**—Defense advanced research projects agency.

**Data Access Arrangement (DAA)**—Interface between modem and telephone lines.

**DAS**—*See* dual attached stations.

**Datagrams**—Unacknowledged packets sent on the Internet.

**Data Link**—Communication link established between node points in a network.

**Data-Link protocol**—*See* protocols.

**Data-Linking** or **Control Characters**—Characters used to establish and maintain a communications link between two stations.

**Data Circuit Terminating Equipment (DCE).**

**Data Communications Equipment (DCE)**—Equipment used to place data from the communications terminal onto a medium and return data from the medium to another terminal.

**Datascope**—Test equipment used to monitor a serial data stream.

**Data Terminal Equipment (DTE)**—The hardware responsible for interfacing communications equipment to computer terminals.

**Data Under Voice**—Method that utilizes normally unused band areas for supplemental voice transmissions.

**dB**—Decibel ratio.

**dBm**—Power in decibels referenced to 1 milliwatt (mW)

**DC3**—Device Control character 3.

**DCD**—Data Carrier Detect. Detect received signal after link is established.

**DCE**—*See* Data Communications Equipment.

**DDD**—Direct Distance Dialing.

**DE**—Discard Elgibility. Determines which frames can be discarded to avoid congestion.

**DES**—Data Encryption Standard. 64-bit key encryption method.

**Dedicated or Leased Line**—Private line between two subscribers.

**DEF**—Dual Error Flag.

**Delta Modulation**—Based on a difference in sample levels rather than the levels themselves.

**DFB**—Distributed Feedback Bragg Semiconductor Laser. A semiconductor laser used in multiplexed systems.

**DFWMAC**—Distributed Foundation Wireless Media Access Control. Wireless MAC specification.

**DHCP**—Dynamic Host Configuration Protocol. Application for adding and deleting hosts on an IP network.

**DIBOC**—Distributed input buffering with output control. Small ATM switch output buffer.

**Diffuse**—Infrared connections that allow signal to bounce off wall or object.

**Direct Sequencing**—Mobile transmission that divides and transmits message portions at different frequencies.

**DLE**—Data Link Escape character.

**DM**—Disconnect Mode command.

**DMA**—Direct Memory Access.

**DMT**—Discrete Multitone Technology.

**DNS**—*See* Domain Name Service.

**Domain Name**—Part of an Internet address that is used to identify the user.

**Domain Name Service (DNS)**—Function used to assign Internet addresses to users.

**Domain Top**—Part of Internet address that designates the general group or organization a user belongs to.

**Downlink**—Transmission from satellite to earth.

**DPSK**—Differential Phase Shift Keying.

**DQDB**—Distributed Queued Dual Bus protocol.

**Drum Scanning**—Form of optical scanning that uses a helix groove in a drum to detect reflected light from a document.

**DSA**—Directory System Agent. System-to-system interface.

**DSAP**—Destination Service Access Point. End points in communication link.

**DSL**—Digital Subscriber Line. High-speed digital service for transporting voice and data.

**DSSS**—Direct Sequence Spread Spectrum.

**DSU**—Data Service Units. Equipment to facilitate direct ISP service.

**DTE**—*See* Data Terminal Equipment.

**DTI**—Document Type Identification.

**DTMF**—Dual Tone Multiple Frequency.

**DUA**—Directory User Agent. Interface user to system.

**Dual Attached Station**—Station that is connected to both rings of a network.

**DUV**—*See* Data Under Voice.

**E**

**EBCDIC**—Extended Binary Coded Decimal Interchange Code.

**Echo**—A portion of transmitted signal returned to the transmitter.

**Echo Suppressor**—Circuit that disables amplifiers in repeaters to prevent transmitted signals from being echoed back.

**EFS**—*See* Error-Free Seconds.

**EIA**—Electronic Industries Association.

**ENQ**—Enquire character.

**ENQUIRE**—Early application that facilitated access to the Internet.

**E-mail**—System that allows users to communicate directly with one another through electronic messaging.

**EMI**—ElectroMagnetic Interference. Interference caused by a steady transmission of power.

**EMP**—ElectroMagnetic Pulse. Large induced pulse on a cable.

**Entropy Coding**—Assigns smaller digital codes to smaller changes in samples.

**Envelope Delay Distortion**—Method used to test and measure the variance in propagation delay of different signals within the voice band.

**EOT**—End Of Transmission character.

**ESA**—Extended Service Area. Multiple BSAs.

**ESC**—ESCape character.

**ESS**—Extended Service Set. Multiple BSSs.

**ETB**—End of Transmission Block character.

**ETX**—End of TeXt character.

**Extended ASCII**

**Extranet**—A restricted type of intranet network with external access.

**F**

**FAQ**—*See* Frequently Asked Questions.

**Fabrics**—Switched-fibre channel networks.

**Facsimile (FAX)**—Image detection and transmission.

**FCMOS**—Fast Complementary Metal Oxide Semiconductor.

**FCS**—Frame Check Sequence. Error detection characters.

**FDDI**—*See* Fiber Distributed Data Interface.

**FDDN**—*See* Fiber-Optic Distributed Data Network. Military fiber optic network.

**FDM**—Frequency Division Multiplexing. Multiplexing techniques where channels share frequency in a band.

**FDMA**—Frequency Division Multiple Access.

**FEC**—Forward Error Correction. Method of correction that occurs as messages are received.

**FECN**—Forward Explicit Congestion Notification.

**FHSS**—Frequency Hopping Spread Spectrum.

**Fiber Distributed Data Interface (FDDI)**—A fiber optic network protocol based on dual ring topology.

**Fiber-Optic Distributed Data Network (FDDN)**—An avionics application using FDDI.

**File Transfer Protocol (ftp)**—Protocol that allows the uploading and downloading of files to/from the Internet.

**Flatbed Scanning**—Image is scanned as it is moved across a flat surface.

**Flat Top**—Sample and hold that maintains peak values between samples.

**FM**—Frequency Modulation.

**Focal Point**—The point at which light converges on one side of a lens after passing through that lens.

**Formatting Characters**—Characters responsible for the appearance of text, like line feed, carriage return, etc.

**Frame rely**—Routing protocal that uses variable size packets called frames.

**Framing Bits**—Bits that denote the beginning and end of a character.

**Free Token**—Short packet with no data packets attached to it.

**Frequency Hopping**—Constantly changing carrier frequency during a mobile phone transmission.

**Frequency Shift Keying (FSK)**—Modulation technique in which the frequency of a tone is altered by the state of binary bit data.

**Frequently Asked Questions (FAQ)**—Collection of commonly asked questions and answers on a particular subject.

**FRMR**—FRaMe Reject.

**FSK**—*See* Frequency Shift Keying.

**FT1**—Fractional T1 service. Uses T1 type service to send fewer channels.

**FTAM**—File Transfer And Management.

**FTP**—*See* File Transfer Protocol.

**Full Duplex**—Ability to transmit and receive data simultaneously.

**Full-Full Duplex**—Ability to transmit data to one station and receive data from a different station simultaneously.

**G**

**GAN**—Global Area Network. World wide network.

**Geosynchronous Orbit (GEO)**—A satellite in stationary orbit, synchronized to earth's rotation.

**GOPHER**—Database of web sites.

**GFC**—Generic flow control field. Identifies level of congestion control and cell priority.

**GOSIP**—Government Open Systems Interconnection Profile.

**Graded Index Core**—Core with several layers of material with slightly different refractive indexes.

**Group Address**—In a multipoint system, this is an address used to designate a number of secondaries, but not all secondaries.

**GS**—Group Separator.

**GSMC**—Global System for Mobile Communications. Worldwide mobile phone network.

**Guardband**—A portion of unused bandwidth between channels.

**H**

**Half Duplex**—Ability to send and receive data but not at the same time.

**Hamming Code**—Error-correction method based on the number of logic 1 states in a message.

**Handshake**—Term used to describe control signaling between two units prior to transferring information.

**HDLC**—High-Level Data-Link Control.

**HDSL**—High data-rate Digital Subscriber Line.

**HEC**—Header Error Control. Checks errors in headers.

**HLR**—Home Location Register. Database of subscriber home addresses.

**Home Page**—Opening page of a web site.

**HTML**—*See* HyperText Mark up Language.

**HTTP**—*See* HyperText Transfer Protocol.

**Huffman Code**—Data transmission efficiency code that uses shorter codes for more frequently transmitted bit combinations.

**Hybrid Circuit**—Interface between 2-wire local and 4-wire trunk lines.

**Hyper-Links**—Links within a web page that allows the user to go to another web page or web site.

**HyperText Mark up language (HTML)**—Mark up language used for creating web pages.

**Hypertext Transfer Protocol (http)**—Controls access to web sites.

**I**

**IBM**—International Business Machines.

**IC**—Integrated Circuit.

**Idle Line One**—state of a communication line when active and not handling messages.

**IDSL**—Integrated Digital Subscriber Line.

**IEEE**—Institute of Electrical and Electronics Engineers.

**Incident Angle**—The angle at which a light ray meets a surface.

**Incident Ray**—Light ray that initially strikes a surface.

**Injection Laser Diode**—Semiconductor device, that emits a concentrated cone of light.

**International Standards Organization**—Data networking standardizing committee.

**IMAP**—Internet Mail Access Protocol.

**Impulse Impairment**—Change in a signal's amplitude or phase for a very short time period.

**IMSI**—International Mobile Subscriber Identity. Global mobile telephone smart card system.

**Inherent or White Noise**—Steady low-level noise generated by equipment.

**INTELSAT-International Telecommunications Satellite Organization**—An international commercial satellite organization.

**Internet**—Worldwide communications network.

**Internet Service Provider (ISP)**—An entity that makes access to the Internet available to users.

**Intranet**—a local Internet type of network with no external access.

**IP**—Internet Protocol. Internet network layer protocol.

**Interrupt Request IRQ**—Interrupt ReQuest signal informing the CPU to divert from its current program processing to execute a new program.

**ISDN**—Integrated Services Digital Network.

**ISDN**—Integrated Services Digital Network. Digital communications system that transports voice and digital data.

**ISO**—*See* International Standards Organization.

**Isochronous Service**—Network service that can handle data, voice, video, and other types of services.

**ISP**—*See* Internet Service Provider.

**ITB**—End of Intermediate Transmission Block.

**ITU**—International Telecommunications Union. Standards group for networks.

**Link Access Protocol (LAP)**—ISDN data-link protocol related to HDLC.

**LLC**—Logical Link Control. Exchange of data across a channel.

**Local Loop**—Telephone lines between subscriber and local switch station.

**Logical Unit (LU)**—Virtual node that services user terminals.

**Longitudinal Redundancy Check (LRC)**—Error-correction method that uses parity and bit summing.

**Loop-Back**—Circuit that returns transmitted data to the source for the purposes of testing the line.

**LRC**—*See* Longitudinal Redundancy Check.

**LSB**—Least Significant Bit.

**LU**—*See* Logical Unit.

**J**

**Jitter**—Small constant change in a signal's amplitude or phase.

**Jumbo group**—FDM signal made up of 6 mastergroups.

**L**

**LAN**—Local Area Network.

**LANE**—Local Area Network Emulation.

**LAP**—*See* Link Access Protocol.

**LAPB/LAPD**—Link Access Protocol for B/D channels. ISDN data-link protocol related to HDLC.

**LATA**—Local Access and Transport Area.

**Latency**—Time it takes for a packet to migrate through a network.

**LCI**—Logical Channel Identifer.

**LCGN**—Logical Channel Group Number.

**LCN**—Logical Channel Number.

**LEDs**—Light-Emitting Diodes. Semiconductor diode that emits light when it conducts.

**LEO**—Low Earth Orbit. Lower orbit sattelites.

**Lines**—SONET connection between multiplexers.

**Line Control Unit**—Controls the interface of peripheral devices to the data terminal.

**Line-of-Sight**—Infrared connection that requires the receiver to be exactly aligned with the transmitter.

**M**

**MAC**—Media Access Control. Address management protocol. Data link control across media.

**MAN**—Metropolitan Area Network. Regional network.

**Mark**—Logic 1.

**Marker or Marking State**—Logic 1 state.

**Mastergroup**—An FDM signal containing 10 super-groups (600 voice channels).

**MAU**—Media Access Unit. Controls data exchange on LLC.

**MDBS**—Mobile Data Base Station. Interface between mobile networks and other systems.

**MDIS**—Mobile Data Intermediate Station. User authentication and billing.

**MDLP**—Mobile Data-Link Protocol.

**MDS**—Mobile Data Service. Data switch and private data service.

**MDSL**—Multi-Rate Digital Subscriber Line.

**Media Access Control**—One of the four protocols of the FDDI suite.

**Medium**—The transmission path for data.

**MEO**—Medium Earth Orbit. Lower orbit sattelites.

**Message Switching**—Network switching achieved by storing incoming messages and sending them out as lines become available.

**Message Thread**—Queries and responses about a specific newsgroup topic.

**MFOTS**—Militarized Fiber-Optic TranSmission Network. High-speed military fiber optic network.

**MILNET**—MILitary NETwork.

**MIME**—Multimedia Internet Mail Extension. Multimedia format.

**Modem**—Unit that converts between digital data and analog data.

**MOSAIC**—First graphical Web browser.

**MRS**—Mobile Radio Service. Extends MTS coverage.

**MS**—Mobile Station.

**MSAT**—Mobile SATellite. Integrated services satellite network.

**MSB**—Most Significant Bit.

**MSDSL**—Multi-rate digital subscriber line.

**MSI**—Mobile Subscriber Identity. Used with remote access smart card.

**MSISDN**—Mobile Subscriber Integrated Services Digital Network. Subscriber access member.

**MSS**—Mobile Satellite Service.

**MTS**—Mobile Telephone Service. Handels voice, fax, and data services.

**MTSO**—Mobile Telephone Switching Office. Central office for a mobile phone cell.

**Multicasting**—Sending a message on the Internet to many destinations at one time.

**Multimode Fiber**—Cable with a wide acceptance cone that permits light rays to enter at many different incident angles.

**Multipath Fading**—Attenuation and distortion created by the refraction and reflection of transmitted signals.

**MUX**—multiplexer. Unit or software that allows signals to share a single channel.

**N**

**NAK**—Negative Acknowledge character.

**Natural Sampling**—Sample peaks follow actual sample signal.

**NAU**—Network Addressable Unit. SNA subunit responsible for moving data through the network.

**NDP**—Neighbor Discovery Protocol. Replaces ARP used in IPv4.

**NETBIOS**—NETwork Basic Input Output Services interface.

**Network**—A system of interconnected communication stations.

**Newsgroups**—Internet sites used for sharing information on a particular topic.

**NIC**—Network Interface Card. Interface PC to network.

**N-ISDN**—Narrowband ISDN. Narrowband ISDN service.

**NL**—New Line.

**Node**—Entry point into a network.

**NNI**—Network-to-Network Interface. Handles ATM transfers between networks.

**Normal**—Line perpendicular to the surface by which all light ray angles are measured.

**Notch Filter**—A narrow bandstop filter.

**NR**—Frame Number expected to be Received next.

**NRZ**—Non-Return-to-Zero.

**NRZ-AMI or NRZI**—Non-Return-to-Zero Alternate Mark Inversion. Digital encoding formats.

**NS**—Frame Number being Sent.

**Nyquist Frequency**—Minimum sampling frequency.

**Nyquist sample rate (SR)**—Minimum sample rate needed to replicate a signal.

**O**

**Open Systems Interconnection (OSI)**—Method by which systems are designed with general specifications to be applicable to any specific protocols and/or physical systems.

**Open Systems Interconnection (OSI) Architecture**—Open systems network model standard designed by the ISO organization.

**Optical Carriers (OC)**—SONET physical system for STS-1 signals.

**Originate Station**—The primary station responsible for originating the call prior to establishing the data link.

**Overhead**—Any nondata bits or characters sent with a transmission.

**OVRN**—Overrun. Error indication of when a computer fails to read incoming data in time.

**P**

**Packet**—A section of a message.

**Packet Switching**—A network switching technique that interleaves smaller message units called packets and sends them to their destinations.

**PAD**—Packet Assembler/Disassembler. Interface non-standard X.25 networks to X.25.

**PAM**—Pulse Amplitude Modulation.

**Parallel Data**—All the bits of a data word transferred at the same time.

**Parity or Vertical Redundancy Check (VRC)**—Error-detection (parity) and error-correction (VRC) techniques based on the odd or even count of logic ones in a transmitted character.

**Payload**—Data unit of a message transmission.

**Path**—SONET end-to-end connection.

**PBX**—Private Branch Exchange. Office telephone switching unit.

**PCM**—Pulse Code Modulation.

**PCN**—Personal Communications Network. Wireless LAN.

**PCS**—Personal Communications System. Wireless mobile telephone network.

**PCU**—Physical Control Unit.

**PDU**—Protocol Data Unit. Field that holds data payloads.

**Peer-to-Peer Bus**—Network topology with no controlling primary station.

**Permanent Virtual Circuit (PVC)**—Logical path through a network that is there once the system or network is engaged and remains there until it is shut down.

**PF**—Poll/final flag.

**PGP**—Pretty Good Privacy. 1,024-bit key encryption standard.

**Phase Lock Loop (PLL)**—Circuit that detects the difference between an incoming tone and a reference tone. It produces an error in DC voltage corresponding to the difference.

**PHY**—PHYsical protocol of the FDDI suite.

**Physical Unit (PU)**—Actual hardware that interfaces to user terminals.

**Pin**—A type of diode that has a large amount of intrinsic material between the P and N channels.

**PMD**—Physical Media-Dependent Protocol. Specifies physical properties of FDDI network.

**Polarity Violation**—Data error control for alternate mark inversion data streams when two consecutive data bits appear with the same polarity.

**Poll**—Message sent by a primary station asking if a secondary station has traffic to send.

**POP**—Post Office Protocol or Point Of Presence.

**POTS**—Plain Old Telephone System.

**PPP**—Point-to-Point Protocol. Low-level data transport protocol.

**Preamble**—Used by synchronous data streams to establish character synchronization.

**Predictive Coding**—Code developed by predicting a sample's amplitude based on the current sample.

**Primary Rate Interface (PRI)**—Defines an ISDN network service consisting of 23 or 30 B channels and one D channel.

**Primary Station**—Controlling station in a network.

**Protocol**—Set of rules for successful communication between two or more nodes in a network.

**Protocol Analyzer**—Test equipment used to monitor network activity.

**PSK**—Phase Shift Keying modulation that uses digital data to alter the phase of a carrier signal.

**PSTN**—Public Switched Telephone Network. Supplier telephone connections to ADSL llines.

**PU**—*See* Physical Unit.

**Pulse Spreading**—Signal distortion caused by different propagation times for each light ray traveling through a cable.

**PVC**—permanent virtual channel. Constantly established communication links.

**Q**

**Quadrature Amplitude Modulation (QAM)**—Phase and amplitude modulation technique.

**Quantization**—Process of approximating sample values by using fixed discrete values.

**Quantization Error or Quantization Noise**—Difference between quantized signal and sample levels.

**QoS**—Quality of Service. Level of data delivery service for reliability and speed.

**QPSK**—Quadrature Phase Shift Keying.


**R**

**RADSL**—Rate Adaptive Digital Subscriber Line.

**Raw Binary Data**—Digital data bits without any interpretation of meaning or use.

**Ray**—Unit of light traveling in a line.

**RBOC**—Regional Bell Operating Company.

**Reflection**—Return of light in a direction opposite of its original course.

**Refraction**—Bending of light ray so that it continues to travel in the same direction but at a slightly different angle.

**Refractive Index**—Value that determines the amount a light ray will be reflected or refracted when it passes from one material to another.

**Repeater**—Long distance line amplifier used to regenerate and reshape degraded signals.

**Remote or Secondary Station**—Non-controlling station in a data link.

**Restricted Token**—Reserved FDDI tokens for higher-priority stations.

**RFC**—Request For Commands. Documents that define parts of the Internet.

**RIM**—Request Initialization Mode.

**RIP**—Routing Information Protocol. Routing protocol used with TCP/IP.

**Ring**—Topology that connects all stations into a ring format. Messages are passed from one station to the next, in an orderly fashion, around the ring.

**RLSD**—Received Line Signal Detect.

**Rotation Time**—Access algorithm for the FDDI network.

**RSA**—Rivest, Shamir, and Adleman.

**RSVP**—Resources reSerVation Protocol.

**RTS**—Request To Send.

**RZ-AMI**—Return-to-Zero Automatic Mark Inverted.


**S**

**SAN**—Storage Area Network.

**SAP**—Service Access Point. Source address.

**SAR**—*See* Segmentation And Reassembly.

**SAS**—*See* Single Attached Stations.

**SASE**—Specific Application Service Element.

**SCPS**—Space Communications Protocol Standards. Set of satellite communications standards.

**SDA**—Station Device Address.

**SDH**—Synchronous Digital Hierarchy. European version of SONET.

**SDLC**—Synchronous Data-Link Control. Bit-oriented data-link protocol.

**SDSL**—Symmetric Digital Subscriber Line.

**Section**—SONET connection between devices.

**Segmentation and Reassembly (SAR)**—ATM process for dividing messages into cells and reassembling them correctly at the destination point.

**SEF**—Single Error Flag.

**Selection**—Message sent by a primary asking a secondary station if it is ready to receive traffic.

**Selection Process**—Primary message inquiring if a secondary is ready to receive traffic.

**Serial Data**—Digital data transferred one bit at a time.

**Serial Interface Adapter (SIA)**—Manchester encoder/decoder chip.

**Serial Line Interface Protocol**—Loss or addition of a bit due to a skip in sampling time.

**Service Node**—Manages ISDN system access control functions.

**Server**—Protocol translation between elements on a network.

**Session**—Functional activity between two network nodes.

**SET**—Secure Electronic Transactions. Authenticates credit cards.

**SGML**—Standard Generalized Mark up Language.

**Shannon Channel Capacity**—Relates data rate to bandwidth and bits per symbol.

**Signal-to-noise ration (SNR)**—A ratio of the level of a signal compared to the level of the noise in the same circuit.

**SIM**—Set Initialization Mode command or Subscriber Identity Module. Global mobile telephone smart card system.

**Simplex**—Transmission of data in one direction only.

**Single attached station**—Station that is normally connected only to the primary ring of a dual ring network or interfaced through a concentrator.

**Single Mode Fiber**—Cable with a narrow acceptance cone that concentrates light travel through the core's center.

**SLIP**—Serial Line Interface Protocol.

**Slope Overload**—Large difference between original signal and replicated delta-modulated signal.

**Smart Card**—Used to authorize access to a mobile phone network.

**SMT**—Station ManagemenT.

**SMTP**—Simple Mail Transfer Protocol. E-mail protocol.

**SNA**—*See* Systems Network Architecture.

**Snell's Law**—Defines the relationship between the ratio of critical and incident angles to the ration of refractive indexes between two meeting surfaces.

**SNR**—*See* Signal-to-Noise Ratio.

**SNRM**—Set Normal Response Mode command.

**SOH**—Start Of Heading.

**SONET**—Synchronous optical network. Fiber optic transport network.

**SPA**—Station Polling Address.

**Space**—Logic 0.

**SPN**—Subscriber's Premises Network. Multiplex broadband ISDN service.

**Spot Beam**—Satellite transmission that covers a given area on the earth.

**SPS**—Symbols Per Second.

**SSA**—Station Selection Address.

**SSAP**—Source Service Access Point. End points in communication link.

**SSL**—Secure Sockets Layer. Security protocol to verify certificate of authority issuer.

**Standing Waves**—Echo created by impedance mismatch between sending and receiving terminations.

**Standing Wave Ratio (SWR)**—Measure of quality of a microwave transmission line.

**Star**—Topology that uses a central controlling station called a hub to direct traffic on the network.

**STATDM**—STATistical Time Division Multiplex.

**Station**—Communications equipment.

**STAT MUX**—STATistical MUltipleXer.

**STDM**—Synchronous Time Division Multiplex.

**Step Index Core**—Fiber core that has only one refractive index.

**STS**—Synchronous Transport Signal. SONET multiplexing function.

**STX**—Start of TeXt.

**Supergroup**—60-channel FDM signal.

**SUSE**—Specific User Service Element.

**SVC**—Switched Virtual Circuit. Communication links that are established when needed.

**SWR**—*See* Standing Wave Radio.

**Symbol**—An electrical parameter used to represent one or more data bits.

**SYN**—SYNchronization character.

**Synchronous Data**—Serial data that requires a synchronizing clock signal between sender and receiver.

**Switched Virtual Circuit**—Virtual path through a network that is created when it is needed as a message is being transmitted.

**Symbol**—A single electrically measurable signal, which represents groups of one or more data bits.

**Symbol Rate**—Rate at which symbols are transmitted in symbols per second SPS.

**Systems Network Architecture (SNA)**—IBM's open systems architecture model for network communications.

**T**

**TA**—*See* Terminal Adapter.

**Tandem Switch**—Switch station that interconnects several local switch stations and/or toll stations.

**TCP**—Transfer Control Protocol.

**TDM**—Time Division Multiplex. Multiplexing technique that assigns channel signals to time slots.

**TDMA**—Time Division Multiple Access.

**TE**—*See* Terminal Equipment.

**TELCO**—TELephone COmpany.

**TEP**—Terminal End Point.

**Terminal Adapter (TA)**—ISDN unit used for converting non-ISDN messages into ISDN format.

**Terminal Equipment (TE)**—ISDN physical user network entry equipment. TE may also be used for Terminal Endpoint.

**Throughput**—Rate at which information is transferred from one point to another.

**TMSI**—Temporary Mobile Subscriber Identification. Actual number on the airwave.

**Token**—Short message that precedes transmission of data packets on a token ring or bus network.

**Toll Station**—Long distance switch station.

**Topology**—Physical composition of a network. Comes in three basic forms: bus, ring, or star (hub).

**TPDU**—*See* Transport Protocol Data Unit. TCP data unit frame.

**Traffic**—Data messages.

**Transfer ID**—XID. Destination address.

**Transmission Valid Time**—Restricts the length of time a FDDI station can send a packet.

**Transponder**—Satellite repeater that receives at one frequency and transmits at another.

**Transport Protocol Data Unit (TPDU)**—TCP data frame.

**Trunk Line**—Long distance telephone lines.

**TTL**—Transistor Transistor Logic.

## U

**UART/USART**—Universal Asynchronous/(Synchronous) Receiver Transmitter.

**UBR**—Unspecified Bit Rate. Defines amount and type of traffic a channel will handle.

**UDP**—User Datagram Protocol. Application for sending unacknowledged messages.

**UI**—Unnumbered Information frame.

**UNI**—User-Network Interface. Handles ATM transfers between user equipment.

**UNICODE**—Current character code used in data communications.

**UNZ/URZ**—Unipolar Non-Return-(return)-to-Zero.

**Universal Resource Locators (URL)**—Internet address.

**Unrestricted token**—FDDI free token that can be captured by any station on the ring.

**URL**—*See* Universal Resource Locator.

**Uplink**—Transmission from Earth to satellite.

**USENET**—Maintains newsgroup servers.

**UTP**—Unshielded Twisted Pair.

## V

**VBR**—Variable Bit Rate. Defines amount and type of traffic a channel will network.

**VCI**—Virtual Channel Identifier. ATM end point address.

**VCO**—Voltage-Controlled Oscillator is a circuit that produces an AC signal whose frequency is determined by the DC voltage at its input.

**VDSL**—Very high rate Digital Subscriber Line.

**VERONICA**—Very Easy Rodent-Oriented Netwide Index to Computerized Archives—Database of GOPHER documents.

**VGM**—Voice Grade Medium. Cable that transports digitized voice.

**Virtual Channel and Path Identifiers**—ATM network and channel addresses.

**Virtual Tributary (VT)**—Manage lower data-rate services on SONET.

**VLR**—Visitor Location Register. Identifies where a user is or has been.

**VPI**—Virtual Path Identifier.

**VRC**—*See* Parity or Vertical Redundancy Check.

**VSAT**—Very Small Aperture Terminals. Limited satellite network.

## W

**WAIS**—Wide Area Information Service. Internet location search engine.

**WAN**—Wide Area Network. Large section of a country network.

**Wander**—Low-frequency jitter in a delta modulator.

**WATS**—Wide Area Telephone Service.

**Wave Guide**—Transmission medium for microwave signals.

**WDM**—Wave Division Multiplexing. Multiplex method to place several fiber optic channels on to a single line.

**Weighted Noise**—Noise generated when a signal is applied to a circuit.

**Wiring Concentrator**—Hub for centralizing a number of single attached stations.

**Words**—Fixed number of data bits.

**WWW**—World Wide Web. Network for exchanging data.

## X

**XID**—eXchange IDentification.

**XML**—Extensible Mark up Language.

# APPENDIX B

# Complete Extended ASCII Code Chart

| HEX | Dec | ASCII | Print | HEX | Dec | ASCII | Print |
|-----|-----|-------|-------|-----|-----|-------|-------|
| 00 | 00 | NULL | (null) | 20 | 32 | Space | ¶ |
| 01 | 01 | SOH | ☺ | 21 | 33 | ! | ! |
| 02 | 02 | STX | ☻ | 22 | 34 | " | " |
| 03 | 03 | ETX | ♥ | 23 | 35 | # | # |
| 04 | 04 | EOT | ♦ | 24 | 36 | $ | $ |
| 05 | 05 | ENQ | ♣ | 25 | 37 | % | % |
| 06 | 06 | ACK | ♠ | 26 | 38 | & | & |
| 07 | 07 | BELL | ● | 27 | 39 | ' | ' |
| 08 | 08 | BKSP | ◘ | 28 | 40 | ( | ( |
| 09 | 09 | HTAB | ○ | 29 | 41 | ) | ) |
| 0A | 10 | LNFD | ◙ | 2A | 42 | * | * |
| 0B | 11 | VTAB | ♂ | 2B | 43 | + | + |
| 0C | 12 | FMFD | ♀ | 2C | 44 | , | , |
| 0D | 13 | CRET | ♪ | 2D | 45 | – | – |
| 0E | 14 | SHOUT | ♫ | 2E | 46 | . | . |
| 0F | 15 | SHIN | ☼ | 2F | 47 | / | / |
| 10 | 16 | DLE | ► | 30 | 48 | 0 | 0 |
| 11 | 17 | DC1 | ◄ | 31 | 49 | 1 | 1 |
| 12 | 18 | DC2 | ↕ | 32 | 50 | 2 | 2 |
| 13 | 19 | DC3 | ‼ | 33 | 51 | 3 | 3 |
| 14 | 20 | DC4 | ¶ | 34 | 52 | 4 | 4 |
| 15 | 21 | NACK | § | 35 | 53 | 5 | 5 |
| 16 | 22 | SYNC | – | 36 | 54 | 6 | 6 |
| 17 | 23 | ETB | ↨ | 37 | 55 | 7 | 7 |
| 18 | 24 | CAN | ↑ | 38 | 56 | 8 | 8 |
| 19 | 25 | ENDM | ↓ | 39 | 57 | 9 | 9 |
| 1A | 26 | SUB | → | 3A | 58 | : | : |
| 1B | 27 | ESC | ← | 3B | 59 | ; | ; |
| 1C | 28 | FLSP | ∟ | 3C | 60 | < | < |
| 1D | 29 | GPSB | ↔ | 3D | 61 | = | = |
| 1E | 30 | RDSP | ▲ | 3E | 62 | > | > |
| 1F | 31 | UNSP | ▼ | 3F | 63 | ? | ? |

| HEX | Dec | ASCII | Print | HEX | Dec | ASCII | Print |
|-----|-----|-------|-------|-----|-----|-------|-------|
| 40 | 64 | @ | @ | 60 | 96 | ` | ` |
| 41 | 65 | A | A | 61 | 97 | a | a |
| 42 | 66 | B | B | 62 | 98 | b | b |
| 43 | 67 | C | C | 63 | 99 | c | c |
| 44 | 68 | D | D | 64 | 100 | d | d |
| 45 | 69 | E | E | 65 | 101 | e | e |
| 46 | 70 | F | F | 66 | 102 | f | f |
| 47 | 71 | G | G | 67 | 103 | g | g |
| 48 | 72 | H | H | 68 | 104 | h | h |
| 49 | 73 | I | I | 69 | 105 | i | i |
| 4A | 74 | J | J | 6A | 106 | j | j |
| 4B | 75 | K | K | 6B | 107 | k | k |
| 4C | 76 | L | L | 6C | 108 | l | l |
| 4D | 77 | M | M | 6D | 109 | m | m |
| 4E | 78 | N | N | 6E | 110 | n | n |
| 4F | 79 | O | O | 6F | 111 | o | o |
| 50 | 80 | P | P | 70 | 112 | p | p |
| 51 | 81 | Q | Q | 71 | 113 | q | q |
| 52 | 82 | R | R | 72 | 114 | r | r |
| 53 | 83 | S | S | 73 | 115 | s | s |
| 54 | 84 | T | T | 74 | 116 | t | t |
| 55 | 85 | U | U | 75 | 117 | u | u |
| 56 | 86 | V | V | 76 | 118 | v | v |
| 57 | 87 | W | W | 77 | 119 | w | w |
| 58 | 88 | X | X | 78 | 120 | x | x |
| 59 | 89 | Y | Y | 79 | 121 | y | y |
| 5A | 90 | Z | Z | 7A | 122 | z | z |
| 5B | 91 | [ | [ | 7B | 123 | { | { |
| 5C | 92 | \ | \ | 7C | 124 | \| | \| |
| 5D | 93 | ] | ] | 7D | 125 | } | } |
| 5E | 94 | ^ | ^ | 7E | 126 | ~ | ~ |
| 5F | 95 | _ | _ | 7F | 127 | DEL | ⌂ |

| HEX | Dec | ASCII | HEX | Dec | Print · | HEX | Dec | ASCII |
|-----|-----|-------|-----|-----|---------|-----|-----|-------|
| 80 | 128 | ç | AB | 171 | ½ | | | |
| 81 | 129 | ü | AC | 172 | ¼ | D6 | 214 | ╥ |
| 82 | 130 | ē | AD | 173 | ¡ | D7 | 215 | ╫ |
| 83 | 131 | â | AE | 174 | << | D8 | 216 | ╪ |
| 84 | 132 | ä | AF | 175 | >> | D9 | 217 | ┘ |
| 85 | 133 | ā | BO | 176 | ░ | DA | 218 | ┌ |
| 86 | 134 | à | B1 | 177 | ▒ | DB | 219 | █ |
| 87 | 135 | ç | B2 | 178 | ▓ | DC | 220 | ▄ |
| 88 | 136 | ê | B3 | 179 | │ | DD | 221 | ▌ |
| 89 | 137 | ë | B4 | 180 | ┤ | DE | 222 | ▐ |
| 8A | 138 | ē | B5 | 181 | ╡ | DF | 223 | ▀ |
| 8B | 139 | ï | B6 | 182 | ╢ | E0 | 224 | ∂ |
| 8C | 140 | î | B7 | 183 | ╖ | E1 | 225 | β |
| 8D | 141 | ī | B8 | 184 | ╕ | E2 | 226 | τ |
| 8E | 142 | Ä | B9 | 185 | ╣ | E3 | 227 | π |
| 8F | 143 | À | BA | 186 | ║ | E4 | 228 | Σ |
| 90 | 144 | Ē | BB | 187 | ╗ | E5 | 229 | σ |
| 91 | 145 | æ | BC | 188 | ╝ | E6 | 230 | μ |
| 92 | 146 | Æ | BD | 189 | ╜ | E7 | 231 | Υ |
| 93 | 147 | ô | BE | 190 | ╛ | E8 | 232 | Φ |
| 94 | 148 | ö | BF | 191 | ┐ | E9 | 233 | θ |
| 95 | 149 | ō | C0 | 192 | └ | EA | 234 | Ω |
| 96 | 150 | û | C1 | 193 | ┴ | EB | 235 | δ |
| 97 | 151 | ū | C2 | 194 | ┬ | EC | 236 | ∞ |
| 98 | 152 | ÿ | C3 | 195 | ├ | ED | 237 | φ |
| 99 | 153 | ö | C4 | 196 | ─ | EE | 238 | ∈ |
| 9A | 154 | ü | C5 | 197 | ┼ | EF | 239 | η |
| 9B | 155 | ¢ | C6 | 198 | ╞ | F0 | 240 | ≡ |
| 9C | 156 | £ | C7 | 199 | ╟ | F1 | 241 | ± |
| 9D | 157 | ¥ | C8 | 200 | ╚ | F2 | 242 | ≥ |
| 9E | 158 | ₧ | C9 | 201 | ╔ | F3 | 243 | ≤ |
| 9F | 159 | ƒ | CA | 202 | ╩ | F4 | 244 | ⌠ |
| A0 | 160 | ā | CB | 203 | ╦ | F5 | 245 | ⌡ |
| A1 | 161 | ī | CC | 204 | ╠ | F6 | 246 | ÷ |
| A2 | 162 | σ | CD | 205 | ═ | F7 | 247 | ≈ |
| A3 | 163 | ū | CE | 206 | ╬ | F8 | 248 | ° |
| A4 | 164 | ñ | CF | 207 | ╧ | F9 | 249 | • |
| A5 | 165 | Ñ | D0 | 208 | ╨ | FA | 250 | · |
| A6 | 166 | a̲ | D1 | 209 | ╤ | FB | 251 | √ |
| A7 | 167 | o̲ | D2 | 210 | ╥ | FC | 252 | n |
| A8 | 168 | ¿ | D3 | 211 | ╙ | FD | 253 | z |
| A9 | 169 | ⌐ | D4 | 212 | ╘ | FE | 254 | • |
| AA | 170 | ¬ | D5 | 213 | ╒ | FF | 255 | blank |

# APPENDIX **C**

# Network Timeline

1835—Morse invents telegraph—first data communications equipment.

1849—First telegraph printer.

1850—Western Union.

1860—First "high"-speed printers at 15 bps.

1861—First coast-to-coast telegraph.

1874—Thomas Edison develops method for full duplex signaling on a single wire. This leads to the development of teletype systems.

1876—Bell invents telephone—Communications.

1877—Bell Telephone Company.

1881—First long distance trunk lines between Boston, MA, and Providence, Rhode Island.

1885—AT&T.

1891—Alexander Graham Bell sent signal 200 meters using sunlight as carrier.

1898—Marconi wireless telegraphy.

1910—Lightguide—Hollow tube internally coated with highly reflective material.

1921—First AM radio broadcasts—KDKA, Pittsburgh.

1931—Teletype at 100 baud rate.

1934—FCC formed.

1936—FM radio began.

1940—Bell Labs connects TTY machine to electronic calculator, forming the beginning of an electronic computer.

—ENIAC, MARK I, and II computers.

1947—Transistor invented.

1950—Time division multiplexing (TDM) applied to telephony.

1951—UNIVAC computer by Remington Rand.

1953—First IBM Computer.

1956—Optical fibers.

1957—Sputnik 1.

—Creation of Advanced Research Projects Agency (ARPA).

1958—First coast-to-coast microwave radio link in Canada.

—Explorer 1—the space race is on.

—MIT develops first time sharing for computers.

1959—Integrated Circuits arrive.

—First Mini-computer—PDP-1.

—Short-lived SCORE communication satellite.

1960—Ruby Rod Laser invented by ATT Bell Labs.

1961—Gas laser.

—T1 lines.

—Leonard Kleinrock develops packet switching.

1962—Satellite communications begin.

—COMSAT organized.

—Solid-state laser.

—Pulse code modulation (PCM).

1964—IBM 360 mainframe.

1965—First commercial communications satellite—Intelsat 1 or Early Bird.

—Theodore Nelson coins the term hypertext.

—Larry Roberts and Thomas Merrill develop first WAN.

1966—Fiber optic cable.

1967—Token Ring.

—Lawrence Roberts paper on ARPANET.

1968—FCC Rule 67 allowing anyone to connect to AT&T phone system.

—ARPANET commissioned—UCLA first host.

1969—Brian W. Kernighan and Dennis M. Ritchie develop UNIX. Advanced Research Projects Agency and Bolt, Baraulk, and Newman (BBN) develop a packet-switched network, called ARPANET to interconnect sites for the purpose of:

a)  remote login (TELNET)

b)  file transfer

c)  remote printing

It opens at 50 Kbps with four hosts: UCLA, UCSB, University of Utah, and Stanford Research Institute.

—Data access arrangement—to allow equipment to be connected to the telephone company lines.

—Leased lines allowed by FCC.

—MCI microwave link between Chicago and St. Louis.

1970—Large-scale integration ICs—the microprocessor.

—Corning Glass Works produce fiber optic cables.

—10 ARPANET hosts.

—Robert Metcalfe and CSMA/CD.

—Denis Ritchie and Kenneth Thompson develop UNIX.

—First automatic teller machine (ATM).

1971—INTEL 4004 microprocessor.

—23 ARPANET hosts.

—8″ Floppy Disk (IBM).

1972—INTEL 8008 microprocessor.

—Roy Tomlinson puts first E-mail program on ARPANET.

—37 ARPANET sites.

—HP3S Hand Held Calculator.

1973—Robert Metcalfe (Xerox) and David Boggs: Ethernet at 2.944 Mbps.

—TCP/IP proposed as standard for ARPANET.

—File transfer protocol (FTP).

—First international ARPANET hosts: University College of London and Royal Radar Establishment of Norway.

—Dataphone digital service.

—Packet communications.

1974—Vint Cerf and Bob Kahn: Protocol for Packet Network intercommunication, which becomes TCP/IP in 1976.

—IBM SNA: system network architecture.

—Domain name service (DNS) for ARPANET addressing.

—CP/M disk operating system.

1975—ALTAIR 8800 first desktop PC.

—Bill Gates and Paul Allen (from Dartmouth College) form Microsoft to develop BASIC for a new PC.

—Satellite Business Systems (SBS) formed.

—Laser printer.

—TELNET.

1976—Ethernet LAN.

—Whitfield Diffle and Martin Hellman: public key cryptography.

—Steve Jobs and Stephen Wozniak found Apple.

—X.25 packet switch standard released by CCITT (now ITU).

—First e-mail is sent to Queen Elizabeth of England.

—CRAY 1 supercomputer.

1977—Datapoint's ARCnet 2.5 Mbps PC LAN.

—User DATAGRAM protocol.

—IBM DES encryption with 8-bit key.

—Short haul fiber cables.

—TRS-80 first personal computer with keyboard and CRT screen by Radio Shack.

—111 ARPANET hosts.

—Hayes 300-bps Modem.

—Commodore PET PC.

1978—ISO ratifies X.25 specification.

—Commercial e-mail over ARPANET.

—Open Systems Integration (OSI) model.

—First long-haul digital radio system by Bell Northern Labs.

—APPLE II PC.

—51/4≤ disk.

—8086/8088 Microprocessor by Intel.

1979—Xerox Ethernet LAN.

—USNET (NetNews) newsgroup on Internet.

—CompuServe begins as Micronet.

—Onyx System releases first commercial UNIX system.

—Visicalc spreadsheet.

—Wordstar word processor.

—68000 microprocessor by Motorola.

1980—ENQUIRE program to access information from ARPANET.

—10-Mbps Ethernet standard.

—Seagate hard drive.

—PC DOS selected for IBM PC.

1981—IBM PC.
   —Hayes Smartmodem.
   —10-Mbps Ethernet.
   —Multimode fiber cable.
   —Sir Clive Sinclair introduces Sinclair ZX81 Computer.
   —Osborn Portable Computer.
   —BITNET—because it's time network.
   —CSNET—computer screen network.
1982—Compaq and Sun founded.
   —IBM token ring.
   —Breakup of AT&T—baby Bells.
   —First VSAT service.
   —First Novell PC LAN.
   —Internet term coined.
   —3-COM ships Ethernet adapter boards.
   —Adobe Systems founded.
1983—Apple Lisa with first GUI.
   —500 Internet sites.
   —Military part of Internet splinters off to form MILNET.
   —TCP/IP adopted as Internet standard protocol.
   —CD-ROM.
   —Tandy 100 Laptop.
   —Compaq IBM Compartibles.
   —Routers.
   —Cellular phones.
   —First computer virus infection.
   —IEEE approves 802.3 for thick coax.
   —Radia Perlman: 802.1 Spanning Tree Algorithm.
   —Lotus 123 Spreadsheet.
1984—1,000 Internet Users.
   —Macintosh GUI.
   —Domain Name Service (DNS).
   —MCI Mail to supply e-mail services.
   —IBM PC Network LAN.
   —William Gibson coins the term cyberspace.
   —Netware 1.0.
   —C++ first release.
   —Ethernet bridge.
   —31/2≤ 720K floppy disk.
1985—Windows 1.0—Microsoft/IBM answer to Apple's GUI.
   —Netware 286.
   —Apple talk.

   —IBM 4-Mbps token ring.
   —IEEE 802.3 approves thin coax for Ethernet.
   —Prelude: first ATM network at 280 Mbps.
   —2400-bps modem.
   —Adobe PageMaker.
1986—Network General: first LAN protocol analyzer, the Sniffer.
   —First ISDN service from Illinois Bell for MacDonalds.
   —National Science Foundation (NSF) develops own IP type network, linking 5 centers but could not afford to link to all universities. This leads to first regional Internet services provider (ISP) to supply Internet connections to a user.
1987—9600-bps modem.
   —10,000 Internet hosts.
   —Quattro Spreadsheet.
   —IBM PS/2.
   —National Science Foundation develops NSFNET enabling organizations to access Internet without meeting defense-related criteria.
   —ANSI ratifies FDDI standard.
   —First Ethernet Hub.
1988—10BaseT.
   —16-Mbps IBM Token Ring.
   —Department of Defense adopts OSI, claims TCP/IP is interim protocol.
   —Internet WORM virus knocks out 6,000 users—results in formation of CERT (Computer Emergency Response Team).
   —Ethernet on UTP cable.
   —Steve Jobs—NExT Computer.
   —88000 RISC Microprocessor.
1989—Netware 386.
   —T-3 Muxes.
   —Lotus Notes.
   —100,000 Internet hosts.
   —Marcus Ranum: firewalls.
   —SNMP.
1990—IBM/Motorola form ARDIS.
   —Frame relay.
   —Ethernet switches.
   —FDDI on UTP.
   —McGill University introduces ARCHIE as FTP search engine.

—Commercial dial-up Internet access.

—ARPANET officially decommissioned and renamed Internet.

—300,000 Internet users.

—Tim Bemers and Robert Callian: World Wide Web (WWW).

—Windows 3.0.

1991—Lotus buys cc:Mail for e-mail service.

—Motorola's Altair—first wireless LAN.

—ATM: asynchronous transfer mode.

—Windows NT.

—RAID: redundant array of interfaced drives.

—Paul Linder and Mark McCahill at University of Michigan introduce GOPHER.

—WAIS: Wide Area Information Services.

—AMD 386 microprocessor gives INTEL competition.

1992—1,000,000 Internet users.

—HTML, (HyperText Mark up Language) for web building introduced.

1993—Netware 4.0.

—Marc Andressen and Eric Bina: MOSAIC, first graphical browser: 40,000–50,000 users in first year.

—Clipper Chip.

—Token ring switch.

—VERONICA added as search engine.

—Ethernet on fiber-optic cable.

—Pentium microprocessor.

—MOSAIC, first graphic browser.

1994—28.8-Kbps V.34 modems.

—Fast Ethernet.

—Microsoft NT Server 3.5.

—Department of Defense drops OSI-only requirement.

—Marc Andreesen and Jim Clark—Netscape Browser.

1995—Windows 95.

—SATAN: Security Analysis Tool for Auditing Networks.

—6,000,000 Internet users.

—Sun introduces JAVA.

1996—5.4% of US population uses the net.

—Wave division multiplexing.

—IP switch.

—Gigabit Ethernet.

—9,500,000 Internet hosts.

—Apple IMAC Computer.

1997—11% of US population (1% of world population) account for 30–50% traffic on local telephone carriers to use the Internet.

—70,000 discussion groups on the Internet.

—16,500,000 Internet users.

—150 countries connected via the Internet.

—Telecommunications Reform Act.

—622 Mbps packet over SONET.

—Supreme Court declares Communications Decency Act of 1996 unconstitutional.

—Browser wars heat up—free Netscape 3.0 and Internet Explorer 3.0.

—56K Modems.

—Cable modems.

—Iridium and Teledesic low earth orbit (LEO) satellite systems proposed, funded, and begun.

—DHTML (dynamic HTML) and XML (extensive mark up languages for handling 3D animations and virtual reality for web pages..

1998—Windows 98.

—Netscape 4.0 and Internet Explorer 4.0: both will need some fixes later in the year.

—Dragon Software introduces voice recognition packages that allow you to control your PC and dictate text. Also allows correction by voice commands. Hello HAL! (2001: A Space Odyssey).

—E-commerce: electronic business over the Internet.

—Many "doomsday" voices speak out about Y2K problems.

1999—Pentium III breaking 500-MHz clock barrier.

—Internet Explorer 5.

—America Online acquires Netscape.

—Gigabit Ethernet on UTP cable.

—Y2K overhype—how easy it is to panic people!

# APPENDIX **D**

# Facsimile

---

## D.1 INTRODUCTION TO FACSIMILE

**Facsimile (FAX)** is an old technology that has come of age. It involves a process of digitizing documents so that they can be sent through a network or through the public telephone system to be exactly reproduced at the receiving end. The quality of the reproduction depends on many factors, including the efficiency of the facsimile (fax) equipment, and the dependability of the network used to transmit the digital information. Today there are stand-alone fax machines that are interfaced to personal computers (PC) and the telephone lines, fax cards that plug inside of a PC; and other devices that include technological improvements over the original fax machines and systems.

---

## D.2 TRANSFERRING ENTIRE DOCUMENTS

The idea of scanning a full document and transferring it, in its entirety, from one location to another is not new. As early as 1842, a patent for such a scheme was awarded to Alexander Bain. Although Bain's machine had little resemblance to facsimile machines of today, the basic concepts of optical scanning that he introduced are in use today. Practical methods of transmitting pictures over long distances were used during World War II and after by major newspaper wire services to carry daily photographic records of the events of the day between Europe, Asia, and America. Successful transfer of an image from source to destination requires several functional elements, including optical scanning, data conversion for transmission and reception, and the ability to produce a hard copy of the image at the receiving end.

The European and Asian communities adopted fax for the purpose of exchanging business documents as well as news reports in great numbers following the Second World War. For some unknown reason, the United States was slow in adopting the technology. It wasn't until personal computers were firmly established as a facet of American life that facsimile use began to take hold in the States. However, once the idea of faxing all kinds of information became popular in the United States, it took off like the wind. It has only been the more amazing acceptance of the Internet for sharing information that has stopped the excessive use of faxing.

## D.3 SCANNING METHODS

**drum scanning**
form of optical scanning that uses a helix groove in a drum to detect reflected light from a document.

**flatbed scanning**
image is scanned as it is moved across a flat surface.

**frequency modulation (FM)**

The main purpose of scanning a document is to translate that document into a digital form so that it can be sent, using data communications and networking technology, to a remote site. Different intensities of light reflected from the documents page generate an analog signal, which is either used to modulate an analog fax transmitter or digitized and sent to a fax modem to be modulated and transmitted through the telephone lines to an end user. A number of methods have been used to optically scan a document. For fax operation, two basic methods are used, **drum scanning** and **flatbed scanning.** With a drum scanner, a cylindrical drum, 19 inches in length, is coated with a black opaque finish. A helix groove is cut from one end of the drum to the other as illustrated in Figure D-1. Mirrors and lens are placed at the drum's center to direct the light entering through the groove toward a photoelectric cell, which converts the light energy to electrical current.

The drum is rotated at 120 revolutions per minute (rpm). Light projected onto the document face is reflected and directed through a lens into the helix groove as the drum rotates. The document is fed through the scanner by means of rollers driven at 60 cycles per second as shown in Figure D-2. The combination of the drum rotation and document feed causes the reflected light to represent a single line of the document. Since the movement is constant, there is no overlapping between lines. The received light that has been directed through the drum's center is converted to electrical current, amplified, and caused to modulate a 2,400-Hz carrier. For an analog system, **frequency modulation (FM)** is used,

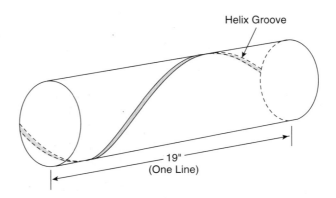

**FIGURE D-1** Drum Scanner

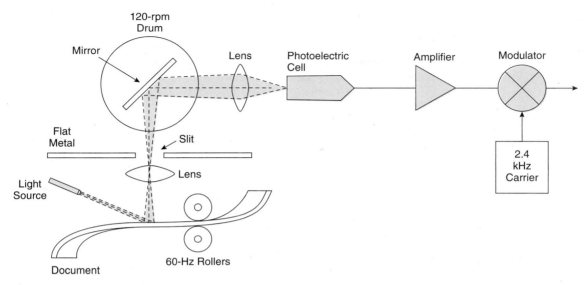

**FIGURE D-2** Facsimile Drum Scanner

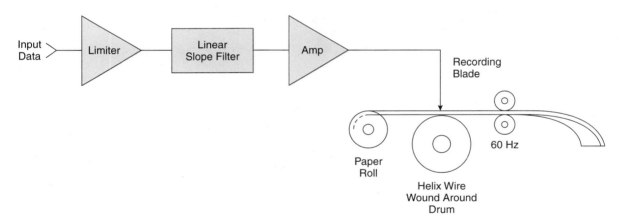

**FIGURE D-3** Drum Printer

producing a range of frequencies between 1,500 and 2,300 Hz (lower sideband), depending on the strength of the light.

The earlier analog fax machines transferred a document at a rate of 96 lines per minute using the drum scanner method. At the receiver side (Figure D-3), the frequency modulated signal is demodulated and filtered to assure that the 1,500–2,300 Hz signals alone are passed through the receiver. A linear slope filter is employed to pass the 1,500-Hz signal at a strength ten times that of the 2,300-Hz signal. The lower frequency represents black and the higher, white. Frequencies in between are varying shades of gray. The stronger signal is required to assure that the black area will be imprinted on the hard copy at the printer. Once again, the hard copy paper is pulled through the printer at a sixty cycle per second rate. A nonconductive drum with a wire wound in helix fashion from end to end, rotates at 120 rpm. A recording blade, which is stationary, is positioned on one side of the paper, as shown in Figure D-3. A voltage between V (white) and 60V (black) is placed between the helix wire and the blade. The level of the voltage depends on the strength of the signal from the slope filter. The hard copy paper is made of a special electrosensitive material. The stronger the voltage between the wire and blade, the darker the print becomes. The rotation of the helix on the drum assures a left-to-right traveling contact point across the paper. The forward motion of the rollers feeds the paper at the 96 lines per minute rate of the scanner.

Drum scanners were developed for analog fax machines, but few of that type are in use today. With the advent of digital fax machines, a second method called flatbed scanning has evolved. The scanned document is pushed or pulled across an aperture slot in a flatbed metal plate (Figure D-4). A light source is projected at an angle through the slot and is reflected from the document's surface back through the slot straight down to a lens surface. The reflected light is directed by mirrors and lenses to a **charge coupled device (CCD),** which produces an electrical current in proportion to the amount of light applied to it. Since the scanner is stationary, the number of moving parts is greatly reduced. The analog electrical information is converted into a digital code and sent to its destination. At the destination, the digital codes are reassembled into data required for any number of different printers or terminal displays.

> **charged coupled device (CCD)**
> converts light to electrical energy.

### Section D.3 Review Question

1. Which scanning method
   a. Is used for analog fax machines?
   b. Uses frequency modulation?
   c. Has a motionless bed?
   d. Projects light through a helix?
   e. Uses a recording blade and charge coupled devices?

## D.4 FAX STANDARDS

The International Telecommunications Union (ITU) standards for facsimile are known as *Recommendations T.2 and T.3* for analog fax systems, and *T.4 and T.30* for digital systems. T.2 or Group I specifications use frequency modulation and a document rate of 6 minutes per page. The resolution of the copy is 96 lines per inch, with a scan rate of 180 lines per minute. The other analog recommendation, T.3 or Group II, uses **amplitude modulation (AM),** decreasing the document rate to 3 minutes per page. Resolution remains the same, but the scan rate is doubled to 360 lines per minute.

> **amplitude modulation (AM)**

Both analog methods are considered slow when compared to digital faxes. At 3 or 6 minutes per page, the time to reproduce several pages would tie up a data system for a

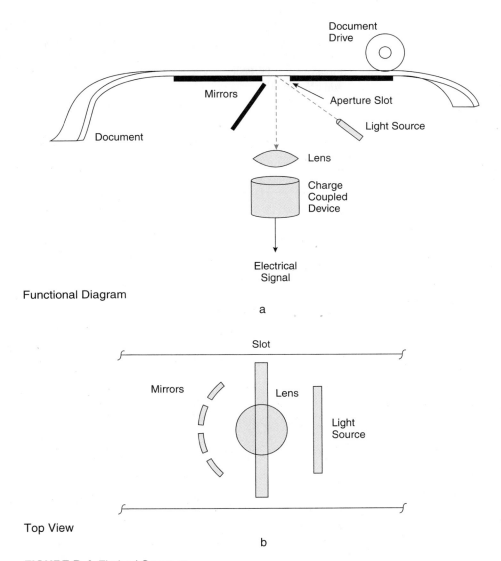

Functional Diagram

a

Top View

b

**FIGURE D-4** Flatbed Scanner

considerable length of time. Digital fax systems were developed to decrease document time while increasing resolution. The idea is to convert the analog signals representing the strength of the reflected light into digital pulses. A digital logic level of 1 represents black and 0 represents white. Thus a train of digital pulses is created representing the light level of a sampled spot. A specific set of consecutive 1s and 0s determines each gray level of scale between black and white. The method used creates different code lengths (number of bits per sample), each with varying numbers of 1s and 0s.

**Huffman code**
data transmission efficiency
code that uses shorter codes for
more frequently transmitted bit
combinations.

To increase the rate of transmitting this information, the digital pulse trains are encoded using a modified **Huffman code** (Table D-1). The purpose of using the code is to compress the actual number bits sent from the source to the destination. Notice on Table D-1 that for most of the conversion, there are fewer Huffman code bits than the original count of white and black bits.

## TABLE D-1

### Huffman Code Set

| Number of Bits | White Bits | | Black Bits | |
| --- | --- | --- | --- | --- |
| | Huffman Code (In Hex) | Number of Code Bits* | Huffman Code (In Hex) | Number of Code Bits* |
| 0 | 35 | 8 | 037 | 10 |
| 1 | 07 | 6 | 2 | 3 |
| 2 | 7 | 4 | 3 | 2 |
| 3 | 8 | 4 | 2 | 2 |
| 4 | B | 4 | 3 | 3 |
| 5 | C | 4 | 3 | 4 |
| 6 | E | 4 | 2 | 4 |
| 7 | F | 4 | 2 | 5 |
| 8 | 13 | 5 | 05 | 6 |
| 9 | 14 | 5 | 04 | 6 |
| 10 | 07 | 5 | | |
| 11 | 08 | 5 | 05 | 7 |
| 12 | 08 | 6 | 07 | 7 |
| 13 | 03 | 6 | 04 | 8 |
| 14 | 34 | 6 | 07 | 8 |
| 15 | 35 | 6 | 018 | 9 |
| 16 | 2A | 6 | 017 | 10 |
| 17 | 2B | 6 | 018 | 10 |
| 18 | 27 | 7 | 008 | 10 |
| 19 | 0C | 7 | 067 | 11 |
| 20 | 08 | 7 | 068 | 11 |
| 21 | 17 | 7 | 06C | 11 |
| 22 | 03 | 7 | 037 | 11 |
| 23 | 04 | 7 | 028 | 11 |
| 24 | 28 | 7 | 017 | 11 |
| 25 | 2B | 7 | 018 | 11 |
| 26 | 13 | 7 | 0CA | 12 |
| 27 | 24 | 7 | 0CB | 12 |
| 28 | 18 | 7 | 0CC | 12 |
| 29 | 02 | 8 | 0CD | 12 |
| 30 | 03 | 8 | | |

*(Continued)*

**TABLE D-1**

**Huffman Code Set**—*Continued*

| Number of Bits | White Bits | | Black Bits | |
| --- | --- | --- | --- | --- |
| | Huffman Code (In Hex) | Number of Code Bits* | Huffman Code (In Hex) | Number of Code Bits* |
| 31 | 1A | 8 | 069 | 12 |
| 32 | 1B | 8 | 06A | 12 |
| 33 | 12 | 8 | 06B | 12 |
| 34 | 13 | 8 | 0D2 | 12 |
| 35 | 14 | 8 | 0D3 | 12 |
| 36 | 15 | 8 | 0D4 | 12 |
| 37 | 16 | 8 | 0D5 | 12 |
| 38 | 17 | 8 | 0D6 | 12 |
| 39 | 28 | 8 | 0D7 | 12 |
| 40 | 29 | 8 | 06C | 12 |
| 41 | 2A | 8 | 06D | 12 |
| 42 | 2B | 8 | 0DA | 12 |
| 43 | 2C | 8 | 0DB | 12 |
| 44 | 2D | 8 | 054 | 12 |
| 45 | 04 | 8 | 055 | 12 |
| 46 | 05 | 8 | 056 | 12 |
| 47 | 0A | 8 | 057 | 12 |
| 48 | 0B | 8 | 064 | 12 |
| 49 | 52 | 8 | 065 | 12 |
| 50 | 53 | 8 | 052 | 12 |
| 51 | 54 | 8 | 053 | 12 |
| 52 | 55 | 8 | 024 | 12 |
| 53 | 24 | 8 | 037 | 12 |
| 54 | 25 | 8 | 038 | 12 |
| 55 | 58 | 8 | 027 | 12 |
| 56 | 59 | 8 | 028 | 12 |
| 57 | 5A | 8 | 058 | 12 |
| 58 | 5B | 8 | 059 | 12 |
| 59 | 4A | 8 | 02B | 12 |
| 60 | 4B | 8 | 02C | 12 |
| 61 | 32 | 8 | 05A | 12 |
| 62 | 33 | 8 | 066 | 12 |
| 63 | 34 | 8 | 067 | 12 |
| 64 | 1 | 5 | 00F | 10 |
| 128 | 12 | 5 | 0C8 | 12 |
| 192 | 17 | 6 | 0C9 | 12 |
| 256 | 37 | 7 | 05B | 12 |
| 320 | 36 | 8 | 033 | 12 |

*(Continued)*

## TABLE D-1

**Huffman Code Set—*Continued***

| Number of Bits | White Bits | | Black Bits | |
| --- | --- | --- | --- | --- |
| | Huffman Code (In Hex) | Number of Code Bits* | Huffman Code (In Hex) | Number of Code Bits* |
| 384 | 37 | 8 | 034 | 12 |
| 448 | 64 | 8 | 035 | 12 |
| 512 | 65 | 8 | 006C | 13 |
| 576 | 68 | 8 | 006D | 13 |
| 640 | 67 | 8 | 004A | 13 |
| 704 | 0CC | 9 | 004B | 13 |
| 768 | 0CD | 9 | 004C | 13 |
| 832 | 0D2 | 9 | 004D | 13 |
| 893 | 0D3 | 9 | 0072 | 13 |
| 960 | 0D4 | 9 | 0073 | 13 |
| 1024 | 0D5 | 9 | 0074 | 13 |
| 1088 | 0D6 | 9 | 0075 | 13 |
| 1152 | 0D7 | 9 | 0076 | 13 |
| 1216 | 0D8 | 9 | 0077 | 13 |
| 1280 | 0D9 | 9 | 0052 | 13 |
| 1344 | 0DA | 9 | 0053 | 13 |
| 1408 | 0DB | 9 | 0054 | 13 |
| 1472 | 098 | 9 | 0055 | 13 |
| 1536 | 099 | 9 | 005A | 13 |
| 1600 | 09A | 9 | 005B | 13 |
| 1664 | 18 | 6 | 0064 | 13 |
| 1728 | 09B | 9 | 0065 | 13 |
| SYNC | 001 | 11 | 001 | 11 |

*The number of Huffman code bits plus the code itself yield all unique combinations. Thus, the codes for 1,2, and 10 white bits and 12 black bits are all different. They are 000111, 0111, 00111 and 000111, respectively.

The basic principal behind the Huffman code is that certain symbols or combinations of symbols appear in text more frequently than others. For instance, the letters E and T are used more frequently than Q and F. The combination of letters TH appears most frequently of random letter pairs. The word THE is the most frequently used word. The process is to assign the shortest code to the most frequently used symbols. A message coded in a Huffman code would then contain fewer digital bits than any compression code that uses a fixed number of bits for every symbol.

To illustrate how a Huffman code is developed, assume that a Huffman code is to be developed to represent the vowels A, E, I, O, U, and Y as the only symbols. A statistical analysis is run on a reasonably large sampling of text. Arbitrarily, and just for the pur-

E 35%    O 25%    I 15%    A 15%    U 8%    Y 2%

Symbols by Percentiles

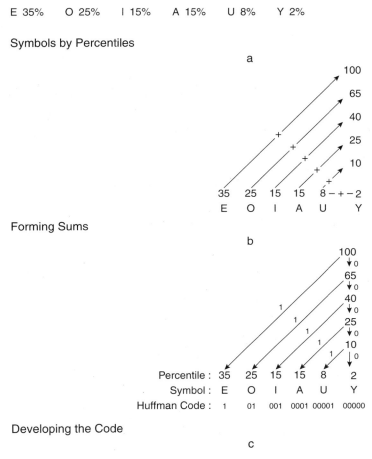

a

Forming Sums

b

Developing the Code

c

**FIGURE D-5** Huffman Code Example for Vowels

poses of example, assume the frequency of the appearance of these vowels in the sampled text is E = 35% of the time, O = 25%, A = 15%, U = 8% and Y = 2%. Diagramically, start by listing the vowels horizontally, sequentially in descending order at the bottom of a page as shown in Figure D-5a. Add the two lowest percentages (U and Y) and place this sum above the Y symbol. Add the next two lowest percentage (A) and place this sum above the last sum. Continue with this until you reach 100% (Figure D-5b). Draw a line from the 100% to the highest percentage (E). Add another line from the next lowest sum to the next lowest percentage (O). Continue the process to the letters I, A, U, and Y. Finally, draw connecting lines between each sum so the diagram looks like Figure D-5c. Label each slanted line with a 1 and each vertical line section with a 0. To create the Huffman code, start at 100% and trace your way to each letter. Write down the binary code for each symbol. Note that for this simple example, as the percentage decrease, the binary code gets larger, with the letter E having only one bit (1) and the letter Y having

5 bits (00001). In reality, there would be at least two symbols with one bit codes (1 and 0) and four with 2 bits (00, 01, 10, and 11) and so on. This example is just here to give you a feel for how the Huffman code is developed. Huffman codes can be used with any type of database—it does not have to be text. For facsimile use, varying levels of shade from black to white, having been encoded into digital data can be compressed using a Huffman scheme.

Recommendation T.4, or Group III, produces documents at the rate of one minute per page using Huffman code compressed digital data streams. 1,728 points are scanned per line, resulting in a horizontal resolution of 204 lines per inch and a vertical resolution of 98 lines per inch. The transmission rate is specified at 4800 bps but can be optionally operated at 9600 bps given a fairly clean telephone line. Phase shift keying (PSK) and Quadrature Amplitude Modulation (QAM) techniques are used to transmit and receive the digital data. A secondary handshake channel used for establishment of a communications dialogue (requests and acknowledgements) operates at 300 bps using frequency shift keying (FSK) modulation. Recommendation T.30 defines the handshake procedures to be used by fax systems.

### Section D.4 Review Questions

1. Compare the document rates for the ITU fax recommendations.
2. What is the purpose of the Huffman code?
3. What modulation techniques are used by each fax recommendation?

## D.5 FAX SYSTEMS

An overall block layout for a fax system, which employs the used of an undedicated telephone line, is shown in Figure D-6. The Voice/Fax MUX block discerns between a fax signal and regular voice call on the telephone line. In this block, voice communications are directed to a regular telephone handset, while fax data are sent to a fax modem. The modem receives and demodulates the fax data, which are decoded and processed by the processor block and sent to a specific peripheral, a printer to produce a hard copy of the document or a terminal for video display. Fax data can be entered into the system through a scanner of from data previously scanned and stored on disk or in a computer's memory. The data are encoded and sent to the fax modem for modulation and transmission on the telephone line.

Figure D-7 shows a Group III fax transmitter. A document is scanned and sampled to produce a 1728-bit-per-line digital stream. The stream is applied to a binary counter, which counts the bits representing black and the bits representing white. The count is used as an address to the Huffman code Read Only Memory (ROM). The data output of

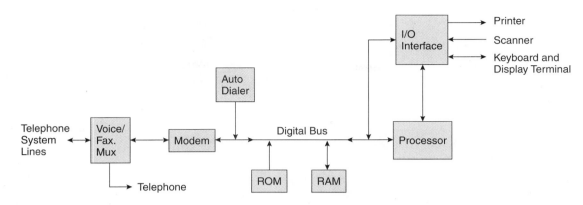

**FIGURE D-6** Facsimile System

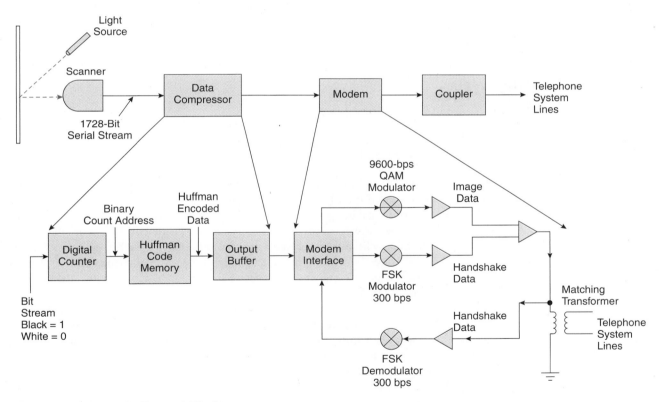

**FIGURE D-7** Facsimile Transmit Blocks

the ROM is the compressed Huffman code, which is used to modulate a QAM modem. Any handshake control data are sent to a FSK modem using an 1,650-Hz tone for a logic 1 and an 1,850-Hz tone for a logic 0. The combined QAM and FSK signals are applied to the telephone lines and sent to their destination.

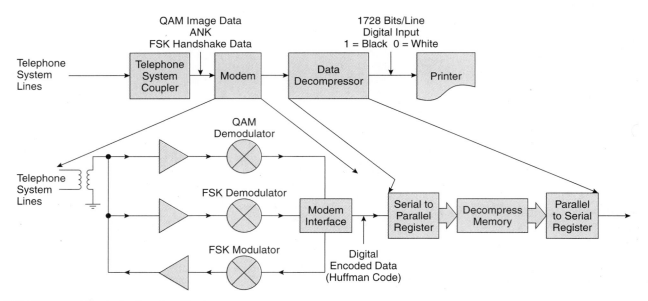

**FIGURE D-8** Facsimile Receive Blocks

Received data are demodulated using a QAM demodulator for the fax information and FSK demodulator for the control data. Figure D-8 shows the fax receiver blocks. The digital data from the QAM demodulator are converted from a serial stream to parallel data and are used to address a Huffman decompressor (decoder) ROM. Parallel data from the ROM is converted back to a serial stream to duplicate the information generated by the scanner at the transmitting site. This digital information will be decoded further as required by the printer or video terminal that reproduces the actual document or image.

A fax machine whose sole purpose is to transmit and receive fax data uses the decoded ROM data directly to reproduce a hard copy of the scanned document. Fax printers used with these machines come in three types—electrothermal, electrostatic, and thermal. An electrothermal printer uses a dry process with a two-layer, white-over-black conductive paper. Black bits cause a large voltage to be applied, which burns off the white upper layer to reveal the black layer underneath. The closer to gray, the less voltage is applied and the more white remains, reducing the intensity of the black area. Electrostatic printers use a special coated paper that is electrically chargeable. Black bits cause the coating to be charged and attract a black toner. The paper is then passed over a heater, where the toner particles are permanently bonded to the paper. Thermal paper printers actually cause black areas to be burned into the paper by the application of a voltage to a stylus. The darker the image the higher the stylus voltage and the more the area is burned.

Facsimile is a method that has been in use for a long time. Its slow acceptance in usage worldwide was a result of the cumbersome size and process employed by earlier machines. With today's technology, small, compact fax machines at affordable prices are in use everywhere. Additionally, fax plug-in cards for IBM, Macintosh, and other personal computers have brought fax capabilities to every application. Fax servers have been added to local area networks to allow many users at a business site to have direct access to fax services through their networked terminal or workstation.

---

### Section D.5  Review Questions

1. Which fax system block separates voice and fax calls?
2. How are scanned digital data translated to Huffman compressed data in a basic fax system?
3. Describe the different methods applied by fax printers to recreate the document.

---

## SUMMARY

Facsimile is an old technology brought up-to-date for the transfer of documents, electronically, over public and private communications links. Fax is incorporated in all types of systems and has become, along with the Internet, an integral part of our lives.

---

## ANSWERS TO REVIEW QUESTIONS

**Section D.2**
1. 1842
2. World War II

**Section D.3**
1a., b., d., and e.—Drum. c.—flatbed.

**Section D.4**
1. T.2—6 minutes/page.
   T.3—3 minute/page.
   T.4—1 minute/page.
2. Reduce the number of bits to be transmitted (compression code).
3. T.2—FM.
   T.3—AM.
   T.4—PSK and QAM.

**Section D.5**
1. VOICE/FAX block.
2. Counter counts number of black bits and number of white bits. Count is used to form an address to the Huffman Code ROM. Data from the ROM is the compressed Huffman code.
3. Electrothermal, electrostatic, thermal printers.

# APPENDIX E

# Answer Key to Odd-Numbered Questions

---

## CHAPTER 1

### Section 1.1

1. Some possibilities: Morse code, telephone, computers, satellites, fiber optics, token ring, Ethernet, ARPANET, Internet, Microprocessor, Microsoft, Novell, Datagrams, IBM PC, modems, AT & T breakup, Windows, ISDN

3. Microsoft, Novell, IBM, Banyon, Motorola, others

### Section 1.2

5. All are part of DTE except b) modem, which is DCE.

7. Primary station controls the data link, determining when a secondary can communicate with it.

### Section 1.3

9. a) simplex   b)   full duplex   c) half duplex
   d) full-full duplex

### Section 1.4

11. Alphanumeric, formatting, and data link

13. 
| Character | ASCII | EBCDIC |
|-----------|-------|--------|
| B | 42 | C2 |
| f | 66 | 86 |
| ! | 21 | 5A |
| LF | 0A | 25 |
| STX | 02 | 02 |
| NULL | 00 | 00 |
| 5 | 35 | F5 |
| " | 22 | 6F |
| shift | 0E/0F | 0E/0F |
| EOT | 04 | 37 |

### Section 1.5

15. Parallel system, number of transmissions = 256/8 = 32
    Transmission time = 32 x 1/1200 = 26.7 mS
    Serial system, transmission time = $256 \times 1/1200$ = 213 mS

### Section 1.6

17. Synchronous data does not include framing overhead bits that asynchronous does. Asynchronous data does not require a synchronized clock between transmitter and receiver. Synchronous data is more efficient because of the lower overhead.

19.  0 1 0 0 1 1 0 0 0 1 0 0 1 1 0 0 1 1 0 0 0 1 1 1 0 0 0 1 0 0 0 1 1 0 0
    -- SYNC -------- SYNC -------- T ----------- h -----------
    1 0 0 0 1 0 1 0 0 0 1 0 1 0 0 0 0 0 0 1 0 1 1 0 0 0 1 0 1 1 0 0 1 0
    i ------------ s --------- s p a c e --------- t ------------- i --
    0 0 1 0 0 1 0 1 0 0 1 1 0 1 0 0 0 0 1 1 1 1 1 0 1 1 0
    ---------—m ------------- e ---------- ? -----

21.  Efficiency = 256/258 * 100% = 99.2%

## Section 1.7

23.  Error detection
25.  Figure 1–K1

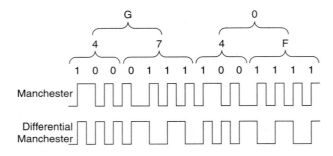

## Design Problems

1.  Any form of code that uses 3 ones and 4 zeros for each character. Each code is unique. Error detection is accomplished by counting the number of ones per character and as long as they are all three, there is no error.

# CHAPTER 2

## Section 2.2

1.  Local loop or local switch station

3.  Toll station connects trunk lines. Tandem station connects multiple toll and/or local switch stations.
5.  Regenerate and reshape long distance signals.

## Section 2.3

7.  Circuits to carry calls are determined at the time the call is placed.
9.  1,004 Hz
11.  The basic, unconditioned specifications for attenuation/gain and propagation delay variance
13.  Leased, dedicated, or private lines
15.  The carrier signal is modulated by a low-frequency sine wave. The time between peaks of the envelope created by the modulated wave form are measured to determine the relative delay of the carrier signal.
17.  Signal-to-noise ratio and harmonic distortion
19.  135 db

## Section 2.4

21.  Impulse gain hit ± 3db for maximum of 4 mS
    Impulse phase hit of no more than 20° shift for a maximum of 4 mS.
23.  External electrical disturbances (storms, ham operators, etc.)
25.  Prevent signals from being echoed back down the line
27.  By sending a 2,010 to 2,240 Hz signal on the line for a minimum of 400 mS
29.  Amplitude and phase
31.  a) interference
    b) crosstalk
    c) white noise
    d) attenuation/gain variance
    e) standing wave or echo
    f) interference
    g) interference

# CHAPTER 3

## Section 3.2

1. Parity
3. Exclusive OR
5. Overhead
7. Automatically requests a retransmission of a message when it detects that an error has occurred
9. Can only correct a single bit error
11. LRC bits are the result of applying an exclusive OR function to each of the same bit positions of each character in a message.
13. A one in the receive LRC result indicates a error has occurred at that bit position in a bad character.
15.

| Character | LSB | MSB | VRC |
|-----------|------|------|-----|
| T | 0010 | 1010 | 1 |
| h | 0001 | 0110 | 1 |
| e | 1010 | 0110 | 0 |
| Space | 0000 | 0100 | 1 |
| Y | 1001 | 1010 | 0 |
| e | 1010 | 0110 | 0 |
| l | 0011 | 0110 | 0 |
| l | 0011 | 0110 | 0 |
| o | 1111 | 0110 | 0 |
| w | 1110 | 1110 | 0 |
| Space | 0000 | 0100 | 1 |
| B | 0100 | 0010 | 0 |
| r | 0100 | 1110 | 0 |
| i | 1001 | 0110 | 0 |
| c | 1100 | 0110 | 0 |
| k | 1101 | 0110 | 1 |
| Space | 0000 | 0100 | 1 |
| R | 0100 | 1010 | 1 |
| o | 1111 | 0110 | 0 |
| a | 1000 | 0110 | 1 |
| d | 0010 | 0110 | 1 |
| LRC | 0010 | 1100 | 1 = 134H |

## Section 3.3

17. 10011)1001100110010111001101110100111110010000
    10011
    00011001
     10011
     10100
     10011
      11111
      10011
      11001
      10011
      10100
      10011
       11101
       10011
       11100
       10011
        11111
        10011
        11001
        10011
        10101
        10011
         11001
         10011
         10100
         10011
          11101
          10011
          11101
          10011
           11100
           10011
           11101
           10011
            11011
            10011
            10000
            10011
             11000
             10011
    CRC =   1011

19. Figure Key 3–2

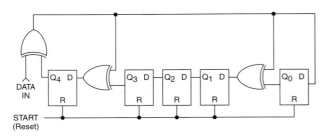

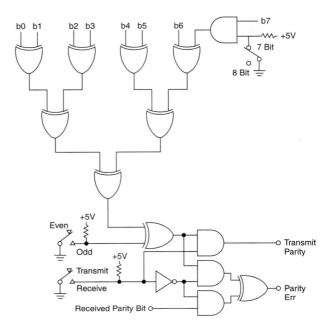

21.

| | | |
|---|---|---|
| m | 0110 | 1101 |
| m | 0110 | 1101 |
| i | 0110 | 1001 |
| l | 0110 | 1100 |
| l | 0110 | 1100 |
| e | 0110 | 0101 |
| r | 0111 | 0010 |
| @ | 0100 | 0000 |
| d | 0110 | 0100 |
| e | 0110 | 0101 |
| v | 0111 | 0110 |
| r | 0111 | 0010 |
| y | 0111 | 1001 |
| - | 0010 | 1101 |
| p | 0111 | 0000 |
| h | 0110 | 1000 |
| x | 0111 | 1000 |
| . | 0010 | 1110 |
| e | 0110 | 0101 |
| d | 0110 | 0100 |
| u | 0111 | 0101 |

Checksum = 0101 0101 = 55H

23. 8 bits
25. $H_5 . . . H_0 =$
27. In-line monitoring
   Loop-back test

## Design Problems

1. Figure 3–K3

# CHAPTER 4

## Section 4.2

1. A topology is the description of how a network is functionally formed.
3. All nodes on a bus topology have equal access to the medium. The main drawback is contention for the use of the bus.
5. A poll is a message sent by the primary inquiring if a particular secondary node has any traffic to send.
7. Since all nodes can communicate directly on a peer bus, there is no delay in having to go through a primary station first.
9. Each station must check the bus for inactivity before attempting to access the bus.
11. A ring network overcomes access contention by passing all messages from one node onto the next node in a continuous ring pattern until they return to the originating node.
13. A hub has the ability to switch by-pass a failed node and allow the remaining nodes to continue communicating in a ring fashion.

15. Free Token
17. Because messages are passed from station to station in a ring fashion. Each message passes through the switched connection made by the hub.

## Section 4.3

19. Large number of interconnections that have to be made to connect every node to each other.
21. Because messages are sent and received without any delay other than the electrical propagation delay of the medium used.
23. When using message switching, entire messages are stored at the switching station, which holds them until the destination node is ready to receive them. For packet switching, all messages are divided into smaller units called packets, which are interleaved with each other. These packets are then sent and directed to their intended receiver nodes. This allows portions of messages to be constantly received by a number of receiving nodes.
25. Circuit switch. Message switching is also known as store and hold.

## Section 4.4

27. LAN—operated in a single building.
    MAN—covers a geographical area like a city or county.
    WAN—covers a large portion of the country (could also be fully countrywide).

## Section 4.5

29. An open system is a communication blueprint that allows any type of protocols to be used.
31. Payload is the data portion of a message. It may contain pure data or the full message stream of an upper level protocol.
33. The physical layer specifies the hardware, medium, electrical, mechanical, and functional specifications for a network.

35. Physical layer addresses are the exact physical address of a device or node in a network. At the data-link layer, addresses may specify either a physical address or a logical (virtual) address.
37. The media access control layer specifies address management functions at the data-link level.
39. Connectionless circuits establish connections as they are needed. Connection-oriented circuits are established and maintained even if there is no traffic to be sent on them.
41. Network Layer
43. Transport Layer
45. Type A systems can operate given a certain level of circuit failure. Type B cannot tolerate circuit failures.
47. File management, connection and disconnection of nodes, binding process names to network addresses, permits multiple applications to share a virtual circuit, user to network or central host interface, fault recovery
49. To provide a user access to the network
51. X.500 Global Directory Service

## Section 4.6

53. None—physical specifications are left to existing physical standards. Does specify types of physical nodes used by the SNA model.
55. Path control layer
57. Bracketing
59. Programs that manage the network and establish and control interconnections between nodes.
61. Logical unit
63. Manage and interface several peripheral devices to the SNA network.
65. Maintain control over the peripheral nodes connected to them.

## Section 4.7

67. The explicit route is the actual physical route taken by a message to go from one node to another.
69. Connectionless or bandwidth-on-demand service

# CHAPTER 5

## Section 5.2

1. An UART's main function is serial/parallel data conversion.
   Additional functions include adding framing and parity bits, setting the data rate, checking incoming data for parity, framing and overrun errors.
3. The number of character bits, number of stop bits, and parity are optional. One start bit is required for all asynchronous data characters.
5. The ratio between the clocks and the data rate are to aid the receiver in correctly shifting in data from the received data line.

## Section 5.3

7. MARK: −5 to −15v SPACE: 5 to 15v
9. a) RTS—request to send sent by the UART signals the MODEM that the UART wants to send and receive data.
   b) CTS—clear to send sent in return by the modem informs the UART that the modem is ready to accept and send data.
11. Modem sends data set ready (DSR) to the UART to signify that it is electrically connected and ready. DTR is a similar signal called data terminal ready returned by the UART.
13. 50 feet 2,500 pf
15. Common, transmit and receive data, RTS, CTS, DSR, DTR, CD

## Section 5.5

17. Common line noise rejection

## Section 5.6

19. MARK—data logic 1; SPACE—data logic 0
21. VCO—convert digital data into tones
    RTS/CTS circuit—handshaking to DTE

DSR/DTR generation and sensing—establish electrical compatibility with DTE.
23. MARK tone, logic 1
25. The originate modem is used by the station that initiates a call to establish a data link. The answer modem is used by the receiving station. Electrically, the two modems differ in the frequencies used for MARK and SPACE tones. The originate modem's transmit frequency pair is the answer's receive frequency pair. Likewise, the frequencies used by the answer modem's transmitter are the same as the originate's receiver.
27. Control and acknowledge data

## Section 5.7

29. Balanced modulator
31. Compares symbol phases to determine the binary value of the current bit.
33. PSK is affected by phase shifts and hits, FSK is not.
35. Tribit
37. Two
39. 16
41. 1050 sps
43. 90°
45. 13500 bps
47. a) 2400 sps
    b) 64 symbols
49. 28 Kbps
51. V.8 handshaking
53. Upstream: 28.8 Kbps, downstream: 56 Kbps
55. PCM

## Research Assignment

5.1 Lots of modems are manufactured with new ones announced every month. The headache is to find any indepth technical material on any of them. Specifications and the like are supplied by the manufacturers and magazine articles show applications and tout each modem's abilities. The challenge is gather the technical background on a selected modem. The depth of the assignment should reflect the availability of the material.

# CHAPTER 6

## Section 6.2

1. Full duplex originate and answer mark and space signals share the telephone line bandwidth.
3. By mixing each 4-kHz channel at a different carrier frequency.
5. 92–96 kHz
7. Mastergroup and jumbo group.
9. 0–564 kHz
11. For STDM, time slots are assigned whether they are used or not, for STAT MUX, time slots are assigned as needed.

## Section 6.3

13. Natural samples follow the actual signal amplitudes. Flat-top samples retain the same amplitude throughout the sample period.
15. 2,760 Hz

## Section 6.4

17. Discrete levels are easier to encode.
19. .598 mV

## Section 6.5

21. Successive Approximations
23. Shorter binary codes lead to shorter messages.

## Section 6.6

25. A single bit is sent for each sample time. A 1 indicates an increase in sample amplitude, a 0 is a decrease in sample amplitude.
27. When a large change in sample amplitudes is experienced

## Section 6.7

29. E1 uses 30 channels and operates at 2.048 Mbps.
31. Framing

33. Superframe: 288 extended superframe: 576
35. Avoid 8 consecutive 0s.
37. Two consecutive logic 1s with the same voltage polarity
39. Slip occurs when a sampling error occurs. A bit "slips" out of place and is misinterpreted as a data or framing error or a missing bit.
41. Data, carrier loss, bipolar violations, framing type, framing error, AMI or B8ZS coding, CRC errors

## Section 6.8

43. Slope overload occurs when there is a large difference in the original signal and the replicated delta modulated signal caused by a large change in neighboring samples.
45. $\mu$ law: 1.023 V     A law: 0.955 V
47. 100101101000 error of 5 from original 365 = 1.37%

## Section 6.9

49. 64 Kbps—4096 Mbps
51. R2 = 2.4R1

## Design Problems

The ability to complete the design problem lab projects will depend on the facilities, availability of components, and time allocated to lab work. These projects are representative of the theory discussed in the chapter.

# CHAPTER 7

## Section 7.2

1. A protocol is a set of rules for successful data communications.
3. Poll—inquires if a station has data to send. Selection—inquires if a station is ready to receive data.

## Section 7.3

5. A secondary is required to respond to polling and selection messages from the primary. In some data-link protocols, the secondary is required to respond to all of the primary messages.

7. By use of the start framing bit in each asynchronous character.

9. a) DC3 indicates a polling message.
   b) / ACK—secondary is ready to receive messages.
   c) //—secondary is in the local mode (off-line).
   d) * *—secondary is in the send mode—response to a selection.

11. When secondary has no message but must respond to primary, it sends a handshake message.

## Section 7.4

13. Anywhere in the message.

15. 0111111000111010101110011111011000011110011111101100001111110

17. More than 15 consecutive ones indicates an idle line condition between transmission of messages.

19. Supervisory

21. By a zero in the least significant bit of the control field.

23. This is more detailed here than in the text. Since the address field is a byte, there are 256 possible address combinations. However, one is used as a broadcast address (FFH) and one for test purposes (00H), so there are actually 254 maximum possible stations that could be identified with individual addresses and no group addresses in the SDLC protocol.

25. a) yes
    b) resend frames 2 & 3

27. The secondary sends unnumbered request messages.

29. An SDLC SNRM does not initiate the secondary's initialization program, SIM does.

31. HDLC has a larger address field, longer NS and NR numbers, additional supervisory code, and allows for a larger data field.

33. 127 consecutive frames for HDLC

35. HDLC includes a selective frame reject supervisor frame that SDLC does not have. HDLC can call for the retransmission of a single frame by using this supervisor message.

37.  7E   3C  D4  F9  9D  4E  C8  C4  D3  C3  40  89
    flag  address- contr.   H   D   L   C  sp   i

    A2  40  81  40  94  96  99  85  40  A5  85  99  A2
     s  sp   a  sp   m   o   r   e  sp   v   e   r   s

    81  A3  89  93  85  40  D7  99  96  A3  96  83  96
     a   t   i   l   e  sp   P   r   o   c   o   c   o

    93  48  xx   xx  7E
     l   .   CRC    flag

39. The asynchronous response mode allows a secondary to respond to a primary without receiving a polling message.

41. For the asynchronous protocol, 11 bits are required per character (8 data, 2 framing, 1 parity). There are a total of 542 characters in the transmission (512 + 30 graphic characters) for a total of 542 x 11 = 5962 bits. Actual data bits are 512 x 8 = 4096 bits. The efficiency = 4096/5,962 x 100% = 68.7%. BISYNC doesn't require framing or parity bits, but does require a 5-character preamble and two additional SYNC characters per block for a total of 7 × 3 = 21 characters for the message. Add 6 bytes for 3 CRC-16 characters brings the overhead total to 27. Total bytes for the message are 512 + 27 + 30 = 569. Efficiency is then 512/569 x 100% = 90%.
   SDLC overhead includes 2 flags, 1 control field and 2 CRC bytes per frame for a total of 15 bytes. There are no graphics characters for a total of 512 + 557 bytes. The efficiency of the transmission is 512/557 x 100% = 97.2%.

## Section 7.5

43. Network lines uninterrupted, analyzer embedded in system, readily available, custom configurable, easily upgraded

# CHAPTER 8

## Section 8.1

1. a) 802.2
   b) 802.6
   c) 802.3
   d) 802.1
   e) 802.5
   f) 802.11
   g) 802.1z
   h) 802.14
   i) 802.9a
   j) 802.2
   k) 802.3

## Section 8.2

3. LAN—one room or building
   MAN—several sites within a limited region
5. How stations access media.

## Section 8.3

7. Data rate, coax cable, length of segments, number of nodes attached to single segment, repeaters, minimum and maximum distances between nodes
9. 100
11. Bus, star, and ring
13. HDLC
15. 10 Mbps
17. 100 Mbps, use of fiber and copper cables
19. CSMA/CA uses collision avoidance instead of collision detection. This is achieved by using a central controlling station that detects when multiple stations want to access the bus and then resolves the contention issue before access is granted.
21. CSMA/CD over fiber optics and copper cable.

## Section 8.4

23. Free token

## Section 8.5

25. Removes message from data stream
27. If a station fails, the entire network is down

## Section 8.6

29. Bridges connect between two similar networks.
31. A router directs traffic through numerous similar and/or dissimilar networks.

## Section 8.7

33. IEEE 802.6
35. Request and acknowledge slots

## Section 8.8

37. 4,096 channels
39. X.25 Flow control frames do not use the poll/final flag.
Frame reject format is not used with X.25 flow control.

# CHAPTER 9

## Section 9.2

1. Integrated Services Digital Network
3. Transfer information in digital form. Regular phone system uses analog.
5. Bearer data
7. 2 B channels + 1 D channel
9. 23 B channels + 1 D channel 1.544 Mbps
11. Convert non-ISDN data into ISDN format.
13. R— non-ISDN data input
    S— ISDN formatted data to network
    T—physical interface between customer and common carrier
15. Physical interface between central office transceivers and local loop lines
17. a) A channel
    b) H4 channel
    c) E channel
    d) H11 channel

19. Network Termination
21. Framing
23. Undefined
25. Link-access protocol (for D and for B channels)
27. Entry point into network
29. Command for traffic from the network, response for traffic to the network
31. Set normal response mode
33. Identifies the protocol format for the payload
35. Setup, pure data, acknowledge message
37. Automatic number identification
39. First national implementation of ISDN

## Section 9.3

41. SPN—multiplexes incoming data and transfers them to the broadband node
    BTI—user interface to broadband services
43. Bearer services, supplementary services, teleservices
45. User information, flow control, error recovery
47. Manages network resources
    Provides maintenance and coordination between all BISDN functions
49. Manages the majority of control functions for system access.
    Call processing, administrative functions, switching and maintenance functions

## Section 9.4

51. 802.9
53. HDLC
55. Frame control

## Section 9.5

57. Digital subscriber line—voice and data services

## Section 9.6

59. Circuit switch
61. Time slot as acknowledgment

## Section 9.7

63. 53 bytes
65. VCI identifies a single virtual channel in an ATM network.
67. Permanent virtual channel is a path that is established before data is sent on it. Switched virtual channels are established on a need basis as the first packet is sent.
69. UNIs interface users to the network.
71. Constant bit rate carries the same level of traffic continually. Variable bit rates carry traffic that varies in volume
73. Perform error detection on the header, determine if its cell can be dropped during congestion, help to locate cells within a data stream.
75. Combined responsibilities of the convergence and physical medium dependent sublayers. Essentially to assure that the physical plant can handle the ATM transfers.
77. Fill ATM time slots with idle cells when a cell is expected but none are available.
79. Cell loss priority and explicit forward congestion control
81. Burst traffic occurs when traffic is intermittent. That is, there are periods of heavy traffic flow and periods of no traffic flow.
83. Bandwidth on demand means that a portion of the bandwidth is assigned use only when it is needed and is freed up when data transfers using it are complete. This saves valuable bandwidth resources and makes them available when needed.
85. Ratio of high-priority cells to those that can be discarded.
87. places data into a common format and maps cells onto the transmission medium
89. AAL 0
91. Class A uses constant bit-rate data and class B uses variable bit-rate data.
93. Address resolution protocol
95. Networks, subnetworks, and end system nodes
97. AAL 5
99. Fabric blocking occurs when the capacity of the ATM switch is less than the sum of all its inputs.

101. Output buffering used in small ATM switches that handle light traffic. Distributed input buffering places buffers at each input port—cells are sent through only when output bandwidth is available.
103. Available bit-rate traffic

## Section 9.7

105. Frame relay handles variable size packets, ATM uses a fixed cell-size packet.
107. HDLC

## CHAPTER 10

## Section 10.1

1. ARPANET
3. An information-sharing collection of networks

## Section 10.2

5. A data packet that is sent without expecting acknowledgement
7. TCP/IP
9. MOSAIC

## Section 10.3

11. GOPHER—menu-driven index of databases
    ARCHIE—databases of ftp sites
    WAIS—text indexing for large text files
    VERONICA—text in a GOPHER menu
    WHOIS—allows users to query compatible servers for information about Internet users.
13. They are bounced back to the sender.
15. User agent (UA)
17. Enterprise-wide directory service of e-mail and other addresses
19. Article

21. Home page
23. FAQ—frequently asked questions

## Section 10.4

25. Serial line interface protocol (SLIP) and point-to-point protocol (PPP)
27. Bearer handles data and delta (D) handles control information.
29. ISDN bridge/router
31. Filters cable signal to extract Internet signal.

## Section 10.5

33. Domain Name Service
35. Domain top is a central classification for the type of web site.
37. Class A
39. Network and host numbers
41. URL with all ones (255.255.255.255)
43. Does not pass multicast messages to subnetwork where there aren't any members of the group.
45. Reduce traffic

## Section 10.6

47. Authentication and cryptography
49. A public key is maintained on a server and is used to encrypt messages. A private key is maintained by a user and is used to decrypt messages.
51. A brute force attack is a method to test cryptographic keys by testing various key combinations using several computers and mathematical algorithms.
53. DES has a small key size, which was cracked in 3 hours using a brute force attack. This weakness was resolved using triple DES that included a longer key code and three encoding processes.
55. 112-bits

57. Skipjack is an encrypting algorithm used for the Clipper Chip.
59. It allows access to encrypting keys to authorized government agencies.

## Section 10.7

61. SIM is a smart card.
63. Hash value is a scrambled message comprised of a key and file data.
65. Small block of data that contains a public key and endorsement by a certificate of authority.
67. Class 1—public key bound to e-mail, no validation
    Class 2—issues certificate based on consumer database information such as credit ratings
    Class 3—requires in-person appearance or notarized application
69. Secure Electronic Transfer (SET)
71. A web of trust is a group of related users that hold an individuals' public key.
73. The downside of web of trust is that the extended use of a public key may get into hands of people far removed from the original individual.

## Section 10.8

75. Screened host—routes Internet requests to a bastion host that determines if the request should be sent onto the Internet or blocked.
    Screened Subnet firewall does the same thing for multiple networks by using a network of screened hosts.
    Dual homed gateway firewall does the blocking at the applications level, determining which accesses to allow for a particular application already under use.
77. Application level proxies—change IP addresses and check validation of users

## Section 10.9

79. An intranet is a "baby Internet" that uses similar functions but is limited in access to an enclosed network.

## Section 10.10

81. Transmission control protocol/Internet protocol
83. Delivery of messages with the same or different, but connected networks.
85. Class 0—simple class, single connection with no error checking
    Class 1—basic recovery class adds network fault recovery to class 0 service.
    Class 2—multiplexing class allows multiple connections to be multiplexed.
    Class 3—combines classes 1 and 2.
    Class 4—adds error detection and correction to the services already provided by the other classes
87. SYN and ACK
89. 128 bits IPv4 is 32 bits

---

# CHAPTER 11

## Section 11.1

1. Light is faster and is not affected by most electromagnetic disturbances.

## Section 11.2

3. Reflection occurs when a light ray "bounces" off a surface and returns toward its source.
5. Air
7. An incident ray is one that enters a fiber core.

## Section 11.3

9. The core and cladding differ by the amount of doping (impurities) each has, resulting in different refractive indexes.
11. Light rays that enter at different angles are reflected against the cladding each time they reach the cladding wall. This cause some rays to travel longer distances to transverse the core than others that enter at an angle closer to the centerline. Since the rays arrive at different times at the end of the cable, the resulting signal is spread.
13. Single mode generally used by lasers and edge-emitting, light-emitting diodes, concentrates the

light at the core's center. All the energy at the receiver arrives at the same time, reducing pulse spread effects. Multimode cores used primarily with surface-emitting diodes allow light rays to enter at numerous incident angles as long as they fall within the acceptance cone. However, pulse spreading is a factor to be considered when using multimode fiber cables.

15. Critical angle = 49.25°
    Brewster angle = 37.15°
17. Refracted
19. Acceptance angle = 35.54°
21. 1.292

## Section 11.4

23. Light-emitting diodes are less costly and have a wider cone of emitted light than laser diodes.
25. Edge-emitting, light-emitting diodes concentrate light into a tighter dispersion pattern. They require more precise coupling alignment than do surface-emitting diodes.

## Section 11.5

27. Light energy applied to the window of an optical detector causes electron and hole flow (current) to occur. The intensity difference of logic 0 and 1 light levels determines the difference in current in the detector.

## Section 11.6

29. Misalignment of splices and connectors.
31. Lateral or axial; longitudinal or space gap; angular
33. Bend radius = 150 times the diameter of the cable.
35. Minimize the number of bends and make them as loose as possible.
37. 48.75 mm
39. Impurity and air bubbles

## Section 11.8

41. Frame formation and error detection, protocol translation between nonnetwork protocols and FDDI.

43. PMD—specifies physical properties associated with the fiber in use, including bandwidth and waveform characteristics.
    PHY—actual physical interface into the network.
45. Wiring concentrator is a dual-attached device used to connect single attached stations to the FDDI ring.
47. When a station fails, the traffic on the primary ring is shuttled to the secondary ring at the bad station. The traffic then travels in the opposite direction until it reaches the same station, where it is returned to the primary ring to continue onto the next station in the ring.
49. A polarity violation occurs when using automatic mark or space inversion and there are two consecutive positive or negative level bits.
51. Polarity violations are used in beginning and ending frame delimiters.

## Section 11.9

53. Isochronous service supplies integrated data, voice and video information.
55. Isochronous
57. Packet switching and circuit switching
59. Basic mode supports packet-switched data.
    Hybrid mode supports circuit-switched isochronous data and packet data.

## Section 11.10

61. Channel rings called loops and arbitration
63. L port with the lowest addresses—usually host stations.
65. Class 1— acknowledged connection service
    Class 2—switched frame service
    Class 3—one-to-many datagram service
    Class 4—connection-based service
67. Connection-based service: class 1 and 4
69. Class 4

## Section 11.11

71. Synchronous optical transport network
73. 51.48 Mbps data rate service

75. Regenerate attenuated optical signals—extend path length
77. Add and drop channels without demultiplexing entire STS signal.
79. Line, section, and path headers, payload

---

# CHAPTER 12

## Section 12.2

1. A wave guide acts as a resonant cavity to the transmitted frequency, passing that signal from transmitter to radiating antenna.
3. A standing wave is a reflected signal due to impedance mismatch of improperly tuned wave guide.
5. Line of sight
7. Multipath fading describes the effect of reflected transmitted signals by the Earth's surface. Refracted and reflected signals arrive with the transmitted signal tend to add in phase cancellation of the transmitted signal power.

## Section 12.3

9. Intermittent use
   Flexible communication times
   Mobility
11. Mobile telephone switching office (MTSO)
13. No health risks have yet been identified.
15. Analog and digital
17. Frequency modulation
19. Authentication, frequency hopping, and direct sequencing
21. Mobile station (MS)
23. Roaming
25. Base transceiver station (BTS)

## Section 12.4

27. Set of parameters that define the behavior of wireless connections.

29. Narrowband channel at 900 MHz
31. Code division multiple access
    Cellular digital multiple access

## Section 12.5

33. 802.11
35. Diffuse—allows signals to be bounced off a wall or object
    Line of sight—transmitter and receiver must be exactly aligned.
37. Spread spectrum
39. a) Single cell within the infrastructure-based network
    b) Grouping of a number of BSAs
    c) A group of stations associated with the same access point
    d) Multiple BSSs connected via a distribution system

## Section 12.6

41. 30-kHz radio frequency channels
    Packet data
    Connectionless network
    Full duplex using channel pairs
43. Channel sniffing
45. Mobile data intermediate station
47. By transmitting a data signal, the voice system thinks a channel is being tampered with, sealing the channel to voice use and leaving it available to data use.

## Section 12.7

49. Repeater—receive, regenerate, and retransmit signal
51. Stationary orbit that is synchronized to Earth's rotation. Main advantage is that these satellites are easy to track and target for transmissions.
53. Low propagation time is the chief advantage. The need for numerous LEO satellite for complete cover is the main disadvantage
55. Four

57. Solar cells
59. Reduces risk of interference between signals. Allows for multiple receptions and transmissions
61. Federal Communications Commission
63. ECHO
65. International Telecommunications Satellite Organization (INTELSAT)
67. Supervisory burst is used for establishing symbol and burst timing. Closing bursts terminate transmission and report status about transmission capacity and possible detected errors.
69. Increased channel capacity by allowing transmission of two signals simultaneously. Radiated power is divided between both signals.
71. To maximize the gathering of low-power signals
73. Space Communications Protocol Standards (SCPS)
75. How satellites interface with land-based networks
77. GEO
79. QPSK and TDMA
81. 66 plus six back up
83. Mobile subscriber integrated services digital network (MSISDN) number
85. McGraw Cellular and Microsoft

87. Law enforcement
Public safety
Fleet management
Airline reservations
Federal aviation administration
Fishing fleets, barges and tugs
Trucking and rail freight
AMTRACK
Point-to-point telephone links

# APPENDIX D

1. Facsimile is the sending and receiving of digitized images. In analog form, it has been around since WWII (1940s).
3. Frequency modulation
5. Flatbed scanners are used for digital fax machines and have fewer moving parts than analog drum scanners.
7. The purpose of the Huffman code is to reduce the number of bits sent for a given pattern of data.

# INDEX

## A

acceptance angle and cone, 395
access unit, 315
adaptive delta modulation, 204
adaptive pulse code modulation 201
address tables, 286
advanced communications technology satellite (ACTS), 458
advanced radio data information service (ARDIS), 443
air interface, 437
A-law, 216
alternate mark inversion, 211
American National Standards Institute (ANSI), 108, 411
amplitude modulation, 490
analog to digital converter, 190, 199
antennas, 454
ASCII, 231
asymmetric digital subscriber line, 316
asynchronous communications interface adapter (ACIA), 129
asynchronous data link protocol, 231, 233, 252
asynchronous transfer mode, (ATM) 292, 324, 420, 443, 458
    address resolution protocol, 332, 334
    ATM adaption layers, 328, 331
    bit rate types, 329
    burst, 329
    cell loss priority, 326
    cell relay, 324
    convergence layer, 331
    generic flow control, 326
    header error control, 326
    LAN emulation (LANE), 335
    network to network interface, 325
    permanent virtual channel, 329
    quality of service, 328
    segmentation and reassembly, 331
    switched virtual channel, 329
    switches, 336
    user network interface, 325
    virtual channel identifier, 326
    virtual path identifier, 326
attachment unit interface, 283
authentication, 367
automatic request for retransmission (ARQ), 26, 68

## B

backbone, 266, 274, 420
bandwidth, 39, 50
bandwidth on demand, 329, 420
baseband, 180, 267
basic service area, 441
baud rate, 2
beacon token, 416
binary synchronous control protocol, 232
bipolar violation, 208, 210
BISYNC data-link protocol, 232, 235, 250, 252
bit-error rate, 83
bits per second (bps), 2, 16
block check character, 233, 236
breakout box, 143
Brewster's angle, 390
bridge, 285, 320
broadband, 180, 267
broadcast address, 230
broadcast storm, 287

## C

CCITT, 108, 140
cells, 434
cellular digital multiple access (CDMA), 438
cellular digital packet network, 442
cellular telephone, 434
certificate of authority, 369
channel sealing, 442
character codes, 2, 12, 131
    alphanumeric, 2, 13
    ASCII, 13, 131
    Baudot, 12
    EBCDIC, 13, 131
    UNICODE, 13
charge coupled device, 490
checksum, 78
channel, 180, 190
circuit switching, 103
circular polarization, 456
cladding, 392

Clipper chip, 367
cluster, 448
C-message bandwidth, 39
coax cable, 265
CODEC, 217, 320
code division multiple access (CDMA), 435
collision, 93
companding, 213
COMSAT, 450
contention, 92
continuous wave keying (CWK), 2
crosstalk, 54
cryptography, 365
    asymmetrical (public) key, 366
    brute force attack, 366
    DES, 366
    plain text attack, 366
    RSA public key signature, 369
    symmetrical (private) key, 365, 366
    triple DES, 366
CSMA/CA, 273, 439
CSMA/CD, 261
cyclic redundancy check (CRC) 73, 250

**D**
data types
    asynchronous, 18
    differential Manchester, 23
    forms, 20, 22, 23
    Manchester, 23, 264, 278
    NRZ-AMI, 412
    synchronous, 17
data communication equipment (DCE), 6, 137
data scope, 136
data terminal equipment (DTE), 6, 137
decibel (dB), 42
dedicated lines, 40
delta modulation, 202
differential pulse code modulation, 201
diffuse infrared, 440
digital signature, 369
digital subscriber line, 315
digital to analog converter, 190, 199
direct distance dialing (DDD), 40
direct memory access, 285
direct sequencing, 436
distributed feedback Bragg laser, 409
distributed queued dual bus, 291
DOMSAT, 451
downlink, 445, 449
drum scanning, 488
dual tone multiple frequency touchtone pad, 31

**E**
E1 carrier—see T1 carrier
EBCDIC, 231
echo, 55
echo plex, 152
Electronic Industries Association (EIA), 108
envelope delay distortion, 44
entropy encoding, 201
error-free seconds, 84
Ethernet, 5, 261
Ethernet frame, 269
extended superframe, 206

**F**
fabrics, 419
facsimile (FAX), 487
fast Ethernet, 273
fault recovery, 275
FDDI-II, 418
fiber distributed data interface (FDDI), 5, 335, 411, 415
    dual attached stations, 412
    MAC, 412, 441
    physical layer (PHY), 412
    single attached station, 412
    station management protocol, 414
    wiring concentrator, 412
fiber optic cable, 265, 388, 394
fiber optic data distributed network (FDDN), 417
fibre channel, 419
firewalls, 371
    address spoofing, 373
    dual-homed gateway, 372
    filtering, 373
    screened host, 372
    screened subnet, 372
    stateful inspection, 373
flatbed scanning, 488
focal point, 395
forward error correction, 72
fractional T1, 209
frame relay, 337
    discard eligibility, 338
    explicit congestion notification, 338
    frame format, 338
framing bits, 18
free token, 276
frequency hopping, 435
frequency modulation, 435, 488
full duplex, 12, 152

## G

gateway, 288
geostationary or geosynchronous orbit (GEO), 445, 452
gigabit Ethernet, 272, 274
global system for mobile communications, 436
graded index, 398
group address, 229
Government OSI Profile (GOSIP), 113


## H

half duplex, 12, 152
Hamming code, 79
harmonic conditioning, 48
hash value, 369
HDLC—see SDLC
Huffman Code, 492
hybrid circuit, 37


## I

IEEE 802, 258, 420
IEEE 802.9, 314
IEEE 802.11, 439
IEEE 802.16, 460
impulse gain/phase hit, 53
infrared, 440
injection laser diode, 392, 402
Institute of Electrical and Electronic Engineers (IEEE), 108, 257
INTELSAT, 450
International Standards Organization (ISO), 108
International Telecommunications Union (ITU), 108, 140, 293
Internet, 6, 347
    Adobe Acrobat Reader, 352
    ARCHIE, 355
    ARPANET, 348
    cable modem
    CERN, 349
    datagrams, 348
    domain name service, 348, 361
    e-mail, 348, 353
    ENQUIRE, 348
    Eudora Light, 354
    extranets, 374
    FAQ, 358
    GOPHER, 356
    home page, 357
    HTML, 349
    http, 361
    intranets, 374
    IP 348, 375, 420
    IP address, 362
    IPv6, 376
    ISP, 358
    JUGHEAD, 356
    MIME, 351
    Mosaic, 350
    multicasting, 363
    news groups, 355
    PPP, 359
    POP, 355
    request for commands, 357
    SGML, 349
    SLIP, 359
    SMTP, 355
    TCP, 379
    TELENET, 357
    URL 361, 375
    VERONICA, 356
    WAIS, 356
    WWW, 349
    X.400, X.500, 355
interrupt request, 136
IRIDIUM, 448, 452, 458
ISDN, 5, 250, 302, 359, 458
    automatic number identification, 309
    basic rate interface, 302, 307
    bearer (B) channel, 302
    BISDN, 311
    channels, 305
    command/response, 308
    delta channel, 302
    exchange termination, 305
    frame format, 306
    hierarchy, 306
    line termination, 305
    link access protocol, 307
    network termination, 304
    N-ISDN, 311
    primary rate interface, 302, 307
    service access point, 308
    service node, 313
    subscriber's premises network, 311
    terminal adapter, 303, 310
    terminal end point, 308
    terminal equipment, 303
    transfer ID (XID), 308
isochronous service, 418


## J

jitter, 56

**L**

laser, 401
latency, 325
leased lines, 40
light emitting diode, 392, 400
line conditioning, 42
line control unit (LCU), 6
line of sight infrared, 440
local access and transport areas (LATA), 37
local loop, 30, 34
logical link control, 259, 270
loop back, 3, 145
longitudinal redundancy check (LRC) and VRC, 250
low earth orbit (LEO), 446, 452, 458

**M**

mark, 2, 129
media access control, 259, 260, 276, 315
media access unit, 259
medium, 8
medium earth orbit (MEO), 446, 452
message switching, 103
microprocessor, 5
microwave, 432
militarized fiber optic transmission system (MFOTS), 417
mobile satellite (MSAT), 459
mobile telephone switching office, 434
modem, 4, 8
    8PSK, 163
    ASK, 153
    balanced modulator, 154
    clock recovery, 161
    constellation (phaser or vector) diagram, 160, 165, 167
    differential PSK, 156
    FSK, 127, 150, 169, 181, 496
    phase lock loop, 152
    PSK, 153, 162, 169, 496
    QAM, 154, 166, 169, 496, 498
    QPSK, 158, 161, 458
    V.34 and V.90, 170
    voltage controlled oscillator, 150
Morse code, 2
multimode fiber, 397
multipath fading, 433
multiplexing, 180
    FDM, 180, 181, 316
    FDMA, 452
    groups, 182
    guardband, 181
    jumbo group, 185
    mastergroup, 185
    stat-mux, 189

supergroup, 185
TDM, 187, 316
TDMA, 435, 453, 458
WDM, 409

**N**

network, 89, 110, 259
    CAN, 106, 259
    GAN, 106, 259
    LAN, 105, 259, 261, 312, 320, 335, 439
    MAN, 106, 259, 291, 460
    WAN, 106, 259, 310
network interface card, 283
nodes, 6, 89, 262, 319
notch filter, 50
null modem, 143

**O**

Open Systems Interconnection (OSI) Model, 109, 115, 326, 412, 419, 457
    application layer, 113, 262, 292, 293
    data-link control layer, 109, 268, 293, 307, 420
    logical link control (LLC), 110
    media access control (MAC), 110
    network layer, 110, 293, 420, 458
    payload, 108
    physical layer, 109, 127
    presentation layer, 112
    sessions layer, 112
    transport layer, 111, 420, 458
optical carrier, 421
overrun (OVRN), 136

**P**

packet, 104, 188, 270, 293, 420
packet switch, 104
parity, 62, 250
payload, 230
permanent virtual circuit, 121, 293
personal communications network, 437
personal communications system, 437
photodetector, 402
photodiode, 402
phototransister, 402
PIN diode, 404
plain old telephone system (POTS), 181, 316
point to point protocol, 318
polarity violation, 22
poll, 91, 227
preamble, 19, 269

predictive coding, 201
pretty good privacy, 366, 371
private branch exchange, 318
protocol, 8, 89, 227
protocol analyzer, 249
protocol data unit, 315
pulse amplitude modulation, 191
pulse code modulation (PCM), 171, 198
pulse spreading, 44, 396

R

reflection, 389
refraction, 389
refractive index, 390, 398
repeaters, 36
restricted token, 417
rotation time, 417
router, 289
routing information protocol, 289
RS232C, 10, 131, 133, 137, 250, 306
RS422/423, 147
RS449, 145, 250

S

sampling theory, 191
    flat-top sampling, 194
    natural sampling, 194
    Nyquist sampling, 191
    sample and hold, 196
satellite, 443
satellite business system, 448
SDLC/HDLC, 238, 250, 252, 337, 414
    address field, 240, 269
    asynchronous disconnect mode, 248
    asynchronous response mode, 248
    frame check sequence, 247, 338, 414
    frame format, 239
    frame reject format, 246
    HDLC control field, 243
    SDLC control field, 241, 295
    supervisory frame, 244, 294
    unnumbered frames, 244
secure electronic transactions, 370
security, 367
selection, 92, 227
serial interface adapter, 270
shielded twisted pair cable, 265
signal-to-noise ratio (SNR), 48
simplex, 10
single-mode fiber, 397
Skipjack, 367

slip, 210
slope overload, 204
small computer system interface (SCSI), 420
smart card, 367, 436, 459
Snell's law, 390
SONET, 421
space, 2, 129
spanning tree protocol, 287
spot beam, 445, 456
standing wave, 55, 432
station, 6, 10
    primary, secondary, 6, 10
station device address, 233
station polling address, 232
step index, 398
storage area network, 420
subscriber identity module, 436
superframe, 206
supernet, 415
switched virtual circuit, 121, 293
symbol, 3, 16, 153
synchronous digital hierarchy, 421
synchronous transport signal, 421
systems network architecture (SNA), 114, 250, 309, 319
    cluster controller, 117
    logical unit, 117
    network addressable unit, 119
    nodes, 118
    path control network, 120
    physical unit, 117

T

T1 carrier, 206, 208, 322, 423
tandem switch, 35
TCP/IP, 289, 375
TELCO, 30
Teledesic, 459
Teletype/Telex, 2, 13
10BASET cable, 264, 273
throughput, 3
token bus, 274
token ring, IBM 281
toll station, 35
topology, 90, 260
    bus, 90, 261
    dual ring, 96
    hub or star, 93, 273
    mixed, 99
    ring, 95, 279
transparency, 237, 239
transponder, 443
trunk line, 35

**U**

universal asynchronous receiver transmitter (UART), 8, 128
universal synchronous asynchronous receiver transmitter (USART), 8
unshielded twisted pair (UTP), 34, 261, 264, 388
uplink 445, 449
user datagram protocol, 375

**V**

very small aperture terminal (VSAT), 460
virtual tributary, 422
voice band, 181
voice grade medium, 322

**W**

wander, 210
wave guide, 432
web of trust, 371
weighted noise, 50
white noise, 48

**X**

X.25 Protocol, 293, 310, 320
    flow control, 294
    packet assembler/disassembler, 293
    packet frame, 294
    supervisor frame, 294